THE SOCIETY FOR APPLIED BACTERIOLOGY
TECHNICAL SERIES NO 28

Genetic Manipulation

TECHNIQUES AND APPLICATIONS

Edited by

J. M. GRANGE
National Heart and Lung Institute
University of London
London

A. FOX &
N. L. MORGAN
South Bank Polytechnic
London

OXFORD

BLACKWELL SCIENTIFIC PUBLICATIONS

LONDON EDINBURGH BOSTON

MELBOURNE PARIS BERLIN VIENNA

Editorial offices:
Osney Mead, Oxford OX2 0EL
25 John Street, London WC1N 2ES
23 Ainslie Place, Edinburgh EH3 6AJ
3 Cambridge Center, Cambridge,
Massachusetts 02142, USA
54 University Street, Carlton
Victoria 3053, Australia

Other Editorial offices:
Arnette SA
2, rue Casimir-Delavigne
75006 Paris
France

Blackwell Wissenschaft
Meinekestrasse 4
D-1000 Berlin 15
West Germany

Blackwell MZV
Feldgasse 13
A-1238 Wien
Austria

First published 1991

Set by Setrite Typesetters, Hong Kong
Printed and bound in Great Britain
at the Alden Press, Oxford
and bound at the Green Street
Bindery, Oxford

British Library
Cataloguing in Publication Data

Genetic manipulation: techniques and applications.
1. Genetic engineering.
I. Grange, John M. II. Fox, Arnold, *1928–*
III. Morgan, N. L. (Neil L. *1951–* IV. Series
575.10724

ISBN 0-632-02926-9

DISTRIBUTORS

Marston Book Services Ltd
PO Box 87
Oxford OX2 0DT
(*Orders:* Tel: 0865 791155
Fax: 0865 791927
Telex: 837515)

USA
Blackwell Scientific Publications, Inc.
3 Cambridge Center
Cambridge, MA 02142
(*Orders:* Tel: 800 759-6102)

Canada
Oxford University Press
70 Wynford Drive
Don Mills
Ontario M3C 1J9
(*Orders:* Tel: (416) 441–2941)

Australia
Blackwell Scientific Publications
(Australia) Pty Ltd
54 University Street
Carlton, Victoria 3053
(*Orders:* Tel: (03) 347–0300)

Library of Congress
Cataloging in Publication Data

Genetic manipulation: techniques and applications/
edited by J.M. Grange, A. Fox & N.L. Morgan.
p. cm.—(The Society for Applied Bacteriology technical series; no. 28)
"Based on the Society of Applied Bacteriology Autumn Meeting held at the South Bank Polytechnic, London, on 18 October 1989" Pref.
Includes bibliographical references.
Includes index.
ISBN 0-632-02926-9
1. Genetic engineering—Congresses.
2. Biotechnology—Congresses. I. Grange, John M. II. Fox, A. (Arnold), 1927 — .
III. Morgan, N. L. (Neil L.), 1951 — .
IV. Society for Applied Bacteriology. Autumn Meeting (1989: South Bank Polytechnic)
V. Series: Technical series (Society for Applied Bacteriology); no. 28.
[DNLM: 1. Biotechnology—congresses.
2. Genetic Engineering—congresses. W1 S0851F no. 28/TP 248.6 G3283 1989]
TP248.6.G465 1991
660'.65—dc20
DNLM/DLC
for Library of Congress

Contents

Contributors

T. Atkinson, *Division of Biotechnology, Public Health Laboratory Service, Centre for Applied Microbiology and Research, Porton Down, Salisbury, Wiltshire SP4 0JG, UK*

J. Baker, *Division of Biotechnology, Public Health Laboratory Service, Centre for Applied Microbiology and Research, Porton Down, Salisbury, Wiltshire SP4 0JG, UK*

Pamela Bankes, *Campden Food and Drink Research Association, Chipping Campden, Gloucestershire GL55 6LD, UK*

J.G. Banks, *Campden Food and Drink Research Association, Chipping Campden, Gloucestershire GL55 6LD, UK*

A.D.T. Barrett, *Department of Microbiology, University of Surrey, Guildford, Surrey GU2 5XH, UK*

P.J.R. Barton, *Department of Cardiothoracic Surgery, National Heart and Lung Institute, Dovehouse Street, London SW3 6LY, UK*

Sarah Bell, *Molecular Biology Group, Department of Microbiology, University of Surrey, Guildford, Surrey GU2 5XH, UK*

J.E. Benbough, *Division of Biologics, Public Health Laboratory Service, Centre for Applied Microbiology and Research, Porton Down, Salisbury, Wiltshire SP4 0JG, UK*

A.M. Bennett, *Division of Biologics, Public Health Laboratory Service, Centre for Applied Microbiology and Research, Porton Down, Salisbury, Wiltshire SP4 0JG, UK*

R.P. Betts, *Campden Food and Drink Research Association, Chipping Campden, Gloucestershire GL55 6LD, UK*

J.E. Bishop, *Biochemistry Unit, National Heart and Lung Institute, Dovehouse Street, London SW3 6LY, UK*

N.J. Brand, *Department of Cardiothoracic Surgery, National Heart and Lung Institute, Dovehouse Street, London SW3 6LY, UK*

Virginia Bugeja, *Department of Biological Sciences, Hatfield Polytechnic, Hatfield, Hertfordshire AL10 9AB, UK*

J.S. Campa, *Biochemistry Unit, National Heart and Lung Institute, Dovehouse Street, London SW3 6LY, UK*

C.J. Carter, *Life Sciences Division, Amersham International PLC, White Lion Road, Amersham, Bucks HP7 9LL, UK*

S. CHAN, *GENE-TRAK Systems, 31 New York Avenue, Framingham, Massachusetts 01701, USA*

C.L. CLAYTON, *Department of Medical Microbiology, St Bartholomew's Hospital, London EC1A 7BE, UK*

I.F. CONNERTON, *Department of Microbiology, University of Reading, London Road, Reading, Berks RG1 5AQ, UK*

J.W. DALE, *Molecular Biology Group, Department of Microbiology, University of Surrey, Guildford, Surrey GU2 5XH, UK*

J.A. DAVIES, *British Biotechnology Ltd, Watlington Road, Cowley, Oxford OX4 5LY, UK*

LINDA DAVIES, *British Biotechnology Ltd, Watlington Road, Cowley, Oxford OX4 5LY, UK*

L.M. DUNSTER, *Department of Microbiology, University of Surrey, Guildford, Surrey GU2 5XH, UK*

A.J. EASTON, *Department of Biological Sciences, University of Warwick, Coventry CV4 7AL, UK*

GABY GAYER-HERKERT, *Institut fur Microbiologie, Technische Hochschule Darmstadt, Schnittpahnstrasse 10, 6100 Darmstadt, Germany*

C.A. GIBSON, *Department of Microbiology, University of Surrey, Guildford, Surrey GU2 5XH, UK*

E.B. GINGOLD, *Department of Biotechnology, South Bank Polytechnic, London SE1 0AA, UK*

D. HALBERT, *GENE-TRAK Systems, 31 New York Avenue, Framingham, Massachusetts 01701, USA*

P. HAMBLETON, *Division of Biologics, Public Health Laboratory Service, Centre for Applied Microbiology and Research, Porton Down, Salisbury, Wiltshire SP4 0JG, UK*

P.S. HARRIS, *Department of Agriculture and Fisheries for Scotland, Agricultural Scientific Services, East Craigs, Edinburgh EH12 8NJ, UK*

A.J. HAYLE, *Life Sciences Division, Amersham International PLC, White Lion Road, Amersham, Bucks HP7 9LL, UK*

S.E. HILL, *Division of Biologics, Public Health Laboratory Service, Centre for Applied Microbiology and Research, Porton Down, Salisbury, Wiltshire SP4 0JG, UK*

H-Y. HSU, *GENE-TRAK Systems, 31 New York Avenue, Framingham, Massachusetts 01701, USA*

T. JACKSON, *Molecular Biology Group, Department of Microbiology, University of Surrey, Guildford, Surrey GU2 5XH, UK*

C.M. JAMES, *Department of Agriculture and Fisheries for Scotland, Agricultural Scientific Services, East Craigs, Edinburgh EH12 8NJ, UK*

A. JOHNSON, *GENE-TRAK Systems, 31 New York Avenue, Framingham, Massachusetts 01701, USA*

I.D. JOHNSON, *British Biotechnology Ltd, Watlington Road, Cowley, Oxford OX4 5LY, UK*

W. KING, *GENE-TRAK Systems, 31 New York Avenue, Framingham, Massachusetts 01701, USA*

H. KLEANTHOUS, *Department of Medical Microbiology, St Bartholomew's Hospital, London EC1A 7BE, UK*

M.J. KLEINMAN, *Department of Biotechnology, Ngee Ann Polytechnic, Singapore, 2159, Singapore*

A. KNIGHT, *Molecular Biology Group, Department of Microbiology, University of Surrey, Guildford, Surrey GU2 5XH, UK*

H.J. KUTZNER, *Institut fur Microbiologie, Technische Hochschule Darmstadt, Schnittpahnstrasse 10, 6100 Darmstadt, Germany*

G.J. LAURENT, *Biochemistry Unit, National Heart and Lung Institute, Dovehouse Street, London SW3 6LY, UK*

G.E. LYONS, *Department of Molecular Biology, Institut Pasteur, 28 Rue du Dr Roux, 75724, Paris Cedex 15, France*

J.J. MCFADDEN, *Molecular Biology Group, Department of Microbiology, University of Surrey, Guildford, Surrey GU2 5XH, UK*

M.F. MOFFATT, *Department of Microbiology, University of Reading, London Road, Reading, Berks RG1 5AQ, UK*

M. MOZOLA, *GENE-TRAK Systems, 31 New York Avenue, Framingham, Massachusetts 01701, USA*

P. MULLANY, *Department of Medical Microbiology, St Bartholomew's Hospital, London EC1A 7BE, UK*

JANE NEWCOMBE, *Molecular Biology Group, Department of Microbiology, University of Surrey, Guildford, Surrey GU2 5XH, UK*

R.W.A. PARK, *Department of Microbiology, University of Reading, London Road, Reading, Berks RG1 5AQ, UK*

B.H. PARKIN, *Metropolitan Police Forensic Science Laboratory, 109 Lambeth Road, London SE1 7LP, UK*

R. PLANK, *Division of Biotechnology, Public Health Laboratory Service, Centre for Applied Microbiology and Research, Porton Down, Salisbury, Wiltshire SP4 0JG, UK*

S. ROGERS, *Department of Microbiology, University of Reading, London Road, Reading, Berks RG1 5AQ, UK*

P.G. SANDERS, *Molecular Biology Group, Department of Microbiology, University of Surrey, Guildford, Surrey GU2 5XH, UK*

N.A. SAUNDERS, *Division of Microbiological Reagents and Quality Control, Central Public Health Laboratory, 61 Colindale Avenue, London NW9 5HT, UK*

G.E. SCOPES, *Molecular Biology Group, Department of Microbiology, University of Surrey, Guildford, Surrey GU2 5XH, UK*

R.F. SHERWOOD, *Division of Biotechnology, Public Health Laboratory Service, Centre for Applied Microbiology and Research, Porton Down, Wiltshire SP4 0JG, UK*

J.R. STEPHENSON, *Division of Biologics, Centre for Applied Microbiology and Research, Porton Down, Salisbury, Wiltshire SP4 0JG, UK*

SOAD TABAQCHALI, *Department of Medical Microbiology, St Bartholomew's Hospital, London EC1A 7BE, UK*

W.J. VALLINS, *Department of Cardiothoracic Surgery, National Heart and Lung Institute, Dovehouse Street, London SW3 6LY, UK*

A. WARNES, *Division of Biologics, Centre for Applied Microbiology and Research, Porton Down, Salisbury, Wiltshire SP4 0JG, UK*

KAREN E. WEBB, *Host Defence Unit, Department of Thoracic Medicine, National Heart and Lung Institute, Dovehouse Street, London SW3 6LY, UK*

M. WILKS, *Department of Medical Microbiology, St Bartholomew's Hospital, London EC1A 7BE, UK*

R. WILSON, *Host Defence Unit, Department of Thoracic Medicine, National Heart and Lung Institute, Dovehouse Street, London SW3 6LY, UK*

S. WILSON, *GENE-TRAK Systems, 31 New York Avenue, Framingham, Massachusetts 01701, USA*

D.L. WOOD, *The Patent Office, State House, 66–71 High Holborn, London WC1R 4TP, UK*

B. WREN, *Department of Medical Microbiology, St Bartholomew's Hospital, London EC1A 7BE, UK*

M. YACOUB, *Department of Cardiothoracic Surgery, National Heart and Lung Institute, Dovehouse Street, London SW3 6LY, UK*

Preface

Over the last few years the life sciences have been revolutionized by the introduction of techniques for the analysis and cloning of DNA, the insertion of natural or synthesized genetic material into alternative host cells and the isolation of gene products from such modified cells. These procedures are collectively termed genetic manipulation or, more popularly, genetic engineering.

Genetic manipulation is no longer the province of the specialized researcher. The techniques are now finding widespread application in all fields of medicine and biology and the materials required are increasingly available commercially. Nevertheless, application of available techniques to new areas of research is often fraught with unexpected problems and difficulties.

This book is based on a technical demonstration at the Society of Applied Bacteriology Autumn Meeting held at the South Bank Polytechnic, London, on 18 October 1989. A wide range of techniques used in genetic manipulation is presented, commencing with the isolation and analysis of DNA, through the selection and use of cloning vectors and the application of techniques to fields as diverse as plant pathology, forensic science, bacterial taxonomy, cardiac research, diagnostic microbiology, food hygiene and sewage treatment, to problems encountered in large-scale industrial procedures. In addition, chapters on the monitoring of safety in biotechnology plants and on the patenting of the products of genetic manipulation are included. The editors wish to thank all the contributors for their enthusiastic devotion to the demonstration and to the subsequent preparation of their chapters. We also thank the staff and students of the Department of Biotechnology of the South Bank Polytechnic for their help in arranging the demonstration and Dr F. Skinner for his advice and guidance on editorial matters.

J.M. Grange
A. Fox
N.L. Morgan

Extraction, Purification and Assay of DNA

KAREN E. WEBB AND R. WILSON
Host Defence Unit, Department of Thoracic Medicine,
National Heart and Lung Institute, Dovehouse Street,
London SW3 6LY, UK

Recombinant DNA technology depends upon the ability to extract, purify and assay DNA. This can be accomplished relatively easily by following set protocols. This chapter aims to introduce the techniques involved when working with DNA from both prokaryotic and eukaryotic sources.

Isolation of DNA

There are many different methods for isolating DNA from cells, but they all involve broadly similar steps. Bacteria are cultured, harvested by centrifugation and then lysed, releasing their DNA into a crude soup from which it can subsequently be extracted and purified. Eukaryotic cells can be obtained from sources such as blood, cell culture or organ tissue, and again the principle is to release DNA by cell lysis prior to extraction and purification.

Phenol extraction

The most common method of DNA extraction uses phenol, which, with appropriate modifications, can be applied to all situations. The fundamental aim of phenol extraction is to deproteinize an aqueous solution containing DNA. The phenol reagent is mixed with the sample under conditions which favour dissociation of proteins from nucleic acids. After centrifugation, the lower organic phenol phase containing the protein (some of which segregates into the white flocculent interphase) is removed leaving the less dense aqueous phase which contains intact DNA. The easiest way of doing this is to aspirate the upper aqueous layer into a clean tube, leaving the lower phenol layer and the interphase as waste.

The phenol used must first be equilibrated with Tris–HCl, and can then

0–632–02926–9

be kept at 4°C in the dark for up to 1 month. Most commercial liquified phenol (or 250 g of solid phenol dissolved in 100 ml of water at 70°C) can be used, to which 8-hydroxyquinoline is added to a final concentration of 0.1%. This is then extracted several times with an equal volume of 1 M Tris pH 8.0 followed by 0.1 M Tris pH 8.0 until the pH of the aqueous phase is >7.6. It is very important not to use old phenol as the presence of the phenolic oxidation products, which cause brown discoloration, results in cleavage of phosphodiester bonds and cross-linking of DNA strands. 8-Hydroxyquinoline is usually added to the phenol as it delays its oxidation and, due to its bright yellow colour, facilitates the separation of the two phases.

Phenol/chloroform extraction

Chloroform, when used in conjunction with phenol, improves the efficiency of the extraction process by its ability to denature proteins. Its high density makes separation of the two phases easier and it also removes any lipid from the sample. The extraction with phenol and chloroform can either be performed sequentially or as a single extraction step with an equal volume of phenol–chloroform–isoamylalcohol (24:24:1 v/v/v). The isoamylalcohol is added to the chloroform to prevent it from foaming.

Precipitation of DNA

The most versatile method for precipitating DNA is with ethanol which can be used to recover DNA (or RNA) after extraction, to retrieve phenol-extracted products after enzymatic manipulations, or to concentrate the DNA sample. The sample, salt (usually sodium acetate) and ethanol are placed at −20°C or lower. The salt of the nucleic acid precipitates and is sedimented by centrifugation, the ethanol supernatant is removed and the DNA pellet is air dried before resuspension in appropriate buffer. It must be noted that if the sample contains phosphate or 10 mM EDTA, it should not be subjected to ethanol precipitation without prior dialysis, as these materials will be precipitated too. If the sample contains sodium dodecyl sulphate (SDS), this is most effectively removed by making the sample 0.2 M with respect to the salt before the addition of the ethanol, which ensures that the SDS remains soluble.

A very dilute sample can be concentrated by extracting it several times with butanol. The volume of the aqueous DNA phase will be reduced as water partitions into the butanol phase. An ethanol precipitation can be done once the volume of sample has been reduced to the required amount; the procedure also concentrates the salts present in the sample.

An equal volume of isopropanol can be used in the same way as for ethanol precipitation. This minimizes the total volume of the sample as an equal volume of isopropanol is used as opposed to 2.5–3.0 volumes of ethanol. However, an ethanol precipitation step is usually done after such a treatment as the isopropanol is not easily lyophilized due to its relatively low volatility. Another reason for this is due to salt presence which tends to coprecipitate with the DNA.

Spermine precipitation can be useful for the recovery of DNA from dilute solutions due to the selective nature of its action. The procedure yields DNA of relatively high purity, but two sequential precipitations are recommended (Hoopes & McClure 1981).

Isolation of Plasmid DNA from Bacteria

All methods for isolating plasmid DNA from bacteria involve culture of the bacteria, which are then harvested and lysed before purification of the plasmid DNA. The procedure is usually carried out with *Escherichia coli* and exploits the fact that the *E. coli* chromosome is much larger than the plasmid DNA, and also that the larger *E. coli* DNA is extracted as broken, linear molecules whereas the plasmid DNA is obtained as covalently closed circular (CCC) DNA.

Depending on the final use for the DNA, two methods can be employed which differ only very slightly. For high quality DNA, for example to use as a probe template, the large-scale method involving isopycric centrifugation in a solution of caesium chloride is recommended. For many other purposes, including analysis by agarose gel electrophoresis, less pure DNA can be used. The small-scale method has the advantage of being simple and quick, yet it yields DNA which is sufficiently pure to be digested by restriction enzymes or transformed into other cells.

Growth of the bacteria

1 For the large-scale preparation, a 10 ml L-broth culture containing the relevant antibiotics is grown up in a shaking incubator at 37°C overnight from a single colony. Then 1 ml of this culture is used to inoculate 500 ml of L-broth in a 2 litre flask and grown up as above.

2 For the small-scale preparation, a single colony of the bacteria is cultured in 2.5 ml of L-broth in a shaking incubator at 37°C for 18 h, or bacteria can be scraped from the surface of a fresh overnight plate.

It is very important that, if molecular biology grade reagents exist, they *must* be used as DNase and/or RNase contamination may be present in less pure reagents.

Harvesting and lysis of bacteria

The cells are harvested by spinning at 10 000 rpm for 30 min at 4°C for the large-scale preparation or by spinning in a benchtop microcentrifuge for 15 s for the small-scale preparation.

It has been shown that there is a narrow pH range (12.0–12.5) within which denaturation of linear DNA, but not CCC DNA, occurs (Birnboim & Doly 1979). The lysis method described below uses this fact to extract plasmid DNA rapidly from bacterial cells. In the following protocol, values in brackets refer to the volumes used in the small-scale preparation.

Method

1 Resuspend the bacterial cells in 15 ml (100 μl) of solution A (25 mM Tris–HCl pH 8.0, 10 mM EDTA pH 8.0, 50 mM glucose. Make up in batches and autoclave at 10 psi. Add lysozyme just before use by first resuspending the pellet in 7.5 ml of solution A without lysozyme and then adding the final 7.5 ml containing 4 mg/ml lysozyme and mixing thoroughly.
2 Leave for 10–15 (30) min on ice.
3 Add 30 ml (200 μl) of fresh solution B (0.2 M NaOH, 1% sodium dodecyl sulphate (SDS)) which must be made up fresh from stock solutions. Mix well.
4 Leave for 5 min on ice.
5 Add 22.5 ml (150 μl) of 3 M sodium acetate (or potassium acetate) pH 4.8.
6 Leave 10–30 (60) min on ice to allow most of the protein, high molecular weight RNA and chromosomal DNA to precipitate.
7 Centrifuge for 10 min at 13 000 rpm (5 min in a microcentrifuge) at 4°C.
8 Remove 60 ml (400 μl) of the supernatant to a clean tube, being extremely careful not to transfer any of the pellet.
9 Add 120 ml (1 ml) of cold 100% ethanol to the supernatant.
10 Hold at −20°C for 30 min to precipitate the DNA.
11 Centrifuge for 10 min at 12 000 rpm (2 min in a microcentrifuge), at room temperature.
12 Resuspend the pellet in 7 ml (100 μl) of TE (10 mM Tris–HCl, 1 mM EDTA pH 8.0).

The DNA from the small-scale preparation can be used at this stage, whereas the large-scale DNA needs to be purified further.

Purification of DNA

This can be achieved by centrifugation to equilibrium in a caesium chloride–ethidium bromide gradient.

Method

1 For every ml of DNA solution obtained, weigh out exactly 1 g of caesium chloride. (This should be done on foil or in the glassware to be used, *not* in a plastic weighing tray, as much of the compound will stick to it.)
2 Add *exactly* the correct volume of DNA solution and dissolve completely by gentle agitation.
3 Add 0.8 ml ethidium bromide (10 mg/ml in water) for every 10 ml of caesium chloride solution. (From this step onwards the solution container should be wrapped in foil to protect it from light.)
4 Transfer the solution to a polyallomer ultracentrifuge tube and fill to the brim with paraffin oil (at the end, the tubes plus caps must be balanced to within 0.01 g and *no* air bubbles should be present). Crimp the tops.
5 Spin at 55 000 rpm for 22 h, or at 45 000 rpm for 36 h at 21°C.
6 Two bands of DNA should be visible under ordinary light. The upper band consists of linear bacterial DNA and nicked circular plasmid DNA, the lower band consists of CCC DNA. There will also be an RNA pellet in the bottom of the tube.
7 To collect the CCC DNA, the tube should be clamped gently over a long wavelength u.v. source, and a 21-gauge needle carefully inserted into, and left in, the top of the tube in order to facilitate aspiration.
8 Using a second 21-gauge needle and a 5 ml syringe, very carefully insert the needle (flat surface uppermost) into the tube immediately below the CCC DNA band. Gently raise the needle up and aspirate the DNA into the syringe.
9 In order to remove the ethidium bromide from the aspirate, extract repeatedly with an equal volume of sodium chloride-saturated isopropanol until the DNA solution is completely clear.
10 Dilute the solution with 3 volumes of sterile distilled water.
11 Add 3 volumes of ethanol.
12 Hold at −20°C for 30 min.
13 Spin for 15 min at 12 000 rpm at room temperature.
14 Redissolve pellet in 300 μl of TE.

The CCC DNA plasmid DNA is now ready for use and can be stored at −20°C.

A more rapid method is to use 13% polyethylene glycol, molecular weight 8 000 (PEG-8 000), provided that extremely pure plasmid DNA is not required.

1 Add 50 µl of RNase A (1 mg/ml in water boiled for 5 min) and incubate at 37°C for 20 min.
2 Extract twice by adding an equal volume of phenol:chloroform, spinning for 5 min and removing the upper aqueous phase containing the DNA to a clean tube.
3 Add 2 volumes of ethanol and leave at −70°C for 15 min.
4 Centrifuge at 10 000 rpm for 10 min.
5 Resuspend the pellet in 1.6 ml of water, add 0.4 ml of 4 M NaCl and mix well.
6 Add 2 ml of 13% PEG-8 000 and mix.
7 Leave on ice for 1 h.
8 Centrifuge at 10 000 rpm for 10 min.
9 Remove the supernatant and wash the pellet in 70% ethanol.
10 Dissolve the pellet in the appropriate volume of water or TE.

Isolating Genomic DNA From Bacterial Cells

There are numerous methods for isolating the chromosomal DNA from bacterial cells and these vary from species to species. The first method given below has been shown to work with the Gram-negative bacteria *Pseudomonas* while the second method is used for the Gram-positive *Streptococcus*.

Isolating genomic DNA from Gram-negative bacteria

Method

1 Inoculate a single colony of bacteria into 3 ml of nutrient broth and grow to log phase.
2 Pellet the culture in 2 × 1.5 ml Eppendorf tubes.
3 Wash the pellet once in 1.5 ml phosphate buffered saline (PBS), combining both pellets in a single Eppendorf tube.
4 Centrifuge to repellet in a microcentrifuge.
5 Resuspend completely in 250 µl of solution C (0.1 M NaCl, 10 mM Tris–HCl pH 8.0, 10 mM EDTA).
6 Add 25 µl of 10% SDS and heat for 10 min at 65°C.
7 Add 25 µl of 5 mg/ml proteinase-K and incubate overnight at 37°C.
8 Extract twice with an equal volume of Tris–equilibrated phenol.
9 Extract once with an equal volume of water-saturated butanol.
10 Add 10 µl of 3 M sodium acetate pH 5.2.
11 Add 3 volumes of cold 100% ethanol and hold at −70°C for 30 min to precipitate the DNA.

12 Centrifuge for 5 min.
13 Decant the ethanol and air dry the pellet.
14 Resuspend in 200 μl TE plus 2.5 μl of 4 mg/ml RNase. (Boil the RNase solution for 5 min before use to destroy any DNase that may be present).
15 The DNA can now be cut with restriction enzymes or incubated for 1 h at 37°C to allow the RNase to work.

Isolating genomic DNA from Gram-positive bacteria

Method

1 Inoculate a fresh single colony into 400 ml of prewarmed nutrient broth and grow statically overnight at 37°C.
2 Pellet the cells at 10–12 000 rpm for 20 min at 4°C and remove all the supernatant.
3 Resuspend the pellet in 8 ml TES (50 mM Tris–HCl pH 8.0, 25% sucrose, 1 mM EDTA).
4 Add a spatula of lysozyme and transfer to a 36 ml heat-sealing ultra-centrifuge tube. (Do not agitate the cells from this point on so as not to damage the DNA.)
5 Leave on ice for at least 15 min.
6 Add 40 μl of 20 mg/ml proteinase-K, mix well and leave on ice for at least 15 min.
7 Add 0.65 ml solution A (0.4 ml 0.5 M EDTA pH 8.0, 0.25 ml 10% SDS), by shooting it into the tube with a Pasteur pipette.
8 Incubate at 65°C overnight or until the solution clears.
9 Add approximately 30 ml of 1.26 g/ml CsCl in TES with 0.15 ml of 10 mg/ml phenylmethysulphonyl fluoride (PMSF) in ethanol. Fill the tube completely with CsCl/TES (do not use liquid paraffin), balance and seal.
10 Centrifuge at 40 000 rpm at 20°C for 18–20 h overnight.
11 The genomic DNA will form a pale band in the middle of the tube. Remove the band as described for plasmid DNA above.
12 Dialyse the DNA against solution B (10 mM Tris–HCl pH 8.0, 1 mM EDTA) for 24–48 h.

Isolating Mammalian DNA

Mammalian DNA can be isolated from a variety of sources, the choice of which is usually dictated by the availability of material. Methods are described below for isolating DNA from three different mammalian sources.

Isolating DNA from blood (Kunkel *et al.* 1977)

The extraction of mammalian DNA from blood is usually favoured as it does not involve the killing of the animal and good yields of DNA can be obtained from relatively small amounts of blood.

Method

1 Collect 10 ml of whole blood in acid citrate dextrose or sodium heparin (to prevent it from clotting) and keep at 4°C until used, or at −70°C for longer periods.
2 Dilute 1 volume of blood with 9 volumes of cold solution D (0.3 M sucrose, 5 mM $MgCl_2$, 1% Triton X-100, 0.01 M Tris–HCl pH 7.6) and keep at 4°C to lyse the cells.
3 Centrifuge at 10 000 rpm at 4°C for 10 min.
4 Resuspend the nuclear pellet in 4.5 ml of cold solution E (75 mM NaCl, 24 mM EDTA pH 8.0).
5 Add 250 μl of 10% SDS to disrupt the nuclei.
6 Incubate overnight at 37°C with a final concentration of 100 μg/ml proteinase-K.
7 Extract with Tris–equilibrated phenol by adding an equal volume, mixing gently, spinning at 10 000 rpm for 5 min and carefully removing the upper layer.
8 Extract twice with chloroform by adding an equal volume of chloroform: isoamylalcohol (24:1), mixing gently and extract as above.
9 Add 0.5 ml of 3 M acetate pH 5.0 and 11 ml of 100% ethanol at room temperature to precipitate the DNA which can then be spooled onto a glass rod.
10 Dissolve the DNA in 1 ml of TE at 4°C.
This process should yield about 200–500 μg DNA/10 ml blood.

Isolating DNA from whole organs (Herrmann & Frischauf 1987)

DNA can be extracted from various organs including the liver, spleen or kidney. The spleen usually yields plenty of pure DNA. If the liver is being used, the animal should be starved for 24 h prior to killing to deplete glycogen levels.

The tissue is cut into small pieces and dropped into liquid nitrogen, it can then be stored at −80°C for over 1 year. It is very important to be properly prepared by having dry ice available, a precooled pestle and mortar, liquid nitrogen (in a well-rinsed container to avoid DNA contamination), a metal spoon, a 400 ml beaker containing 20 ml of TENS (50 mM Tris–HCl pH 9,

100 mM EDTA, 200 mM NaCl) and DNase-free RNase A (100 μg/ml). All the tools should be sterilized.

Method

1 Pour liquid nitrogen into the mortar and add the frozen tissue pieces.
2 Break the tissue into smaller pieces.
3 Quickly grind the tissue into a powder without further addition of liquid nitrogen.
4 Chill one end of the metal spoon by dipping it in the liquid nitrogen.
5 Distribute the powder in small portions onto the surface of prepared buffer. Allow the powder to spread onto the surface then swirl the beaker a little to submerge the material. Continue until everything is in solution.
6 Transfer the solution into a 50 ml plastic tube and shake repeatedly to obtain a more homogeneous solution.
7 Leave the tube on a rocking platform at 30 rpm for 10 min at room temperature.
8 Add 1 ml of 20% sodium dodecyl sulphate (SDS) and invert the tube twice. Then continue to rock the tube for another 10 min.
9 Add 1 ml of proteinase-K (10 mg/ml in water), invert the tube to mix and incubate at 37°C overnight with rocking. (The solution should appear almost clear after proteinase digestion.)
10 Transfer the solution into a flat-sided glass bottle and add 20 ml Tris–equilibrated phenol. This allows the mixture to cover a large surface and aids the mixing.
11 Rock the bottle for 1–3 h at room temperature.
12 Centrifuge the mixture in a 50 ml plastic tube at 3 000 rpm for 10 min at room temperature.
13 Remove the lower phenol layer by aspiration.
14 Repeat the phenol extraction until the aqueous DNA phase is clear.
15 Transfer the DNA solution to a glass centrifuge tube and spin at 9 000 rpm for 20 min at 25°C (to avoid precipitation of SDS). Protein and undissolved material will stick to the glass.
16 Carefully pour the supernatant into a 50 ml plastic tube and then dialyse against TE (10 mM Tris–HCl pH 8.0, 1 mM EDTA) 1:1 000 v/v, firstly at room temperature (to avoid SDS precipitation) and then at 4°C.
17 Add sodium acetate pH 6.5 to a final concentration of 0.3 M and 0.8 volumes of isopropanol. Repeatedly invert the tube gently.
18 Spool out the precipitated DNA using a glass pipette with a sealed end, and place it into 3 ml of TE per organ of starting material.
19 To dissolve the DNA leave the tube rotating at 20 rpm overnight at room temperature.

Isolating DNA from cell culture cells

The method for isolating DNA from cell culture cells is very similar to that described above for organ tissue. However, Bowtell (1987) described a rapid method to isolate eukaryotic DNA from both cells grown *in vitro*, and those isolated from various tissues by teasing through a stainless-steel sieve. This involved resuspending pelleted cells in 1/10th volume of phosphate buffered saline (PBS) pH 7.3 which is then added to 6 mM guanidine hydrochlorate and 0.1 M sodium acetate pH 5.5 (15 ml/10^7 cells), and mixed on a rotating wheel for a minimum of 1 h. The suspension is then layered under 2.5 volumes of 100% ethanol at room temperature and the DNA is spooled out by means of a hooked glass rod. The DNA is then rinsed in 2 × 5 ml of 70% ethanol and resuspended in TE.

Assay of DNA

Determination of concentration and purity

An u.v. absorbance of 1, measured at 260 nm in a 1 cm path length, indicates an approximate concentration of 50 μg/ml of double-stranded DNA, or 40 μg/ml of RNA or single-stranded DNA. The purity of the sample can be estimated by the ratio of the absorbance at 260 nm to that at 280 nm: values of 1.8–1.9 for DNA and 1.9–2.0 for RNA solutions are acceptable. This is because the presence of proteins, which absorb at 280 nm, decreases the ratio, as does phenol and other likely contaminants.

Gel electrophoresis

Gel electrophoresis rapidly resolves mixtures of single or double-stranded DNA molecules by size. At pH values near neutral, DNA is negatively charged and migrates from the cathode to the anode, the migration rate of the DNA through the gel depends on both the size of the DNA molecules and the concentration of the gel matrix. Linear duplex DNA is believed to migrate in an end-on position at a rate inversely proportional to the $\log_{10}$ of its molecular weight. Non-denaturing polyacrylamide gels can be used for the separation of double-stranded DNA fragments between 6 (20% acrylamide) and 1 000 (3% acrylamide) bp whilst non-denaturing agarose gels can be used for fragments ranging from 70 (3% agarose) to 800 000 (0.1% agarose) bp. Single-stranded DNA can be fractionated by either type of gel electrophoresis by the inclusion of a denaturing agent.

The location of the DNA fragments within the gel can be viewed directly by staining the bands with the fluorescent intercalating dye ethidium bromide

It must be noted that ethidium bromide is a powerful mutagen and gloves must be worn when handling gels or solutions containing it.

For most applications, agarose gel electrophoresis can be used, including the analysis of restriction fragments and the preparation of fragments for cloning. The preferred method is to use a horizontal gel system which has the advantage that low percentages of agarose can be used as the gel is supported from below, the gels can be cast in a variety of sizes, are simple to pour, load and handle, and the apparatus is durable and inexpensive.

Gel-running buffers

The stock and working gel-running buffer solutions are shown in Table 1. Tris–acetate is the most commonly used buffer, although its buffering capacity is low and it tends to become exhausted during extended runs.

The agarose gels are made up in running buffer to the desired concentration and poured into the mould. Ethidium bromide can be added to give a final concentration of 0.5 μg/ml. This allows the progress of the DNA bands through the gel to be monitored during the run. Incorporation of ethidium bromide will, however, affect the conformation of certain DNA molecules, thereby changing their mobilities, and it will also damage DNA in the light. It is therefore usually advisable to stain the gel with a 0.5 μg/ml solution of ethidium bromide (in water or running buffer) after the run has finished.

TABLE 1. *Gel-running buffers*

Buffer	Working solution	Concentrated stock solution (1 litre)
Tris–acetate (TAE)	0.04 M Tris–acetate 0.001 M EDTA	50 ×: 242 g Tris base 57.1 ml glacial acetic acid 100 ml 0.5 M EDTA pH 8.0
Tris–borate (TBE)	0.089 M Tris–borate 0.089 M boric acid 0.002 M EDTA	5 ×: 54 g Tris base 27.5 g boric acid 20 ml of 0.5 M EDTA pH 8.0
Tris–phosphate (TPE)	0.08 M Tris–phosphate 0.002 M EDTA	10 ×: 108 g Tris base 15.5 ml phosphoric acid (85%) (1.679 μg/ml) 40 ml 0.5 M EDTA pH 8.0

Gel-loading buffers

The types of loading buffers used contain one or two tracking dyes, 0.25% bromophenol blue and 0.25% xylene cyanol. These can be in a solution of 40% (w/v) sucrose, 30% glycerol or 15% Ficoll (made up in water) to give a 6 × stock solution which can be stored at 4°C, unless Ficoll is used, when it can be stored at room temperature.

Samples are mixed with the loading buffer to the correct dilution and then pipetted into the wells of the prepared gel which is submerged in the running buffer in the electrophoresis tank. The minimum amount of DNA that can be detected by photography, in a 0.5 cm wide band, after staining with ethidium bromide, is 2 ng. If more than 200 ng of DNA is present in a band this size, trailing and shearing will occur. The ideal amount of DNA from simple populations of DNA fragments is 0.2–0.5 μg per 0.5 cm slot. If, however, the sample consists of a very large number of fragments it is possible to load 5–10 μg per lane.

The gel is usually run until the first dye front (bromophenol blue), has travelled three-quarters of the way through the gel. This will depend, however, on the size of the fragments of interest. The gel is then stained for 45 min in a 0.5 μg/ml solution of ethidium bromide, rinsed in distilled water and photographed under u.v. light.

Recovery of DNA from agarose gels

There are numerous established methods for the recovery of DNA fragments from gels after electrophoresis (Maniatis *et al.* 1982), including electro-elution into dialysis bags and into troughs. The methods involve cutting a trough in the gel directly in front of the band to be eluted. The trough can either be filled with running buffer, or a piece of Whatman 3MM paper together with a piece of single dialysis membrane soaked in buffer can be slotted in to it. In the former case, the gel is then run for a further 2–3 min and the buffer is collected from the trough. The process is repeated until all the DNA in the band has been recovered. If the latter procedure is used, the gel is run until the DNA band has run completely into the membrane. In both cases, the DNA has then to be purified.

All these methods are time consuming, as the DNA has to be purified afterwards, and need constant attention. Recently BIO 101 Inc. have produced a *Geneclean* kit, (available from Stratech Scientific Ltd, 50 Newington Green, London, N16), which provides a quick and simple method for recovering DNA from gels. It involves the binding of the DNA to silica particles in a matrix suspension, the washing of the DNA with a sodium chloride/ethanol/water wash, and the subsequent elution of the DNA from the matrix with TE buffer, water or a low salt buffer of choice. Typical yields of DNA from the use of this kit are greater than 60%, and the resulting DNA can be used immediately in enzyme reactions or other manipulations.

References

Birnboim, H.C. & Doly, J. 1979. A rapid alkaline extraction procedure for screening recombinant plasmid DNA. *Nucleic Acids Research* **7**, 1513–1523.

Bowtell, D.D. 1987. Rapid isolation of eukaryotic DNA. *Analytical Biochemistry* **162**, 463–465.

Herrmann, B.G. & Frischauf, A-M. 1987. Isolation of genomic DNA. In *Methods in Enzymology—Guide to Molecular Cloning*, eds Berger, S.L. & Kimmel, A.R. pp. 180–183. New York & London: Academic Press.

Hoopes, B.C. & McClure, W.R. 1981. Studies on selectivity of DNA precipitation by spermine. *Nucleic Acids Research* **9**, 5493–5504.

Kunkel, L.H., Smith, K.D., Boyer, S.H., Borgaonkar, D.S., Wachtel, S.S., Miller, O.J., Breg, W.R., Jones, H.W. & Rany, J.M. 1977. Analysis of human Y-chromosome-specific reiterated DNA in chromosome variants. *Proceedings of the National Academy of Sciences of the United States of America* **74**, 1245–1249.

Maniatis, T., Fritsch, E.F. & Sambrook, J. 1982. *Molecular Cloning*. Cold Spring Harbor, New York: Cold Spring Harbor Laboratory.

Restriction Mapping of DNA

KAREN E. WEBB AND R. WILSON
Host Defence Unit, Department of Thoracic Medicine,
National Heart and Lung Institute, Dovehouse Street,
London SW3 6LY, UK

Restriction enzyme mapping provides a powerful method to begin the analysis of extracted DNA. This is the first step in a process which may ultimately result in obtaining the entire base-pair sequence of a gene of interest. It involves the construction of a physical map which shows the sites at which defined breaks occur in DNA, thereby allowing their distance apart to be accurately determined. This is achieved by using endonucleases, known as *restriction enzymes*, which recognize rather short, specific sequences of DNA as targets for cleavage. More than 400 different endonucleases have been isolated from bacteria, and their names are derived from the strain of origin.

Restriction Enzymes

Most bacteria have defence systems to guard against invasion by foreign DNA. They contain specific endonucleases (restriction enzymes) that make double-stranded cuts in foreign DNA. The bacteria protect their own DNA from the potentially lethal effect of their restriction enzymes by previously modifying it, usually by an appropriate DNA methylase. This involves methylation of certain bases at a limited number of sequences so that the enzymes no longer recognize them as sites.

Because in practice they are used for the molecular dissection of DNA, the restriction enzymes have been characterized primarily with respect to their recognition sequences. The majority of these enzymes recognize sequences 4–6 nucleotides long, but there are some that recognize sequences of up to 8 nucleotides. The enzyme cuts the DNA each time it recognizes the base-pair sequence so that the smaller the number of nucleotides in the site, the more often the enzyme will tend to cut the DNA. Most of the sites contain a dyad axis of symmetry and the bases within the site are uniquely defined, but

0–632–02926–9

```
(a) ...GAATTC...  EcoRI   ...G3'       5'AATTC...
    ...CTTAAG...  ------> ...CTTAA          G...

(b) ...CTGCAG...  PstI    ...CTGCA3'        5'G...
    ...GACGTC...  ------> ...G          ACGTC...

(c) ...CCCGGG...  SmaI    ...CCC3'      5'GGG...
    ...GGGCCC...  ------> ...GGG          CCC...
```

FIG. 1. The three types of end produced when DNA is cut with a restriction enzyme: (a) a 5′-overhang; (b) a 3′-overhang and (c) a blunt-end.

this is not true in all cases. Restriction enzymes with symmetrical recognition sites generally cleave symmetrically within, or adjacent to, the site, whereas those that recognize asymmetric sites tend to cut at a distance from the actual site.

The nature of the cleavage is also very important as the resulting ends determine the suitability of the fragments for subsequent procedures. All restriction enzymes cut their DNA substrates to produce a 5′−phosphate and a 3′−hydroxyl terminal on each strand at each cut. The breaks can be staggered, generating either 5′−phosphate overhangs (Fig. 1a) or 3′−hydroxyl overhangs (Fig. 1b) on each strand, or the breaks can produce 'blunt' ends. (Fig. 1c).

Digestion of DNA with these enzymes produces fragments of different lengths depending on the distribution of the specific sites within the DNA. The fragments can then be separated by gel electrophoresis. In order to work, the digestion reaction has to be performed in the specific reaction buffer for the enzyme being used. Most companies that supply the enzymes also supply the relevant buffer.

Approaches to Mapping

A number of strategies can be used to construct a physical map of the cleavage sites within the DNA being studied. The techniques most commonly used are:

1 Simultaneous digestion with combinations of restriction enzymes.

2 Sequential digestion of an isolated DNA fragment with a second restriction enzyme.

3 Partial digestion of either labelled, or unlabelled, DNA specifically at one terminus.

4 Partial exonucleolytic digestion of the DNA followed by digestion with a restriction enzyme.

A combination of several of these strategies may be necessary to produce a map which is sufficiently accurate and detailed. We will first describe in general the theory underlying these mapping techniques, before addressing each in more detail at a practical level.

The strategy adopted depends partly on the size of the DNA to be mapped, and partly on the detail required. For example, to map a piece of DNA more than 2–3 kb in length, it is usual to begin by analysing the fragments produced by simultaneous digestion with two restriction enzymes that cleave the DNA infrequently (i.e. enzymes that recognize hexanucleotide sequences). From the size of the fragments produced it is usually possible to deduce the relative locations of at least some of the sites. By doing a number of such double restriction enzyme digests it is possible to increase the number of defined sites, until eventually no ambiguities remain. Figure 2 shows a simple example of this type of deduction which is the basis of restriction mapping. The resolution of such maps depends on the accuracy with which the size of DNA fragments produced can be determined relative to standard size fragments of marker DNA. To construct more accurate maps, it is usually necessary to isolate particular fragments by excision from a gel, and then to cleave these with combinations of enzymes recognizing tetranucleotide sequences, and again to measure the size of the fragments produced by polyacrylamide gel electrophoresis.

Whenever mapping accurately, it is useful to work from a fixed point. Two procedures have been developed to achieve this. The first involves partial digestion of an end-labelled fragment of linear DNA (Smith & Bernstiel 1976). The conditions of the digest are adjusted so that, on average, only one cut per molecule occurs, and in this way a ladder of DNA fragments is generated. The sizes of these fragments corresponds to the distance between the labelled end of the DNA and the given restriction sites, whereas the difference in the size between each fragment in the ladder corresponds to the distance between one recognition site and the next for each enzyme studied.

The second procedure (Legerski *et al.* 1978) involves the digestion of a DNA fragment, to varying extents, by a double-strand exonuclease such as *Bal*31 (Fig. 3). This enzyme progressively degrades both the 3′ and 5′ strands of DNA. Under suitable conditions, the linear DNA molecule can be digested from both ends in a controlled manner and the rate of digestion can be calculated. A reaction with *Bal*31 is commenced, samples are withdrawn sequentially and inactivated with EGTA (as the enzyme is absolutely dependent on calcium). After digestion of these samples with the restriction enzyme of interest, restriction fragments can be seen to disappear in a defined order, i.e. the order in which they occur in the DNA molecule. By using DNA consisting of vector sequences at one end (for which the restriction map is known), and the unmapped sequence at the other, it is possible to distinguish fragments from

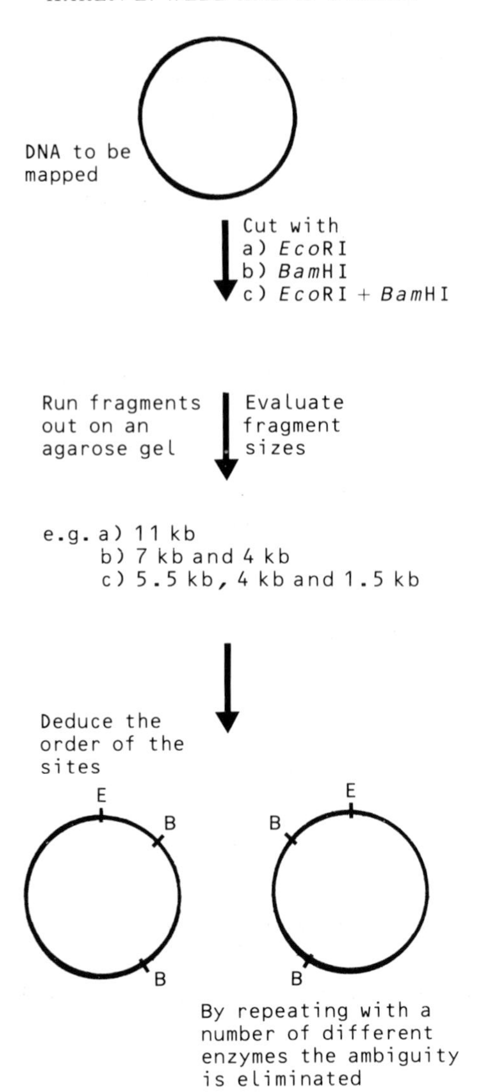

FIG. 2. A very simple example of the basis of restriction mapping.

the two ends and to deduce the order in which the fragments occur in the unmapped DNA.

Having described the theory behind restriction mapping, we will now give a more detailed practical description of how to perform each of these procedures.

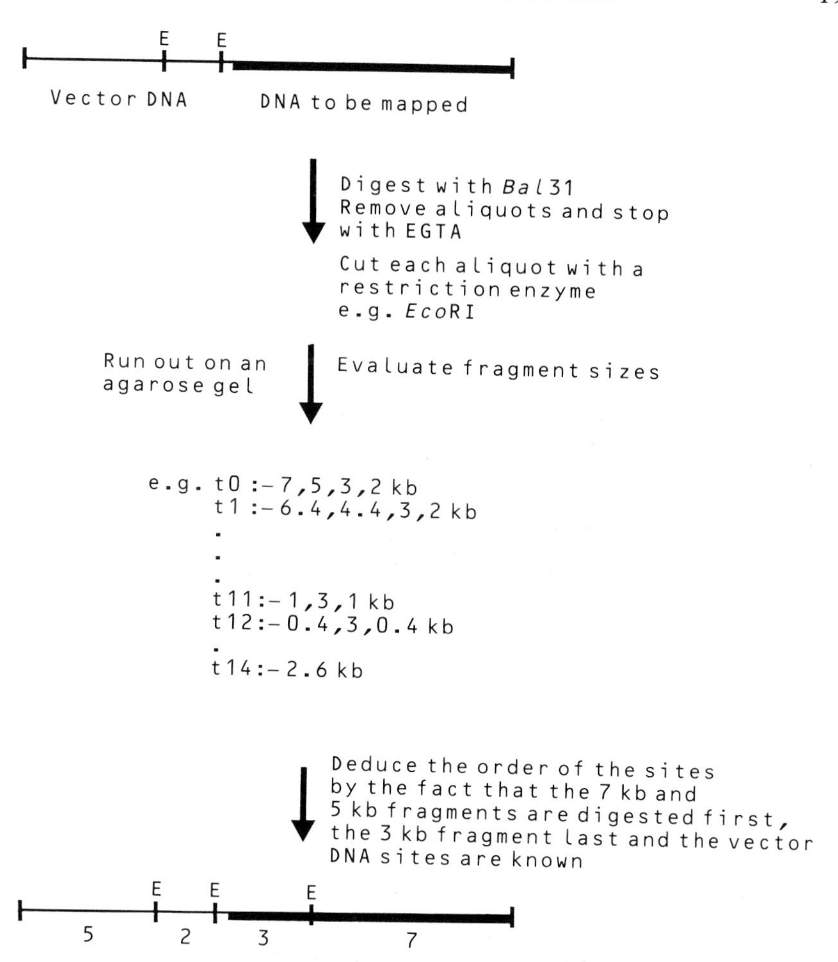

FIG. 3. A simple example of the way the exonuclease *Bal*31 can be used to obtain a restriction map.

Mapping by single and multiple digestions

Method

1 Individually digest 5 μg of DNA with a range of restriction enzymes expected to cut only a few times within DNA.
2 Analyse 0.5 μg of each digested sample by electrophoresis through a 0.9% agarose gel. Calculate the size of the fragments produced using standard marker fragments (usually commercially available bacteriophage DNA cut

with an enzyme producing fragments of known size). This is achieved by plotting a graph of log (size of known fragments) against distance travelled and by reading off the size of the unknown fragments from the distance they have travelled.

3 Cut 0.5 μg of each primary digest with a series of additional enzymes. (The two enzymes may be used simultaneously if their buffers are compatible, otherwise it is necessary to purify the DNA from the first digest by phenol/chloroform extraction and ethanol precipitation.)

4 Repeat step 2. Calculate the size of the fragments produced by this second digestion. With the information thus obtained it is possible to build up a restriction map in the way described in Fig. 2.

Sequential digestion

There are a number of methods that can be used to excise DNA fragments from an agarose gel after electrophoresis. The simplest is to use low melting point (LMT) agarose. This allows the gel containing the DNA fragment of interest to be excised with a razor blade and melted at 65°C without denaturing the DNA. The sample can then be cooled to 37°C without the agarose setting and incubated with a second enzyme.

1 Digest the DNA with an enzyme and run through a LMT agarose gel. (Avoid using buffers containing phosphate as they alter the melting characteristics of the LMT agarose).

2 After staining the gel with ethidium bromide in the usual way, examine the fragments under long wavelength u.v. radiation (short wavelength u.v. radiation will damage the DNA). Using a razor blade, cut out the desired band in as thin a slice as possible. Photograph the gel. The band should be excised prior to photography in order to keep the time that the DNA is exposed to u.v. light in the presence of ethidium bromide to a minimum.

3 Place the gel slice in a preweighed tube and add at least 10 volumes of cold water. Soak the gel slice for 1–2 h at 4°C during which time most of the electrophoresis buffer and free ethidium bromide will difuse out of the gel. The water can then be decanted off and the slice stored, if required, at 4°C in a tightly sealed tube.

4 Hold the gel slice at 65°C until the agarose melts (2–5 min).

5 Transfer the tube to 37°C and add 0.1 volume of prewarmed restriction buffer.

6 Add nuclease-free bovine serum albumin (BSA) to a final concentration of 0.1%.

7 Add the second restriction enzyme (usually two or three times more enzyme is needed with LMT agarose than under normal conditions). Incubate at 37°C for 1 h.

8 Heat the restriction mix to 65°C for 5 min and load into a normal agarose gel in the usual way. Allow the LMT agarose to harden in the wells before starting the run. Remember to adjust the marker DNA so that it contains approximately the same amount of LMT agarose and BSA as the samples.
9 Electrophoresis and analysis of the fragments is performed as above.

Partial digestion of end-labelled DNA

There are a number of ways of end-labelling DNA, but the method described below is rapid and may be carried out immediately after the appropriate enzyme digestion without any further treatment.

The Klenow fragment of *E. coli* polymerase-I carries the 5′,3′ polymerase and 3′,5′ exonuclease activities of the parent holoenzyme, but lacks the 5′,3′ exonuclease. It can therefore be used to fill in the 3′ recessed termini created by digestion of DNA with restriction enzymes. However, when 3′ protruding termini are produced, the more active 3′,5′ exonuclease activity of T4 DNA polymerase is used.

The $(\alpha-{}^{32}P)$dNTP added to the filling in reaction depends on the sequence of the 5′ overhang at the ends of the DNA. For example, if the enzyme used was *Eco*RI the overhang would be labelled with $[\alpha-{}^{32}P]$dATP.

$$\begin{matrix} \ldots p-Cp-Tp-Tp-Ap^{5'} \\ \ldots -pG- \\ \quad OH^{3'} \end{matrix} \xrightarrow{(\alpha-{}^{32}P)dATP} \begin{matrix} \ldots p-Cp-Tp-Tp-Ap-Ap^{5'} \\ \ldots\ldots -pG-pA^*-Ap^*- \\ \quad OH^{3'} \end{matrix}$$

Blunt-ended fragments are labelled by replacement of the nucleotide at the 3′-hydroxyl terminus in the same way.

$$\begin{matrix} \ldots p-Tp-Cp^{5'} \\ \ldots\ldots -pA-pG- \\ \quad OH^{3'} \end{matrix} \xrightarrow{[\alpha-{}^{32}P]dGTP} \begin{matrix} \ldots p-Tp-Cp^{5'} \\ \ldots\ldots -Ap-pG^*- \\ \quad OH^{3'} \end{matrix}$$

Method

1 Digest up to 1 μg of DNA with the desired restriction enzyme in 25 μl of the relevant buffer.
2 Add 2.0 μCi of the appropriate $[\alpha\text{-}{}^{32}P]$dNTP.
3 Add 1 unit of the Klenow fragment (or T4 polymerase if a 3′ extension is to be filled). Incubate at room temperature for 10 min.
4 The labelled DNA can be separated from the unincorporated dNTPs using the BIO 101 Geneclean kit. (This can be obtained from Stratech Scientific Ltd, 50 Newington Green, London, N16.)
5 Mix approximately:
10 cpm of labelled DNA (dissolved in no more than 8 μl water),

1 μg carrier DNA,

1 μl 10 × restriction buffer,

to 10 μl with water;

1–2 units restriction enzyme.

The carrier DNA used is usually of plasmid or bacteriophage origin and is added to make it easier to control the rate at which the enzyme digests the labelled DNA.

6 Incubate at 37°C withdrawing 1.8 μl samples at 2, 5, 10, 15 and 30 min into 1 μl of 0.5 M EDTA.

At this stage all the samples should be combined in order to enhance the detection of poorly represented partial digestion products.

7 Load 5 μl of the combined reaction into a single lane of a 1.4% agarose gel as usual. Use about 10 cpm of a mixture of end-labelled fragments of pBR322 DNA as a size marker.

8 Run the gel until the bromophenol blue dye front has travelled two-thirds of the length of the gel and then expose it to X-ray film at −70°C.

Digestion of exonuclease treated DNA

It is possible to calculate the time (1 min) needed to remove the desired number of nucleotides by the equation given below:

$$\frac{dM_t}{dT} = \frac{-2V_{max}M_n}{[K_m+(S)_0]}$$

where

M_t = molecular weight of linear duplex DNA after t min of digestion;

V_{max} = maximum reaction velocity (expressed as moles of nucleotide removed/litre/min);

M_n = average molecular weight of sodium mononucleotide (taken as 330 Da);

K_m = Michaelis–Menten constant (expressed as moles of duplex ends of DNA/litre);

$(S)_0$ = moles of duplex DNA ends/litre at t=0 min (remains constant until a significant portion of the molecules have undergone complete degradation).

At an enzyme concentration of 40 units/ml:

$V_{max} = 2.4 \times 10^{-5}$ moles of nucleotide removed/litre/min;

$K_m = 4.9 \times 10^{-9}$ moles of duplex termini/litre

Method

1 Precipitate the DNA with ethanol and then dissolve it in a solution of 500 μg/ml BSA.

2 Add an equal volume of 2 × *Bal*31 buffer (24 mM $CaCl_2$, 24 mM $MgCl_2$, 0.4 M NaCl, 40 mM Tris–HCl (pH 8.0), 2 mM EDTA) and incubate for 30 min at 30°C.
3 Add the appropriate amount of *Bal*31.
4 At the appropriate times remove samples from the reaction and add 0.2 M EGTA pH 8.0, to a final concentration of 20 mM. Store the samples on ice.
5 At this stage two alternative procedures are available:
 (a) Samples may be diluted 3-fold with water in order to lower the NaCl concentration from 0.2 M to 66 mM. After addition of 1/10th volume of restriction enzyme buffer (prepared without NaCl), the restriction enzyme is added and the samples are left to digest. The digested samples are then analysed by electrophoresis as usual;
 (b) This procedure is used if the restriction enzyme of interest is inhibited by 66 mM NaCl. The *Bal*31-digested samples are extracted and ethanol precipitated. After washing with 70% ethanol, the pellets are dissolved in the appropriate buffers and digested as in (a).

References

Legerski, R.J., Hodnett, J.L. & Gray, Jr., H.B. 1978. Extracellular nucleases of *Pseudomonas Bal*31. III. Use of the double-strand deoxyribonuclease activity as the basis of a convenient method for the mapping of fragments of DNA produced by cleavage with restriction enzymes. *Nucleic Acids Research* 5, 1445–1464.

Smith, H.O. & Bernstiel, M.L. 1976. A simple method for DNA restriction site mapping. *Nucleic Acids Research* 3, 2387–2398.

The Construction and Use of Cloning Vectors

P. Mullany, B. Wren, M. Wilks, C.L. Clayton and Soad Tabaqchali
Department of Medical Microbiology, St Bartholomew's Hospital, London EC1A 7BE, UK

A brief review of some of the types of commercially available vectors that can be used for gene cloning will be given. The second part of the chapter will discuss the construction of a cloning vector capable of introducing chimeric DNA into a host strain for which established protocols are not yet developed. The construction of a shuttle vector capable of replication in *E. coli* and *Clostridium perfringens* will be used to illustrate one approach to this problem.

The Commercially Available Vectors

A great variety of vectors are currently supplied by a number of companies. This section illustrates which vectors are most suited for a particular cloning project and helps the reader make a logical choice of cloning vector(s). All the vectors described in this section are for cloning genes into *E. coli*. In a typical gene cloning project the flow of manipulations is as follows:

1 DNA is isolated from the organism of interest (see chapters by Webb & Wilson, this volume).
2 A gene library is constructed by means of a suitable cloning vector (usually a phage or cosmid which can accept large DNA inserts).
3 Screening of the gene library (by a suitable DNA probe or immunoassay).
4 Localization of the gene of interest by subcloning (usually in a plasmid vector).
5 Gene sequencing and *in vitro* mutagenesis (usually by placing the gene in a single-stranded DNA phage vector such as M13).
6 Returning the inactivated gene to the original host using a suitable shuttle vector (see p. 33).

Table 1 shows a number of different cloning vectors which are suitable for constructing gene libraries as they can accept large insert DNA molecules.

Genetic Manipulation
0-632-02926-9

TABLE 1. *Primary gene cloning vectors for making gene libraries*

	Insert size (kb)	Cloning sites	Antibody screening possible	Recombinant detection method	*In vivo* subcloning	Antibiotic resistance markers
λZAPII	0–10	*Eco*RI, *Xho*I, *Spe*I, *Hind*III, *Xba*I, *Not*I, *Bam*HI, *Mbo*I	Yes	Inactivation of *lacZ*	Yes	*amp*
λgt11	0–7	*Eco*RI	Yes	Inactivation of *lacZ*	No	NA
λgt10	0–7	*Eco*RI	No	Recombinants produce clear plaques	No	NA
λEMBL3	9–23	*Bam*HI, *Sal*I, *Eco*RI	No	Growth on P2 lysogen	No	NA
λEMBL4	9–23	As above but in opposite orientation		Growth on P2 lysogen	No	NA
pHC79	30–42	*amp* and *tet* genes	No	Inactivation of one of the antibiotic resistances	No	*amp* and *tet*

NA = not applicable.

λZAPII

λZAPII is an insertion phage vector which can accept up to 10 kb of insert DNA. A major disadvantage of cloning in a phage is the difficulty with which DNA is isolated from phage heads. λZAPII however, overcomes this problem by being rapidly converted into a plasmid when it is introduced into a host strain containing a suitable helper phage. λZAPII also contains the Fl (M13) origin of replication so that, in an appropriate host strain, it will produce single-stranded DNA which will be packaged as a filamentous phage which will bud off from the cell. The DNA from the resultant phage can be prepared by a standard protocol and can act as a template for DNA sequencing and *in vitro* mutagenesis. In its double-stranded plasmid form, the λZAPII chimera can also be used to construct exonuclease III/mung bean unidirectional deletions in the gene of interest.

DNA inserts into λZAPII can be expressed as β-galactosidase fusion proteins which can be detected by antibodies, thereby permitting the library to be screened with a suitable immunoassay. A detailed protocol for the use of λZAPII is provided when purchasing the vector.

λgt11

λgt11 is used to insert DNA of up to 7 kb into the β-galactosidase coding region. Under appropriate conditions, the foreign DNA is expressed as part of a fusion protein. The library can be screened with antibody probes. Furthermore, fusion proteins produced in this system can be conveniently isolated by affinity column chromatography. Columns are available from Promega (Appendix II). These contain monoclonal anti-β-galactosidase antibodies linked to the column.

λEMBL3 and λEMBL4

λEMBL3 and λEMBL4 are replacement vectors. There are inverted restriction enzyme sites of *Eco*RI, *Bam*HI and *Sal*I (the sites are in opposite orientation in each vector) for cloning on either side of a non-essential stuffer region. To aid the selection of recombinants the stuffer region contains genes which prevent the phage growing on certain hosts lysogenic for phage P2. The advantage of the λEMBL phages over the other phage systems mentioned is that they can be used to clone up to 23 kb and, because of the P2 selection mechanism, all plaques represent recombinants.

pHC79

pHC79 is a cosmid, that is a plasmid molecule which contains the *cos* ends (the sites recognized by the packaging system of phage λ). This allows pHC79, when containing an insert of between 30–42 kb, to be packaged efficiently into phage heads and transduced to suitable recipients. Selection for recombinants is by inactivation of one of the antibiotic resistance genes carried on pHC79.

Vectors used for subcloning from the primary gene library (Table 2)

Once the library has been constructed and the gene of interest approximately located, it is desirable to subclone the DNA into vectors in which it is possible to carry out a more detailed analysis. This may include restriction enzyme mapping, DNA sequencing, *in vitro* mutagenesis or enhanced expression of the insert DNA in expression vectors.

pBR322

pBR322 is sometimes called the work horse of molecular biology. This vector is still popular in cloning projects, even though more sophisticated and specialized vectors are available. The reason for this is its simplicity and the availability of its complete nucleotide sequence. Ligating foreign DNA into restriction enzyme sites within either the *amp* or the *tet* gene will inactivate that gene, thus giving an easy system for detecting recombinants.

The pUC family

The pUC family of vectors are extremely useful for subcloning and expressing foreign genes. They consist of a pBR322 ampicillin resistance gene and origin of DNA replication ligated to the *lacZ* region of *E. coli*. A polylinker site (containing an array of unique restriction enzyme sites for cloning) has been introduced into the *lacZ* (alpha complementing region) of each of these plasmids. When introduced into a suitable host strain carrying an F′ plasmid that has the *lacZ* region capable of complementing the alpha region, the plasmids will produce a blue colour on agar plates containing X-gal indicator. Recombinants will not be able to hydrolyse the X-gal as inserts will inactivate the β-galactoside gene and therefore give rise to white colonies.

pBS (+/−)

These plasmids are phagemids, that is, depending on the host strain, they can either replicate as a phage (and be packaged and exported), or replicate as a plasmid. They contain an inducible *lac* promoter and the N-terminal region of β-galactosidase. This allows colour selection of the recombinant, as described for the pUC vectors, and the possibility of the production of a fusion protein.

TABLE 2. *Plasmid vectors used for subcloning*

	Vector size (kb)	Antibiotic resistance	β-galactosidase fusion proteins	Unique cloning sites	M13 origin	Uses
pBR322	4.36	*amp*, *tet*	No	See map (Fig. 1)	No	Subcloning into the antibiotic-resistance genes (this plasmid has been completely sequenced)
pBS(+/−)	3.20	*amp*	Yes	*Hin*dIII, *Sph*I, *Pst*I, *Sal*I, *Xba*I, *Bam*HI, *Xma*I, *Sma*I, *Kpn*I, *Sac*I, *Eco*RI	Yes	Subcloning, transcribing and sequencing of insert DNA
pUC19	2.68	*amp*	Yes	*Hin*dIII, *Sph*I, *Pst*I, *Sal*I, *Xba*I, *Bam*HI, *Xma*I, *Sma*I, *Kpn*I, *Sac*I, *Eco*RI	No	Subcloning into the *lacZ* region. Recombinant selected by white colour on plates containing X-gal. Genes cloned into this vector may be expressed as fusion proteins under the control of the *lac* promoter.
pKK175−6	4.4	*amp*, *tet*	No	*Eco*RV, *Sma*I, *Bam*HI, *Sal*I, *Pst*I, *Hin*dIII	No	Cloning of promoters
pKK223−3	4.5	*amp*	No	*Eco*RI, *Sma*I, *Xma*I *Bam*HI, *Sal*I, *Acc*I, *Hin*cII, *Pst*I, *Hin*dIII	No	Over expression of proteins

pBS + and pBS− have the Fl origin of replication in opposite orientations. Therefore, when the phagemid is used to infect *E. coli* cells that carry an Fl or Ml3 helper phage, single-stranded DNA is produced of either the sense of anti-sense strand of the cloned insert. Because these phagemids are of small size (when compared to M13 phage, which is commonly used as the vector to produce single-stranded DNA), they offer the advantage of being able to clone larger inserts more stably.

pKK175−6

This plasmid is a pBR322 derivative containing a promoter-less tetracycline resistance gene. There is a multiple cloning site proximal to the tetracycline resistance gene. If a promoter sequence is cloned into this site it will drive expression of the *tet* gene, therefore allowing easy cloning of promoters that are active in *E. coli.*

pKK223−3

This vector is used for the over-expression of proteins. It contains a multiple cloning site just after the strong *tac* promoter. Genes containing a ribosome binding site and ATG start codon can be inserted into the *Pst*I or *Hin*dIII sites to be expressed. Expression is controlled as the *tac* promoter is under the control of *lac* repressor and can therefore be induced by isopropyl-β-D thiogalactopyroanoside (IPTG) which is a gratuitous inducer of the *lac* operon. The ribosome binding site on the plasmid can be utilized to express inserts cloned into the *Eco*RI and *Sma*I sites if the start codon of the insert is within 10−15 bp from the ribosome binding site.

The host strains and helper phages required for all the above plasmids are shown in Table 3. An explanation of the mutations is given in Tables 4 and 5.

TABLE 3. *Host strains, helper phages and availability*

Vector	Commonly used host or helper phage	Suppliers
λZAP	Many hosts, but most commonly XL1-blue which enhances the blue colour of non-recombinant phage XLI-blue genotype is *rec*Al, *end*Al, *gyr*A96, *thi*, *hsd*R17, (r*k*- mk+), *sup*/E44 *rel*Al, λ-*lac*- (F, *pro*AB, *lad*q Z Δ M15, Tn10 (*tet*^{R})	Stragene (UK Agents: Northumbria Biologicals Ltd)
	Helper phage is VCS-M13. This phage carries a modification in the M13 origin that allows preferential packaging of the vector	

TABLE 3. (contd.)

Vector	Commonly used host or helper phage	Suppliers
λgt11	Y1088 and other suitable *lac* hosts, the genotype of Y1088 is Δ (*lac*U 169), *sup*E, *sup*F, *hsd*R (r- m+), *met*B, *trp*R, *ton* A21, *pro*C :: Tn 5 (pMC 9) NB pMC 9 is pBR322 *lac* Iq	Available from most suppliers of molecular biological products
λgt10	As above	As above
λEMBL3 and 4	LE392 genotype: F-, *hsd*R514 (rk- mk+) *sup*E44, *sup*F58, *lac*YI or Δ (*lac*l ZY) 6, *gal*K2, *gal*T22, *met*B1, *trp*R55 λ- P2392 genotype: as above but carrying a P2 lysogen. λEMBL phage cannot grow on P2392 when it contains the stuffer region. Only recombinants can grow on this host	Available from most commercial outlets (Appendix I)
pHC79	Any *rec*A strain that will produce the λ receptor strain HB101 is commonly used: genotype of HB101. F-, *hsd* S20 (rB-, mB-), *sup*E44, *ara*14, λ-*gal*K2, *lac*Y1, proA2, *rps*L20, *xyl*-5, *mtl*-1, *rec*A13	Gibco BRL (see Appendix I)
pBR322	Any *E. coli* strain that is sensitive to tetracycline and ampicillin	Most suppliers (see Appendix I)
PBS (+/−)	JM101, JM109, NM522, XL1-blue: genotype of JM101: *sup*E, *thi* Δ (*lac* − *pro* AB) (F′, *tra*D36, *pro*AB, *lac*qZ Δ M15) genotype of JM109: *rec*Al, *end*Al, *gyr*A6l, *thi hsd* Rl7 (rk- mk+) *sup*E44, *rel*Al, λ- Δ (*lac* − *pro* AB) (F′, *tra* D36, *pro*AB, *lac*qZ Δ M15) NM522 genotype: *sup*E, *thi* Δ (*lac* − *pro*AB), Δ *hsd*5 (r- m-) (F′ *pro*AB *lac* qz Δ M15) helper phage VCS-M13	Stratagene (UK Agents: Northumbria Biologicals Ltd)
pUC family	Any *lac* Z-host carrying an episome capable of complementing the *lac* Z alpha region of the pUC plasmids, JM101 and JM109 are commonly used	Most suppliers (see Appendix I)
pKK175−6	Any tetracycline and ampicillin-sensitive strain	Pharmacia
pKK223−3	JM105: genotype *thi*, *rps*L, *end*A, *sbc*B15, *hsd*R4, Δ (*lac* − *pro*AB) (F′, *tra*D36, *lac*qZ, Δ M15, *pro* AB+)	Pharmacia

TABLE 4. *Host mutation descriptions*

Symbol	Phenotypic trait
*ara*C	Arabinase regulatory gene: activator and repressor protein
*ara*D	Arabinose L-ribulose phosphate 4-epimerase
*end*A	DNA specific endonuclease I/Mutation shown to improve plasmid mini-preparation yield and quality
*gal*k	Galactose galactokinase
*gal*U	Galactose: glucose-l-phosphate undylyltransferase
*gyr*A	DNA gyrase, subunit A: resistance or sensitivity to nalidixic acid
*lac*I	Lactose, regulatory gene, repressor protein of *lac* operon
*lac*Y	Lactose: galactosidase permease (M protein)
*lac*Z	Lactose-β-D-galactosidase/LacZ deletions produce white phenotype in the presence of Xgal. Wild type produces blue phenotype
*leu*B	Leucine: β-isopropylmalate dehydrogenase
*rec*A	A general recombination for repair of radiation damage and induction of phage λ/Mutation reduces general recombination to one thousandth its normal level, thus minimizing recombination between endogenous and exogenously added DNA
*rec*BCD	Recombination and repair of radiation damage: exonuclease V sub-unit/Mutation reduces general recombination to a hundredth its normal level
*rel*A	Regulation of RNA synthesis: stringent factor: ATP-GTP 3′-pyrophosphotransferase
*rps*L	Ribosomal protein, small; 30S ribosomal subunit protein S12
*sbc*B	Exonuclease 1; suppressor of *rec*B, *rec*C
*sup*E	Amber suppressor; suppresses amber (UAG) mutations
*sup*F	Amber suppressor; suppresses amber (UAG) mutations
thi-1	Thiamine thiazole requirement
*ton*A	Outer membrane protein receptor for ferrichrome, colicin M, and phages T1, T5 and ϕ80
*ton*B	Uptake of chelated iron and cyanocobalamin; sensitivity to phages T1 and ϕ80 and colicins
tsx	Resistance or sensitivity to phage T6
*xyl*A	Xylose; D-xylose isomerase

TABLE 5. *Mutations in restriction and modification genes*

Symbol	Phenotypic trait
dam	DNA adenine methylase/Mutation hinders *dam* gene product methylation of adenine residues in the recognition sequence 5′−GATC−3′
dcm	DNA cytosine methylase/Mutation hinders *dcm* gene product methylation of cytosine residues in the recognition sequence 5′−C*CAGG−3′ or 5′−C*CTGG−3′ (*methylated)
*hsd*M	Host specificity, host modification, DNA methylase M/Mutation hinders the sequence specific methylation modification function of the *hsd*M gene product
*hsd*R	Host specificity, host restriction, endonuclease R/Mutation hinders *Eco*K and *Eco*B restriction of foreign (unmodified) DNA by host
*hsd*S	Host specificity, specificity determinant for *hsd*M and *hsd*R/Mutation hinders the sequence specific DNA recognition function of the *hsd*S gene product
*mcr*A	*E. coli* restriction system/Mutation hinders wild type restriction of DNA modified by the *Hpa*II methylase
*mcr*B	*E. coli* restriction system/Mutation hinders wild type restriction of DNA modified by the *Hae*III. *Alu*I and *Msp*i methylases

Appendix I gives the address of a number of companies that provide the vectors discussed in this chapter. Appendix II gives the addresses of companies that supply other molecular biology products.

The Construction of Vectors for Use with the Clostridia

When introducing recombinant DNA into a host strain for which no existing recombinant DNA technology exists, a number of problems need to be overcome:

1 Introduction of foreign DNA into the host.
2 A means of detecting the DNA in the new host.
3 The newly introduced DNA must be able to replicate stably in the new host.

The authors have developed a protocol for introducing DNA into *Clostridium difficile* by electroporation. This is a technique whereby transient pores are induced in the cell membrane by passing a very high voltage through a dense cell suspension mixed with the transforming DNA. The DNA enters the cell through the transient pores.

The optimal conditions for electroporation can only be found by experimentation. For the clostridia the methods described here and the published methods (Oultram *et al.* 1988, Kim & Blascheck 1989), together with suggestions in the Bio-Rad manual can be taken as a starting point.

We have used the Bio-Rad gene pulser and pulse controller with the following protocol.

1 Inoculate 1 litre of Wilkins–Chalgren broth with *C. difficile* and incubate anaerobically at 37°C until late log-phase.

2 Resuspend pellet to a final volume of 4 ml in 1 mM HEPES NaOH pH 7 buffer.

3 Wash cells 6 times in 1 mM HEPES NaOH pH 7 buffer in 1.5 ml microcentrifuge tubes and finally resuspended in 1 ml of HEPES to a concentration of 10^{12} colony forming units/ml.

4 Add 100 μg of transforming DNA, to the cells and mix.

5 Add mixture to a sterile electroporation cuvette and give 1 pulse of between 0.5 and 6.25 kV cm^{-1} with the gene pulser apparatus set at 250 μF and the pulse controller at 200 Ω.

6 Immediately after pulsing, determine the percentage kill by removing a sample for a viable count.

7 Add 3 ml of Wilkins–Chalgren broth to the cells and allow to recover in the anaerobic chamber overnight at 37°C before plating onto selective medium.

Vectors used in electrotransformation of the Clostridia

The most useful vector to use to transform an organism with no previously characterized transformation system is one which will replicate in both *E. coli* and the new host(s). This will allow the construction of the vector and cloning of insert DNA to take place in *E. coli* where the sophisticated cloning systems, some of which were described above, can be utilized.

A plasmid constructed by Kim & Blascheck (1989) for *E. coli* and *C. perfringens* illustrates a useful approach to adopt when constructing a shuttle vector (Fig. 1). A cryptic *C. perfringens* plasmid was ligated into the *E. coli* vector pBR322 by digesting with the restriction enzyme *Eco*RV which cuts each plasmid once. The resultant plasmid pAK101 now has two origins of replication, one that is recognized in *E. coli* and the other in *C. perfringens*. A gene encoding chloramphenicol resistance that was expressed in both hosts was subsequently ligated into the plasmid. The resultant plasmid pAK201 encoded resistance to chloramphenicol in both *E. coli* and *C. perfringens*. The plasmid was also maintained stably in both hosts.

Prior to vector construction, where possible, it should be determined whether the antibiotic genes used are functional in both *E. coli* and the clostridial host.

For a theoretical discussion of the construction of vectors, see the recent review by Minton & Oultram (1988).

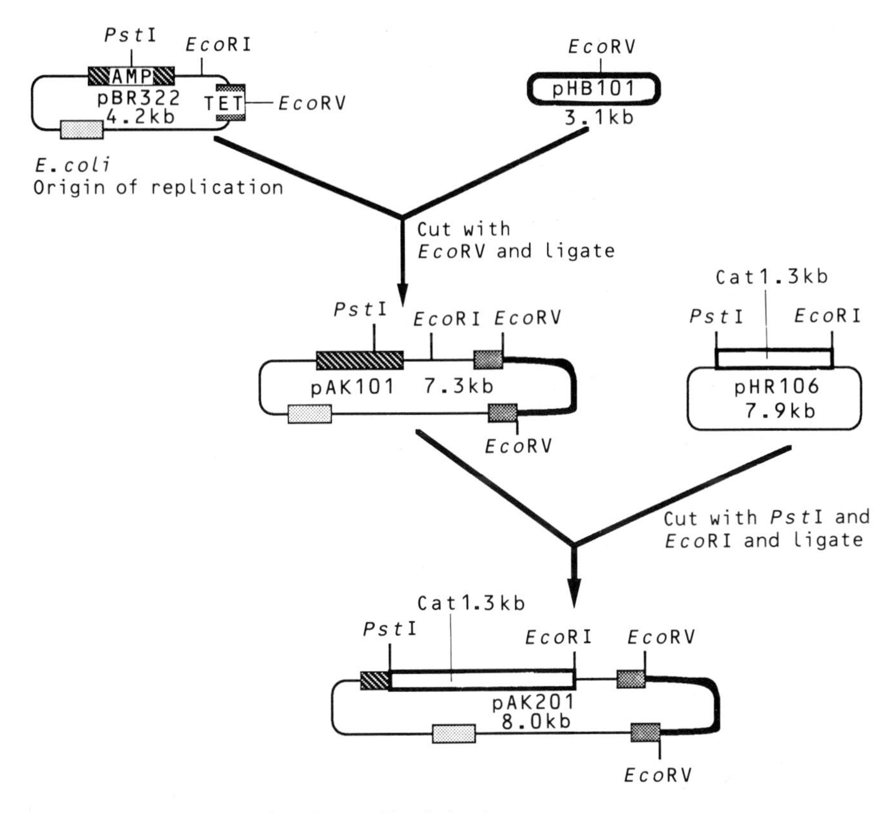

FIG. 1. Construction of a *C. perfringens/E. coli* shuttle vector.

References

KIM, A. & BLASCHECK, H.P. 1989. Construction of an *Escherichia coli—Clostridium perfringens* shuttle vector and plasmid transformation of *Clostridium perfringens*. *Applied and Environmental Microbiology* **55**, 360–365.

MINTON, N.P. & OULTRAM, J.D. 1988. Host: vector systems for gene cloning in *Clostridium*. *Microbiological Sciences* **5**, 310–315.

OULTRAM, J.D., LAUGHLIN, M., SWINFIELD, T-J., BREHEN, J.K., THOMPSON, D.E. & MINTON, N.P. 1988. Introduction of plasmids into whole cells of *Clostridium acetobutylicum* by electroporation. *FEMS Microbiology Letters* **56**, 83–88.

Appendix I: Addresses of Suppliers of Cloning Vectors

Amersham International PLC, UK and Export Sales Office, Lincoln Place, Green End, Aylesbury HP20 2TP. Telephone 0296–395222.

Boehringer, Mannheim House, Bell Lane, Lewes, East Sussex BW7 1LG. Telephone 0273–660444.
Cambridge BioScience, Newton House, 42 Devonshire Road, Cambridge CB1 2BL. Telephone 0223–316855.
Gibco BRL, P.O. Box 35, 3 Washington Road, Paisley PA3 4EP. Telephone 041–889–6100.
International Biotechnologies, 36 Clifton Road, Cambridge CB1 4ZR. Telephone 0223–242813.
Northumbria Biologicals Ltd, South Nelson Industrial Estate, Cramlington NW23 9HL. Telephone 0670–732992.
Pharmacia Ltd, Pharmacia House, Midsummer Boulevard, Milton Keynes MK9 3HP. Telephone 0908–661101.

Appendix II: Addresses of Suppliers of Other Products Useful for Molecular Biology

Bio-Rad Laboratories, Bio-Rad House, Maryland Avenue, Hemel Hempstead, Herts. Telephone 0442–232552.
Promega Biotech, UK Agents: P & S Biochemicals Ltd, 38 Queensland Street, Liverpool L7 3JG. Telephone 051–709–4701.

Adaptor Based cDNA Cloning in the Phage Vectors λgt10 and λgt11

C.J. CARTER AND A.J. HAYLE
Life Sciences Division, Amersham International PLC, White Lion Road, Amersham, Bucks HP7 9LL, UK

The ability to synthesize and clone complementary DNA (cDNA) copies of cellular messenger RNAs (mRNAs) has proved invaluable for analysing the structure, organization and expression of many eukaryotic genes (Jeffreys & Flavell 1979; March *et al.* 1985). The experimental techniques used are, however, complex and involve a series of enzymatic steps and manipulations (Maniatis *et al.* 1982; Perbal 1988). A population of single-stranded mRNA molecules, containing the sequence of interest, is first isolated from an appropriate cell type by established procedures (Perbal 1988). These molecules then act as templates for the synthesis of double-stranded cDNA molecules, which requires a number of sequential enzyme reactions (Boulnois 1987). This cDNA is next converted into a form suitable for cloning, either by attachment of synthetic linkers (Kurtz & Nicodemus 1981) or adaptors (Sartoris *et al.* 1987), or through the addition of homopolymer tails (Villa-Komaroff *et al.* 1978) to its ends. cDNA molecules, modified in this way, can be inserted into a suitable viral or plasmid-based cloning vector to form recombinant molecules, which can then be introduced into suitable host cells and allowed to replicate (Maniatis *et al.* 1982). Each infected cell contains only one recombinant cDNA species, but together these form a 'library' of cDNA clones. The cDNA clone of interest can then be isolated from this library, using an appropriate screening method (Perbal 1988). Since the frequency of occurrence of a particular cDNA clone in the library should reflect the abundance of that sequence in the original mRNA population, construction and screening of very large cDNA libraries (10^5-10^6 clones) will be required to isolate rare mRNA sequences, for example, of $< 0.01\%$ abundance (Williams 1981).

Genetic Manipulation
0-632-02926-9

Experimental Methods

The procedures outlined below permit high efficiency cloning of cDNA molecules (up to 10^8 clones/μg cDNA) by employing an adaptor-based cloning strategy in combination with the phage vectors λgt10 and λgt11. Figure 1 outlines the steps involved in using this approach.

Preparation of cDNA for cloning

Synthesis of cDNA

The cloning strategy described here is designed for cloning double-stranded cDNA molecules containing blunt-ends. The production of blunt-ended cDNA is not covered in detail here, but we can recommend several methods (Perbal 1988; Boulnois 1987; Watson & Jackson 1985) based on that originally described by Gubler & Hoffman (1983). This method involves the use of oligo(dT) or random hexanucleotides to prime first-strand cDNA synthesis from the 3′ to the 5′ end of the mRNA template, in the presence of the enzyme *reverse transcriptase*. The resulting first-strand cDNA is then converted into a double-stranded form by the combined action of RNaseH and DNA Polymerase-I and the ends are finally blunted by T_4 DNA polymerase. This method minimizes the loss of 5′ cDNA sequences which, in earlier methods, resulted from hairpin loop cleavage by S1 nuclease. It therefore produces a higher proportion of longer cDNA molecules.

The cDNA to be cloned should be radioactively labelled, by incorporating ^{32}P-labelled nucleotides, during first-strand synthesis (Boulnois 1987). This allows it to be monitored at various stages of the cloning process, and also provides a means of calculating the amounts of cDNA required for subsequent ligation reactions. This can be achieved by using the following calculation, which makes the assumption that only one labelled dNTP was present during first-strand cDNA synthesis.

If:	Percentage of the labelled dNTP incorporated	$= A\%$
	Amount of the same unlabelled dNTP present	$= B$ nmol
	Amount of this dNTP incorporated into cDNA	$= A\%$ of B nmol
		$= C$ nmol
Then:	total amount of all four dNTPs, incorporated	$= 4 \times C$ nmol
	Average weight of one nmol of dNMP	$= 350$ ng
	Weight of first-strand cDNA synthesized	$= 350 \times 4C$ ng
	Weight of double-stranded cDNA synthesized	$= 700 \times 4C$ ng
If:	Total amount of distintegrations per min (dpm) incorporated into cDNA	$= D$

Then: 1 ng of cDNA should contain $D/28C \times 10^2$ dpm of label

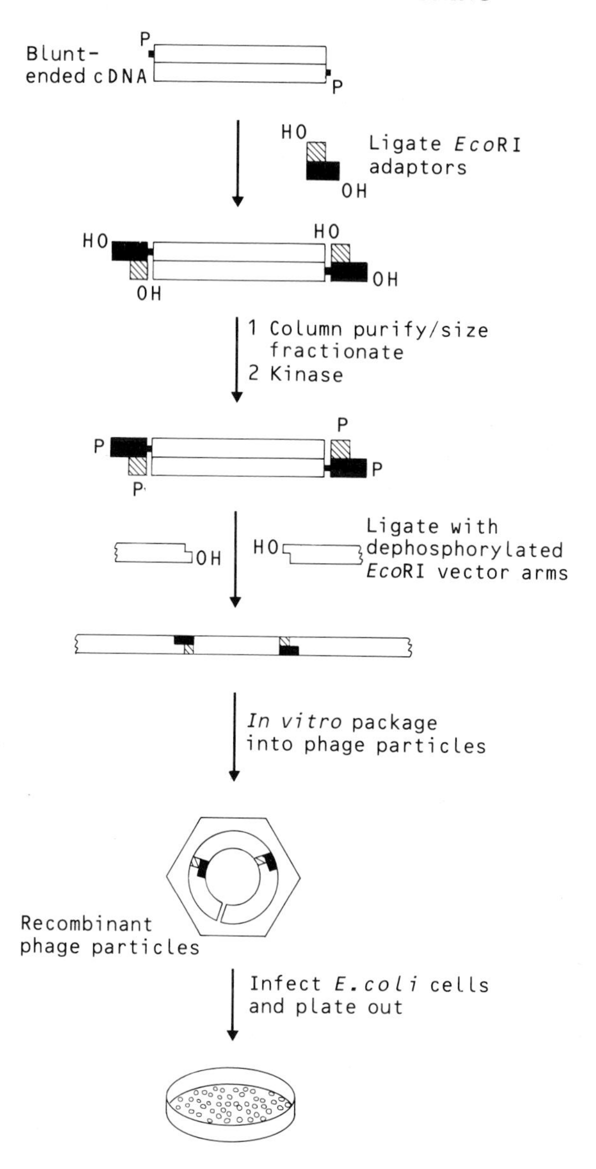

FIG. 1. Flow diagram outlining the steps involved in the adaptor cloning strategy.

cDNAs should be purified by standard methods of phenol extraction and ethanol precipitation (Maniatis *et al.* 1982) but should not be size fractionated, since this is incorporated into a later procedure. The cloning process should not be started with less than 500 ng of purified cDNA, since losses will be experienced at various stages of the procedure, particularly during the precipitation steps. Purified cDNA should be resuspended in TE buffer (10 mM Tris–HCl pH 8, 1 mM EDTA) at a final concentration of 50–100 ng/μl.

Blunt-ended DNA control

A suitable blunt-ended DNA control (for example, M13 RF DNA digested with the restriction enzyme *Hae*III) should be available to take through the whole cloning process in parallel with the cDNA. This will serve as a control cDNA-like molecule to monitor the various steps of the cloning procedure. Ideally, this control DNA should be radioactively labelled before starting the cloning process, and this is most conveniently achieved by replacing the 3′ terminal nucleotide with the corresponding ^{32}P-labelled nucleotide, using the Klenow exchange labelling reaction (Perbal 1988).

Use of oligonucleotide adaptors

The adaptor cloning strategy employs synthetic oligonucleotide adaptors to convert the blunt-ends of the cDNA molecules into cohesive ends, compatible with the cloning site in the vector (Sartoris *et al.* 1987; Perbal 1988). Adaptor molecules are formed by annealing together two short complementary single-stranded oligonucleotides to produce a double-stranded molecule, blunt at one end, for ligation to cDNA, and carrying the appropriate restriction site cohesive end (for example *Eco*RI), for ligation to the vector, at the other (Haymerle *et al.* 1986). Adaptors are preferred to linkers (which carry the restriction site *internally*) since they avoid the need for protective methylation of the cDNA and subsequent digestion with a restriction enzyme to generate the required cohesive end. A typical adaptor molecule, containing an *Eco*RI cohesive terminus, is illustrated in Fig. 2a. It is commercially available from New England Biolab's. Alternatively, if suitable equipment and knowhow are available, it is possible to synthesize adaptor molecules specifically designed for use with a particular cloning vector/system. For example, we have designed and synthesized an adaptor specifically for use with the phage vectors λgt10 and λgt11. This adaptor (see Fig. 2b) contains additional restriction sites which are most useful for assisting in subsequent clone analysis and subcloning procedures.

Addition of adaptors to blunt-ended cDNA

The first step in the cloning procedure involves the addition of a single adaptor molecule to each end of the blunt-ended cDNA. This is a ligation reaction, using the enzyme T_4 DNA ligase, and it requires the 5′-end of the participating DNA strand to carry a phosphate group (Maniatis *et al.* 1982).

```
(a)        EcoRI
     HO  |AA  TTC GAA CCC CTT CG
                 G CTT GGG GAA GC  OH

(b)        EcoRI            BamHI    KpnI  NcoI
     HO  |AA  TTC GAG |GAT CCG GGT AC|C ATG G
                 G CTC CTA G|GC C|CA TGG TAC| C OH
```

FIG. 2. Synthetic oligonucleotides for use in the adaptor cloning strategy. The nucleotide sequences of each adaptor, together with any restriction sites, are shown. (a) New England Biolab's adaptor sequence. (b) Amersham International's adaptor sequence.

The adaptors used in these experiments are deliberately produced without 5′ phosphate groups, to prevent their self-ligation in this reaction (Haymerle *et al.* 1986). The blunt-ended cDNA molecules in the reaction do, however, contain 5′ phosphate groups so that each of their ends can ligate a single adaptor molecule, although covalent linkage can only occur through one of the two adaptor strands. The other adaptor strand should, however, still remain attached through hydrogen-bonding during the subsequent processing steps (Haymerle *et al.* 1986). Self ligation of the blunt-ended cDNA molecules is minimized by using a high molar excess of unphosphorylated adaptors to blunt-ended cDNA in the ligation reaction (Sartoris *et al.* 1987).

The adaptor ligation reaction is performed in a final volume of 20 μl, containing ligation buffer (10 × ligation buffer contains: 500 mM Tris–HCl pH 7.8, 100 mM $MgCl_2$, 10 mM ATP, 100 mM β-mercaptoethanol, 500 μg bovine serum albumin/ml), cDNA *or* control blunt-ended DNA (up to 1 μg), *Eco*RI adaptors (250 pmoles/μg blunt-end DNA) and 5 units of T_4 DNA ligase (Amersham: code no. T-2010). This reaction mixture should be incubated at 15°C for 16–20 h and the reaction is then stopped by adding 2 μl of 0.25 M EDTA. The reaction should *not* be inactivated by heating. If not proceeding directly to the next step, the reaction mix can be stored in this form at −20°C for several days.

Since only the blunt-end of an adaptor can ligate to the blunt-ended cDNA, it will always attach in the same orientation to produce 'adapted' cDNA molecules carrying preformed *Eco*RI cohesive termini. This 'adapted' cDNA is now in a form which, after further purification, can be inserted into suitably prepared cloning vector DNA.

Column purification of 'adapted' cDNA

A purification step is required at this stage to remove the large excess of unreacted adaptor molecules from the now 'adapted' cDNA. This step is necessary to prevent free adaptors from carrying through the subsequent cloning steps and producing 'false recombinants' in the final cDNA library (Perbal 1988). Removal of unwanted free adaptors is most conveniently and efficiently achieved by using a column purification step (Huynh *et al.* 1985). Suitable column materials, effective in separating small DNA molecules (i.e. linkers or adaptors) from larger DNA molecules (i.e. adapted cDNA) are commercially available, e.g. Sepharose CL 4B and Sephacryl S-200 (Pharmacia). Other matrices available, e.g. Biogel A50M (Bio-Rad) and Sephacryl S-1 000 (Pharmacia), are also equally effective for this purpose, but they additionally size-fractionate the cDNA at the same time. This latter procedure, whilst reducing the amount of DNA available for cloning, does ensure that only the larger cDNAs (> 500 bp) are processed further. We routinely use such a size-fractionating column matrix, which is available in prepacked columns, but only as a component of Amersham's cDNA cloning kits. The protocol for column purification presented below relates specifically to the use of Sephacryl S-1 000 and may require modification if an alternative column matrix is employed.

Suitable purification columns can be prepared by pouring about 1 ml of preswollen Sephacryl S-1 000 into a 2 ml plastic syringe, clamped vertically and plugged with siliconized glass wool. As the gel matrix settles, check that no trapped air bubbles are present and ensure that the column is flowing freely. Wash the column, by passing through about 5 bed volumes of elution buffer (TE pH 8). Meanwhile, prepare a rack containing 30 microcentrifuge tubes, suitably labelled, for collecting column fractions, and position this beneath the column. Allow the buffer level on the final wash to just reach the top of the gel and then load all 22 μl of the ligation reaction. Allow this to soak in briefly, add a further 200 μl of TE buffer and then immediately start collecting 3 drop (120 μl) fractions into the labelled collection tubes. As soon as the 200 μl of TE buffer has soaked into the column, completely fill the column with more TE buffer and continue collecting 3 drop fractions up to tube 30. Add more buffer to the column, as required, so that it does not run dry. Use a different column for each separate ligation reaction and perform column separations at room temperature.

Since the adapted cDNA is radioactively labelled with ^{32}P-dNTPs, its distribution within the column fractions can be conveniently detected by Cerenkov counting (Maniatis *et al.* 1982). In order to recover cDNA free from unligated adaptors, pool together consecutive fractions, starting with the first fraction containing radioactivity, until the pooled material comprises no more than 40% of the total radioactivity in all the fractions. This should require pooling about 8 fractions (typically fractions 10–17). Later fractions should not be included within this pool since they will contain increasing amounts of adaptors which, being unlabelled, will not be detectable. Accurately measure the amount of radioactivity present in a 10 μl sample of the pooled material by scintillation counting and determine the amount of cDNA present by means of its specific activity (calculated as described earlier).

Addition of 5′ phosphate groups to 'adapted' cDNA

'Adapted' cDNA molecules, contained in the appropriate pooled column fractions, possess *Eco*RI cohesive termini which carry 5′ hydroxyl groups. These must next be converted to 5′ phosphate groups prior to their ligation with vector molecules, and this can be achieved by incubation with the enzyme T_4 polynucleotide kinase in a suitable buffer containing ATP as phosphate donor (Perbal 1988). The pooled column fractions must first be adjusted by adding 0.2 volumes of 5 × kinase buffer (200 mM Tris–HCl pH 7.6, 50mM $MgCl_2$, 25 mM dithiothreitol, 5 mM ATP). 200 units of T_4 polynucleotide kinase (Amersham: code no. T-2020) per ml of reaction are then added and the reaction mix is incubated at 37°C for 30 min. The reaction mix is then extracted with an equal volume of phenol:chloroform (1:1) twice, followed by with an equal volume of chloroform:isoamylalcohol (24:1) twice, to remove all traces of the kinase. After each extraction, the sample tube should be spun for 1 min in a microcentrifuge, to achieve phase separation. Following centrifugation, remove and discard the lower organic phase and retain the upper aqueous phase.

Recovery of 'adapted' cDNA

The volume of each extracted reaction mix (~1 ml) should next be reduced (to about 350 μl) to allow the purified 'adapted' cDNA to be precipitated in a single microcentrifuge tube. This is most conveniently achieved by further extracting the aqueous sample with butan-1-ol at room temperature (C.R. Mundy, personal communication). This extraction reduces the volume, by removing water into the organic (upper) phase (Wallace 1987), and each volume of butan-1-ol added will extract approximately 0.2 volumes of water. Two extractions, each with 2 volumes of butan-1-ol, should therefore be sufficient to reduce a volume of 1 ml. The volume of butan-1-ol and/or the number of extractions should, of course, be adjusted for different aqueous starting volumes. For each extraction, add the appropriate volume of butan-1-ol to the sample tube, recap the tube and shake the contents vigorously for a few seconds. Allow the phases to separate and then remove and discard the upper butanol layer and repeat the process as required.

The adapted cDNA in ~350 μl is finally precipitated by adding 0.1 volume of 3 M sodium acetate (pH 6.0) and 2.5 volumes of ethanol and leaving overnight at −20°C. The DNA pellet should be collected by centrifugation for 30 min in a microcentrifuge, briefly dried under vacuum and then resuspended in TE buffer (pH 8.0), at a concentration of approximately 20 ng/μl. This purified adapted cDNA is now ready for insertion into a suitable cloning vector and can be stored in this form at −20°C until required.

Choice and preparation of phage vectors for cDNA cloning

Advantages of using phage vectors

The use of λ-phage, rather than plasmid, cloning vectors for constructing cDNA libraries offers two important advantages to the researcher. Firstly, λ-phage DNA can be introduced into *E. coli* cells very efficiently and reproducibly by a technique known as *in vitro* packaging (Hohn 1979). Since high quality λ-*in vitro* packaging extracts are commercially available, this technique generally proves more reliable, efficient and convenient than the alternative of transforming competent cells with plasmids (Huynh *et al.* 1985). The very high cloning efficiencies obtainable with λ-phage vectors are particularly useful for constructing large cDNA libraries for the isolation of cDNA clones for very low abundance mRNA species (Williams 1981). Secondly, λ phage cDNA libraries form plaques which can be grown on plates at much higher densities ($2 \times 10^4 - 6 \times 10^4$ plaques/90 mm plate) than bacterial colonies (5×10^3 colonies/90 mm plate) (Davis *et al.* 1980). The transfer of phage plaques to filter membranes is efficient and sufficiently specific to ensure that plaque filters can be screened at very high plaque densities without significantly increasing the background (Benton & Davis 1977). This ability to perform high density plaque screening enhances the convenience of phage-based systems by reducing the number of plates, and hence filters, that require screening.

Choice of phage vector

The well-known and highly successful phage cloning vectors λgt10 and λgt11 possess several properties which make them ideal for cloning 'adapted' cDNA molecules. Phage libraries prepared with the vector λgt10 form large plaques of uniform size which are ideal for screening with nucleic acid probes (Huynh *et al.* 1985). Phage libraries prepared with the vector λgt11 can be screened with antibody probes since this expression vector produces a fusion peptide partly encoded by the cloned DNA fragment (Young & Davis 1983). The choice between these two cloning vectors will, therefore, be largely determined by the type of probe available for screening the cDNA library once this is produced.

Production of phage vector 'arms'

Both λgt10 and λgt11 carry a unique *Eco*RI cloning site, and cleavage with this restriction enzyme generates two vector 'arms', between which the 'adapted' cDNA can be inserted. The cleaved vector arms possess 5′ phosphate groups at their *Eco*RI-ended termini and can be re-ligated, in the absence of 'adapted' cDNA molecules, to produce functional vector molecules. This will still occur, to some extent, in the presence of 'adapted' cDNA molecules and gives rise to non-recombinant background plaques in the resulting cDNA library. Such backgrounds can be significantly reduced by treating the *Eco*RI-digested vector arms with the enzyme alkaline phosphatase, which converts the 5′ phosphates to 5′ hydroxyl groups (Perbal 1988). Although dephosphorylated vector arms cannot self-ligate, they can ligate with kinased 'adapted' cDNA molecules, carrying 5′ phosphate groups, to produce 'recombinant' molecules.

Preparation of phage vector DNA, production of *Eco*RI-ended vector arms and their subsequent dephosphorylation are all standard procedures and detailed protocols have been described elsewhere (Huynh *et al.* 1985; Davis *et al.* 1980).

Construction of cDNA libraries with λ-phage vectors

The next step in the cloning process involves ligation of the 'adapted' cDNA molecules possessing 5′ phosphorylated *Eco*RI cohesive termini, with *Eco*RI digested, 5′ dephosphorylated vector arms. The efficiency of this reaction will determine the final cloning efficiency and should therefore be optimized. Theoretically, a ratio of 2 vector arms to 1 cDNA molecule should be optimal, but since not every cDNA molecule is ligatable, more cDNA is often required (Maniatis *et al.* 1982). The optimum amount of cDNA will differ between samples but usually lies between 25 and 120 ng per microgram of arms, for a cDNA population of average size range 0.5–5 kb. At least two separate ligation reactions, containing different amounts (usually 25 and 50 ng) of cDNA, should therefore be set up for each cDNA sample.

Use of control ligations

Four additional control ligations should also be included at this stage. These are important for monitoring various steps in the cloning process and will provide additional information necessary to analyse and interpret the results. The ligated controls should comprise:

1 Uncut vector DNA, to measure the efficiency of the *in vitro* packaging reactions.

2 Vector arms *minus* cDNA, to monitor the non-recombinant background.
3 Vector arms *plus* a control *Eco*RI-ended DNA, (e.g. *Eco*RI-digested pAT153), to determine the efficiency of the ligation reaction.
4 Vector arms *plus* the 'adapted' blunt-ended DNA control, to check the performance of the whole cloning process.

These various ligation reactions should be set up as indicated in Table 1 and all reactions should be performed in 1.5 ml microcentrifuge tubes and incubated at 15°C for 16–20 h. The concentration of phage arms is kept at 1 μg/10 μl ligation since this is optimal for the formation of linear concatamers of λ-DNA, which is the most efficient *in vitro* packaging substrate.

Introduction of recombinant DNA into host cells

Following the ligation reaction, recombinant phage DNAs should be assembled into active phage particles, using the highly efficient method of *in vitro* packaging. *In vitro* packaging extracts can either be obtained commercially (from Amersham (code no. N334), Stratagene or Promega Biotec) or prepared by established protocols (Hohn 1979). The infective phage particles that result from the *in vitro* packaging process should contain mostly recombinant

TABLE 1. *Ligation reactions*

Tube	Insert DNA	Undigested Phage DNA	Phage arms (μg)	Ligation buffer (10 ×) (μl)	Water	T4 DNA ligase (units)
1	–	1 μg	–	1	to 10 μl total volume	2.5
2	–	–	1	1	to 10 μl total volume	2.5
3	100 ng *Eco*RI-ended control DNA	–	1	1	to 10 μl total volume	2.5
4	50 ng 'adapted' blunt-end control DNA	–	1	1	to 10 μl total volume	2.5
5	*x* ng 'adapted' cDNA	–	1	1	to 10 μl total volume	2.5
6	*y* ng 'adapted' cDNA	–	1	1	to 10 μl total volume	2.5

phage DNA with integrated cDNA molecules. The total number of active phage particles/ml and the percentage of recombinants present should, however, be determined at this stage by infection of an appropriate bacterial strain with 'packaged' phage particles and counting the number of phage plaques that result after overnight growth on agar plates. This can be achieved by first preparing a series of 10-fold dilutions (10^2-10^7) for each packaged phage reaction. 100 μl of the appropriate dilution (see Table 2) should then be mixed with 100 μl of plating cells (prepared according to Davis *et al.* 1980: for strains see Tables 3 and 4) and incubated at 37°C for 15 min to allow phage adsorption/infection to take place. Each infected cell incubation is then mixed with 4 ml of soft L-agar (0.8%) and plated onto an L-agar plate. These plates are then incubated overnight at 37°C or 43°C for λgt10 or λgt11 infected cells respectively to allow plaque growth to occur. The phage titre/ml packaged phage reaction is then calculated by multiplying the total number of plaques on a plate by the dilution number plated. For example, 200 plaques on the 10^5 dilution plate represents a phage titre of 2×10^7 plaque forming units (pfu)/ml. *In vitro* packaged phage stocks will remain stable for several months when stored at + 4°C.

Identification of recombinants

Both phage vectors have their unique *Eco*RI cloning sites located in a different functional gene and insertion of foreign DNA at this site results in the production of an inactive gene product in each case (Huynh *et al.* 1985). The ability to monitor this loss of activity provides a way of distinguishing between recombinant and non-recombinant phage.

The *Eco*RI site of λgt10 lies within its cI gene, whose normal peptide product, the cI repressor, promotes phage lysogeny in infected cells (Murray *et al.* 1977). In wild type *E. coli* cells, for example, L87 and C600, lysogeny is incomplete and results in the formation of 'turbid' plaques. Complete lysogeny can however occur by infecting *E. coli* cells, for example, NM514 and C600

TABLE 2. *Titration of packaged phage*

In vitro packaged DNA samples	Dilutions to plate	
	L87/Y1090	NM514
λgt10 DNA	10^5, 10^6, 10^7	10^3, 10^4, 10^5
λgt10 arms	10^2, 10^3, 10^4	10^2, 10^3, 10^4
λgt10 arms + control *Eco*RI insert	10^4, 10^5, 10^6	10^4, 10^5, 10^6
λgt10 arms + cDNA or blunt-end control	10^4, 10^5, 10^6	10^4, 10^5, 10^6

TABLE 3. *cDNA cloning—example of results obtained with λgt10*

Tube	λDNA	Insert	L87 titre pfu/μg	NM514 titre pfu/μg arms	Percentage recombinants (on NM514)	Recombinants pfu/μg insert (on NM514)
1	λgt10	—	5.7×10^8	9.8×10^5	—	—
2	λgt10 *Eco*RI arms	—	4.3×10^4	1.5×10^4	—	—
3	λgt10 *Eco*RI arms	100 ng *Eco*RI-ended DNA	3.8×10^7	2.9×10^7	99.95	2.9×10^8
4	λgt10 *Eco*RI arms	50 ng blunt-ended DNA	8.6×10^6	5.9×10^6	99.75	1.2×10^8
5	λgt10 *Eco*RI arms	25 ng cDNA	2.5×10^6	1.4×10^6	98.93	5.5×10^7
6	λgt10 *Eco*RI arms	50 ng cDNA	5.1×10^6	3.2×10^6	99.53	6.4×10^7

TABLE 4. *cDNA cloning—example of results obtained with λgt11*

Tube	λDNA	Insert	Y1090 pfu/μg arms titre	Percentage recombinants	Cloning efficiency pfu/μg insert
1	λgt11	—	3.9×10^8	—	—
2	λgt11 *Eco*RI arms	—	4.4×10^4	—	—
3	λgt11 *Eco*RI arms	100 ng *Eco*RI-ended DNA	3.1×10^7	99.3	3.1×10^8
4	λgt11 *Eco*RI arms	50 ng blunt-ended DNA	9.5×10^6	98.7	1.9×10^8
5	λgt11 *Eco*RI arms	25 ng cDNA	1.8×10^6	97.7	7.2×10^7
6	λgt11 *Eco*RI arms	50 ng cDNA	4.2×10^6	98.2	8.4×10^7

hfl, containing the *h*igh *f*requency of *l*ysogeny (*hfl*+) mutation (Hoyt *et al.* 1982), and no plaque formation is observed. Insertion of foreign DNA into this site results in the production of inactive cI repressor and leads to lytic phage growth with formation of 'clear' plaques on both cell types. These differences provide a clear visual distinction between non-recombinant (turbid plaques) and recombinant (clear plaques) phages on strain L87 (or C600). In contrast, use of NM514 (or C600 *hfl*) cells actually provides a stringent biological selection system in which only recombinant phages form plaques, which are always clear. This ensures a low non-recombinant background even when using vector arms that have not been dephosphorylated. This background can be further reduced by using phage arms lacking 5′ phosphate groups.

The *E. coli* host strains L87 and NM514 are not, at present, commercially available, but comparable cloning efficiencies can be obtained with alternative strains, C600 and $C600_{hfl}$, which are commercially available from Stratagene Cloning Systems, 11099 North Torrey Pines Road, La Jolla, California 92037, USA.

The *Eco*RI site of λgt11 lies within its *lacZ* gene which encodes the enzyme β-galactosidase, expression of which is under *lac* promoter control (Young & Davis 1983). Since this phage is normally grown on *E. coli* strain Y1090 which contains *lac* repressor, the expression of this gene is normally repressed, until a synthetic inducer, such as isopropyl-β-D-thiogalacto-pyranoside (IPTG) (Sigma Chemical Co.), is added (Miller & Reznikoff 1978). Under these conditions, phage plaques producing active β-galactosidase (non-recombinants) can be identified by the ability of this enzyme to form a blue colour when grown in the presence of the chromogenic substrate 5-bromo-4-chloro-3-indolyl-β-D-galactopyranoside (X-gal) (Sigma Chemical Co.). Insertion of foreign DNA into the cloning site leads to the production of an inactive fusion peptide unable to form the blue colour with X-gal.

In the presence of both IPTG and X-gal, non-recombinant phage plaques, expressing active β-galactosidase, appear blue, whilst recombinant phage plaques appear colourless, and this provides a visual means for determining the percentage of recombinants present. Since non-recombinant phages still grow on strain Y1090, it offers no biological selection for λgt11 recombinants and the non-recombinant background can only be reduced by using phage, arms lacking 5′ phosphates. The *E. coli* host strain Y1090 (*hsd*($r_k^- m_k^+$), *lon*$^-$, *Sup*F, pMC9) is commercially available from Pharmacia.

Results of cDNA Library Analysis

Analysis of controls

The results obtained with the various control reactions should be analysed to determine how well the cloning procedure has performed. Cloning efficiencies can be calculated as pfu/μg phage DNA or, where appropriate, pfu/μg insert DNA. Since 1 μg of phage DNA is packaged *in vitro* and diluted out to 0.5 ml, the cloning efficiency (pfu/μg phage DNA) = phage titre/ml × 0.5.

The function of each control, together with the range of cloning efficiencies predicted if the system is working well, is given below.

1 *Whole phage DNA control*: this measures the efficiency of the *in vitro* packaging reaction and should produce a packaging efficiency of $>3 \times 10^8$ pfu/μg phage DNA with either phage vector. With λgt10, the ratio between the titre on L87 (non-selective host) and NM514 (selective host) indicates the degree of biological selection obtained, and should be >100. With λgt11, >99% blue (non-recombinant) plaques should be observed.

2 *Phage arms only control*: since the arms have been dephosphorylated and cannot self-ligate, the packaging efficiency, with either phage in the appropriate cell line, should be $<5 \times 10^4$ pfu/μg phage arms. This should give a direct measure of the non-recombinant background.

3 *Phage arms plus* EcoRI*-ended insert control*: this monitors the efficiency of the ligation reaction. The cloning efficiency should be $>3 \times 10^6$ pfu/μg phage arms, with either phage, and should be significantly greater than that observed with control 2. With λgt10, the ratio of NM514 titres from control 3 and control 2 should be >100. With λgt11, >95% colourless (recombinant) plaques should be observed.

4 *Phage arms plus adapted blunt-ended control*: this monitors the performance of the whole cloning process and should produce cloning efficiencies of $>1 \times 10^6$ pfu/μg phage arms, with either phage.

Analysis of cDNA results

When calculating the results from the cDNA containing reactions, the cloning efficiency/μg cDNA =(number of recombinants × 1 000)/(ng cDNA used). This value should be $>1 \times 10^6$ pfu/μg cDNA.

The results of typical experiments obtained in our laboratory, using the cloning strategy described above to clone cDNA, synthesized from a mouse skeletal muscle mRNA population, in both λgt10 and λgt11, are presented in Table 3 and Table 4 respectively.

Ligation of λgt10 arms with 'adapted' cDNA produced cloning efficiencies, on NM514 cells, of $5.5 \times 10^7 - 6.4 \times 10^7$ pfu/μg cDNA, which are only slightly below that observed for the 'adapted' blunt-end control (1.2×10^8 pfu/μg insert DNA). All cDNA cloning reactions produced an excellent percentage (>98.9%) of recombinants, which reflects the very low background (1.5×10^4 pfu/μg λgt10 arms) on NM514 cells, achieved by a combination of biological selection and dephosphorylation.

Ligation of λgt11 arms with 'adapted' cDNA resulted in cloning efficiencies of $7.2 \times 10^7 - 8.4 \times 10^7$ pfu/μg cDNA, slightly below that (1.9×10^8 pfu/μg insert DNA) observed with the 'adapted' blunt-end control. All cDNA cloning reactions also produced >97% colourless (recombinant) plaques, reflecting the very low background (4.4×10^4 pfu/μg λgt11 arms).

Physical analysis of cDNA libraries

The quality of the cDNA libraries produced using this adaptor cloning strategy can be further assessed by determining the size range of cDNA inserts present in at least 12 clones, picked at random from each library. The physical analysis of these clones requires the small-scale isolation of recombinant phage DNA from high titre phage lysates (>2×10^{10} pfu/ml), prepared most efficiently using a plate-lysate method (Davis *et al.* 1980). The resulting mini-lysate phage DNAs can then be analysed by digestion with suitable restriction enzymes, in order to detect the presence and size of their cDNA inserts. For example, digestion with *Eco*RI restriction enzyme (Amersham: code no. T-1040) releases any insert DNA from the vector arms (32.7 kb and 10.6 kb for λgt10 or 24.1 kb and 19.6 kb for λgt11). Since small cDNA inserts (<1 kb) comprise only a small proportion (<2%) of the λ-DNA isolated, they may prove difficult to visualize on gels. They can, however, be labelled by 'filling in' the cohesive ends of the restricted DNA fragments with an appropriate radioactive nucleotide by means of the Klenow fragment of DNA Polymerase–I (Amersham: code no. T-2141) (Maniatis *et al.* 1982), and subsequently visualized by autoradiography. A typical autoradiograph of *Eco*RI-digested phage DNA, derived from clones picked from the 50 ng cDNA library in λgt10 (Fig. 3), shows that 9 of the 10 clones processed contained cDNA inserts ranging in size from 0.5 kb to 4 kb.

From each set of ligations, the best cDNA library, containing the highest percentage of recombinants with the highest average insert size, should then be screened by an appropriate method to isolate the clones of interest.

Methods of screening phage libraries

The biological selection of recombinant phage combined with large uniform plaque size are two features of cDNA libraries cloned in λgt10 which make them ideal for screening with nucleic acid probes. This technique, known as *hybridization screening*, is well developed and protocols for preparing filter plaque replicas and for performing hybridization reactions have been described elsewhere (Benton & Davis 1977). The exact hybridization protocol will depend on the type of nucleic acid probe being used and the parameters involved have been reviewed (Meinkoth & Wahl 1984).

Since the λgt11 cloning vector is designed to permit controlled expression

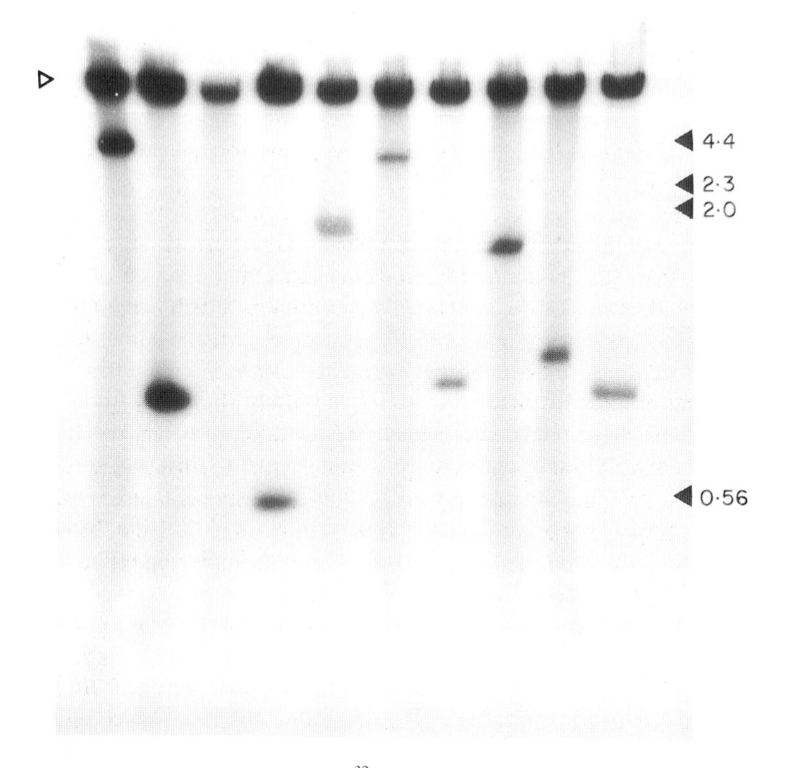

FIG. 3. Autoradiograph of *Eco*RI-digested ^{32}P-labelled λgt11 mini-prep DNA samples. DNA purified from individual recombinant plaques of the 50 ng cDNA library (Table 4) was digested with *Eco*RI and labelled with ^{32}P by means of Klenow fragments as described in the protocols. After electrophoresis through a 1.5% agarose gel, the gel was dried and the autoradiograph was exposed for 2 h. The positions of the molecular weight markers (◀ kb) and the λgt11 *Eco*RI arms (▷) are marked.

of recombinant peptides fused to β-galactosidase, cDNA libraries prepared with this vector are usually screened with a specific antibody probe. This method of screening is known as expression screening (or immunoscreening) and detailed protocols for performing this technique have been published (Huynh *et al.* 1985).

Discussion

Addition of clonable ends to blunt-ended cDNA usually involves the covalent attachment of 5′ phosphorylated linkers. A subsequent restriction digestion is then required to generate a cohesive end and, if possible, the substrate DNA would have been treated previously with an appropriate methylase to prevent this digestion cleaving internal sites. In the studies presented here we have made use of synthetic 'adaptors' carrying a preformed cohesive end, thereby eliminating the need for these two steps and considerably simplifying the procedure.

During ligation with blunt-ended DNA, phosphorylated linkers will also undergo self-ligation to form multimers which are normally removed by the restriction digestion step. This problem can be avoided by using adaptors carrying only one 5′ phosphate, at their blunt-ends (Sartoris *et al.* 1987), and will ensure that only one adaptor molecule adds to each end of the target DNA. Such adaptors, however, can still form dimers, which cannot be cleaved by restriction enzymes and are more difficult than single adaptors to separate away from 'adapted' cDNA by the standard column procedures. Any not removed will compete with the 'adapted' cDNA during subsequent cloning procedures. In the cloning strategy described here, we have used adaptor molecules lacking 5′ phosphate groups at both ends, to ensure single adaptor ligation whilst avoiding dimer formation. Ligation of such adaptors with blunt-ended cDNA, carrying 5′ phosphates, results in covalent linkage through only one of the adaptor-strands, whilst the other strand remains annealed throughout the subsequent cloning procedures.

Removal of excess unreacted adaptor molecules is achieved by a column purification procedure. The column used by us additionally size-fractionates the 'adapted' cDNA to ensure that only the larger molecules (>500 bp) are retained for cloning. We also include a novel butanol extraction step to concentrate the size-fractionated DNA, allowing it to be recovered in a single microcentrifuge tube.

The novel features described above have been combined with advantages offered by phage cDNA cloning vectors to develop a unique cDNA cloning strategy capable of producing representative cDNA libraries at high efficiency (up to 10^8 clones/μg cDNA) and of sufficient size and complexity to permit the isolation of cDNA clones of low abundance ($<0.01\%$) mRNA species from a typical eukaryotic cell. Finally, the success of each step of the cloning process can be closely monitored by including the various control reactions indicated.

Acknowledgements

The authors wish to express their gratitude to Drs Chris Mundy, Martin Cunningham and Kevin McFarthing for helpful advice and discussion during the preparation of this chapter, and to Jane Palmer and Kath Cooper for their experimental contributions. We would also like to thank Mrs Molly Holdsworth for her efforts in typing this manuscript and finally Amersham International PLC for permission to publish this work.

References

BENTON, W.D. & DAVIS, R.W. 1977. Screening λgt recombinant clones by hybridization to single plaques *in situ*. *Science* **196**, 180–182.

BOULNOIS, G.J. 1987. *Gene Cloning and Analysis: A Laboratory Guide*. Oxford: Blackwell Scientific Publications.

DAVIS, R.W., BOTSTEIN, D. & ROTH, J.R. 1980. *Advanced Bacterial Genetics: A Manual for Genetic Engineering*. Cold Spring Harbor. New York: Cold Spring Harbor Laboratory.

GUBLER, U. & HOFFMAN, B.J. 1983. A simple and very efficient method for generating cDNA libraries. *Gene* **25**, 263–269.

HAYMERLE, H., HERTZ, J., BRESSAN, G.M., FRANT, R. & STANLEY, K.K. 1986. Efficient construction of cDNA libraries in plasmid expression vectors using an adaptor strategy. *Nucleic Acids Research* **14**, 8615–8624.

HOHN, B. 1979. *In vitro* packaging of λ and cosmid DNA. *Methods in Enzymology* **68**, 299–309.

HOYT, M.A., KNIGHT, D.M., DAS, A., MILLER, H.I. & ECHOLS, H. 1982. Control of phage development of stability and synthesis of CII protein: Role of the viral cIII and host hflA, and himA and himD genes. *Cell* **31**, 565–573.

HUYNH, T.V., YOUNG, R.A. & DAVIS, R.W. 1985. Constructing and screening cDNA libraries in λgt10 and λgt11. In *DNA Cloning, a Practical Approach*. Vol 1. ed. Glover, D. pp. 49–78. Oxford: IRL Press.

JEFFREYS, A.J. & FLAVELL, R.A. 1979. Rabbit Beta-globin gene contains a large insert in the coding sequence. *Cell* **12**, 1079–1108.

KURTZ, D.T. & NICODEMUS, C.F. 1981. Cloning of 2U globulin using a high efficiency technique for the cloning of trace messenger RNAs. *Gene* **13**, 145–152.

MANIATIS, T., FRITSCH, E.F. & SAMBROOK, J. 1982. *Molecular cloning: A Laboratory Manual*. Cold Spring Harbor, New York: Cold Spring Harbor Laboratory.

MARCH, C.J., MOSELEY, B., LARSEN, A., CERETTI, D.P., BRAEDT, G., PRICE, V., GILLIS, S., HENNEY, C.S., KRONHEIM, S.R., GRUBSTEIN, K., CONLON, P.J., HOPP, T.P. & COSMAN, D. 1985. Cloning, sequence and expression of two distinct human interleukin-I complementary DNAs. *Nature* **315**, 641–647.

MEINKOTH, J. & WAHL, G. 1984. Hybridization of nucleic acids immobilized on solid supports. *Analytical Biochemistry* **138**, 267–284.

MILLER, J.H. & REZNIKOFF, W.S. 1978. In *The Operon*, Cold Spring Harbor, New York: Cold Spring Harbor Laboratory.

MURRAY, N.E., BRAMMAR, W.J. & MURRAY, K. 1977. Lambdoid phages that simplify the recovery of *in vitro* recombinants. *Molecular and General Genetics* **150**, 53–61.

PERBAL, B. 1988. *A Practical Guide to Molecular Cloning*. New York: Wiley Interscience.

SARTORIS, S., COHEN, E.B. & LEE, J.S. 1987. A rapid and improved method for generating cDNA libraries in plasmid and phage lambda vectors. *Gene* **56**, 301–307.

VILLA-KOMAROFF, L., EFSTRATIADIS, A., BROOM, S., LOMEDICO, P., TIZARD, R., NABER, S., CHICK, W.L. & GILBERT, W. 1978. Bacterial clones synthesizing pro-insulin. *Proceedings of the National Academy of Sciences of the United States of America* **75**, 3727–3731.

WALLACE, D.M. 1987. Precipitation of nucleic acids. *Methods in Enzymology* **152**, 41–48.

WATSON, C.J. & JACKSON, J.F. 1985. An alternative procedure for the synthesis of double-stranded cDNA for cloning in phage and plasmid vectors. In *DNA Cloning, a Practical Approach* Vol. 1 ed. Glover, D. pp. 79–88. Oxford: IRL Press.

WILLIAMS, J.G. 1981. The preparation and screening of a cDNA clone bank. In *Genetic Engineering* Vol. 1 ed. R. Williamson. pp. 1–59, New York & London: Academic Press.

YOUNG, R.A. & DAVIS, R.W. 1983. Efficient isolation of genes by using antibody probes. *Proceedings of the National Academy of Sciences of the United States of America* **80**, 1194–1198.

Plasmid Profiling and DNA/DNA Hybridization for Distinguishing Between Mesophilic *Aeromonas* Bacteria

S. Rogers, M.F. Moffatt,
I.F. Connerton and R.W.A. Park
Department of Microbiology, University of Reading, London Road, Reading, Berks RG1 5AQ, UK

After many years of using morphological, biochemical and physiological methods for discriminating between different bacterial species and their strains, new immunological and molecular techniques have been introduced to classify and identify micro-organisms at a fundamental level. The analysis of DNA relatedness is probably the most fundamental way of assessing the similarity of two bacterial isolates. This analysis may be represented, for instance, by DNA/DNA hybridization, DNA sequencing, guanosine:cytosine ratios or 16S rRNA relatedness. Plasmid profiles of bacterial strains can provide an epidemiological marker for tracking the progress of resistance/virulence factors and serologically similar biovars. DNA/DNA hybridization and plasmid profiling will be discussed in this chapter.

DNA/DNA Hybridization

DNA/DNA hybridization has proved a popular technique amongst bacterial taxonomists for grouping bacteria within a genus or species. Bacteria that are difficult to distinguish phenotypically may be separated on the basis of differences in the degree of homology between their nucleotide sequence. Hybridization refers to the passive association of two single-stranded DNA or RNA molecules by molecular fit and charge attraction. The higher the degree of similarity between the nucleotide bases, the greater the degree of hybridization. A certain amount of 'mis-matching' will occur when segments of similarity exist interspersed with areas that are quite dissimilar or when energetically unfavourable charge attractions occur. The hybridization in these areas will be less stable and under appropriate experimental conditions non-specific 'mismatching' can be controlled.

Genetic Manipulation
0-632-02926-9

Classical hybridization studies have utilized techniques such as thermal elution chromatography from hydroxyapatite columns (Britten & Kohne 1966; Brenner *et al.* 1969). DNA from reference strains of the bacteria under study were labelled with ^{32}P phosphorus. The DNA was then mixed with unlabelled DNA from the homologous strain or from other strains of interest. A large excess of the unlabelled DNA ensures good hybridization of labelled with unlabelled DNA. The DNA solution (buffered) was then heat-denatured and incubated for approximately 16 h at a selected temperature. A self hybridization control of labelled homologous DNA was included. After incubation and dilution, the DNA mixture was passed through a column of equilibrated hydroxyapatite. Double-stranded DNA adheres to hydroxyapatite and the single-stranded unhybridized DNA can then be eluted. Double-stranded molecules can be eluted by raising the temperature. Relatedness is expressed by normalizing DNA bound in the heterologous reaction to that bound in the homologous control, allowing for the error in the labelled DNA only reaction. *Yersinia enterocolitica* and *Yersinia pseudotuberculosis* were characterized by DNA/DNA hybridization using thermal elution (Brenner *et al.* 1976). The classification of *Escherichia vulneris* into the Enterobacteriaceae also utilized data gained from hydroxyapatite chromatography (Brenner *et al.* 1982). Farmer *et al.* (1988) characterized *Vibrio metchnikovii* and *Vibrio gazogenes* by comparing phenotypic data with genotypic data from bound DNA.

Liquid hybridization techniques are time-consuming and limited in the number of samples easily handled at one time. Membrane hybridization is a faster method, allowing many samples to be processed together, and minimizing the chance of reannealing of DNA strands from the same preparation. Sample DNA is bound to a membrane, usually of nitrocellulose, although nylon membrane is popular as it is possible to re-hybridize to the filters numerous times. Labelled single-stranded reference DNA is supplied in a hybridization solution. Membrane filter technology was developed in the 1960s by workers such as Gillespie & Spiegelman (1965) and Denhardt (1966). The genetic relationship amongst the genera *Beneckea* and *Photobacterium* was studied by Reichelt *et al.* (1976). Totten *et al.* (1983) used membrane *in vitro* hybridization to detect *Neisseria gonorrhoeae* by probing with the gonococcal cryptic plasmid. The same workers studied relationships amongst 'new' species of *Campylobacter* and *C. jejuni* using 'taxonomic spot blotting' (Fennell *et al.* 1984; Totten *et al.* 1985, 1987).

DNA/DNA hybridization for classifying Aeromonas

The genus *Aeromonas* divides into two main groups. The non-motile psychrophilic aeromonads fall into one species named *Aeromonas salmonicida* whilst three motile, mesophilic species are recognized by Popoff (1984). These are

A. hydrophila, *A. sobria* and *A. caviae*. Three other species have been suggested, *A. media*, *A. veronii* and *A. schubertii* (Allen *et al.* 1983; Hickman-Brenner *et al.* 1987, 1988). *A. salmonicida* is a recognized fish pathogen (Herman 1968). The mesophilic species produce a variety of extracellular substances which may have a role in virulence. These include proteolytic enzymes, haemolysins, cytotoxins and enterotoxins. They have been isolated from patients suffering from diarrhoeal diseases (Rosner 1964; Chatterjee & Neogy 1972; Champsaur *et al.* 1982; Burke *et al.* 1983; Abeyta *et al.* 1986). *A. hydrophila* and *A. sobria* are the species which are most commonly associated with human disease. There is an increasing awareness of the pathogenic potential of *Aeromonas* spp. Their occurrence in the water supply and foods (Burke *et al.* 1984; Abeyta & Wekell 1988) is being studied. The taxonomy of the mesophilic *Aeromonas* is not clear, with many isolates being atypical of the recognized or proposed species and with recent evidence that phenotypic tests do not distinguish between genotypically different isolates. In order to attempt a redefinition of the species, new systems for isolation and identification of *Aeromonas* species have been developed. Generally, isolates are differentiated by biochemical tests such as those developed by Popoff (1984). Other recent techniques have included immunofluorescence (Fliermans & Hazen 1979), esterase typing (Picard & Goullet 1987), serotyping (Fricker 1987) and electrophoretic profiling (Millership & Want 1989). Genetic typing techniques came to the fore with the MacInnes *et al.* (1979) paper on 'Deoxyribonucleic acid relationships among members of the genus *Aeromonas*'. Using S1 nuclease analysis, they showed that the genus could be divided into two groups, homogeneous and non-motile or non-homogeneous and motile. Internal homology groups were not identified. Popoff *et al.* (1981) investigated DNA sequence relationships by means of hybridization methods. Eleven hybridization groups were found. Hickman-Brenner *et al.* (1987, 1988) used DNA–DNA hybridization and hydroxyapatite separation to investigate proposed new species of *Aeromonas*. Phenotypic characterizations of 189 faecal isolates were compared with DNA relatedness by Kuijper *et al.* (1989). Their phenotypic characterizations were found insufficiently specific to distinguish between some hybridization groups. All these workers used hydroxyapatite procedures. By contrast, this chapter deals with the use of 'taxonomic spot blotting' to discriminate between *Aeromonas* isolates (Sladen *et al.* 1988).

Plasmid Profiling

Plasmid profiling provides another means of distinguishing between bacterial isolates. It offers a relatively fast and easy procedure in preference to methods such as biochemical, phage and sero-typing. The latter two methods require the production and maintenance of large numbers of phage stocks and antisera.

Plasmids are small, autonomously replicating, nucleic acid elements. Occasionally they may integrate into the chromosome. Not essential for bacterial growth, they may code for characters that can bestow an advantage on the strain for survival under certain conditions. Replicative and transfer functions such as pili genes, if present, are located on the plasmid. Factors such as antibiotic resistance, adhesins, toxins and degradative abilities may also be plasmid-coded. Multiple plasmids of the same or different molecular size are commonly found in the same cell. The 'pattern' of different plasmids found on a separating gel after electrophoresis may be used as an epidemiological marker for bacterial isolates. The ease with which some plasmids can transfer means that isolates from a discrete environment may carry plasmids that are identical or similar (Datta & Nugent 1984). Plasmid profiling has found a place in the study of hospital-acquired infection. In 1980, profiling revealed that a large (106 kb) plasmid carrying multiple antibiotic resistance was responsible for a recalcitrant *Klebsiella pneumoniae* infection in a neonatal unit (Markowitz *et al.* 1980). The spread of *Campylobacter jejuni* has been mapped by profiling (Tenover *et al.* 1984). Schemes for typing *Salmonella* spp. have been proposed (Bezanson *et al.* 1983; Gotuzzo *et al.* 1987). Poh *et al.* (1988) compared plasmid profiling with serotyping and pyocin typing for *Pseudomonas aeruginosa*, finding profiling to be a good confirmatory technique.

Plasmid profiling of Aeromonas

The epidemiology of *Aeromonas* spp. has become increasingly important with growing awareness of the pathogenicity of some strains. Chang & Bolton (1987) published antibiograms and some details of *Aeromonas* plasmids for 75 isolates. Intestinal infections of *Aeromonas* in the USA were examined by Holmberg *et al* (1986). An 85 MDa plasmid occurred in strains from two unrelated patients living 40 miles apart. *Aeromonas* isolated from the patient's children carried the same plasmid. The occurrence of plasmid profiles in *Aeromonas* spp. was studied and methods developed to interpret plasmid patterns.

Methods

DNA/DNA filter hybridization steps

Isolation and growth of bacteria

Aeromonas spp. were isolated from faeces, food or water by enrichment in alkaline peptone water followed by plating on to a selective medium (Aeromonas medium, Difco). Cultures were kept frozen at −70°C in 15% glycerol with nutrient broth (NB:no 2, Oxoid). For reference DNA production, strains were grown overnight in liquid culture (500 ml NB:no 2 in 2 litre shake flasks at 37°C, 200 rpm for 24 h) or on nutrient agar (Oxoid) at 37°C for 24 h.

Harvest of cells

Cells were harvested from liquid by centrifugation (13 000 ***g***) for 15 min. Growth on plates was scraped off into buffer before washing. Cell pellets were washed twice with phosphate buffered saline (PBS).

Breakage of cell walls and lysis

Pellets were resuspended in 18 ml of sterile Tris–EDTA (TE) buffer. Lysis was achieved by incubating the suspension with lysozyme (1 mg/ml, Sigma) at 35°C for 30 min followed by the addition of 1% sodium dodecyl sulphate (SDS) at 55°C until lysis was evident. The lysate was diluted with an equal volume of TE and incubated overnight with 50 μg/ml proteinase-K (Sigma) at 37°C.

Preparation of reference DNA

The proteinaceous debris was extracted with an equal volume of TE–equilibrated phenol. Phases were separated by centrifugation at 16 000 ***g*** for 3 min. The upper, aqueous phase was carefully drawn off leaving behind the interface and lower phenolic liquor. The phenolic layer may be back extracted with an equal volume of TE, the aqueous layers being combined. An equal volume of phenol:chloroform (50:50) was added, mixed and the extraction repeated. A final extraction with chloroform or ether removed remaining phenolic and protein traces, the aqueous phase being retained. Phenol extractions were repeated until no white debris was visible at the interface. *Aeromonas* DNA was precipitated by adding ice cold ethanol after increasing the salt concentration to approximately 0.3 M with sodium acetate (3 M stock). Precipitation was for 2 h at −20°C. DNA was concentrated by centrifugation at 13 000 ***g***. The pellet was re-dissolved in TE and incubated at 37°C with 50 μg/ml DNase free RNase (boiled) for 30 min followed by digestion with pronase (50 μ/ml) for a further 30 min. After a further chloroform extraction and ethanol precipitation, the DNA pellet was dissolved in TE and the concentration and purity determined by absorption at 260 nm and 280 nm.

Labelling of reference DNA

Radiolabelling of the reference DNA was accomplished by 'nick translation' (Maniatis *et al.* 1975) using ^{32}P-labelled deoxyadenosine triphosphate. In brief 1–2 μg of purified genomic DNA was added to a reaction containing 2.5 μl of 10 × nick translation buffer, 2 μl deoxycytosine triphosphate, 2 μl deoxythymidine triphosphate, 2 μl deoxyguanosine triphosphate (each 0.1 mg/ml stocks), 4 μl ^{32}P-labelled deoxyadenosine triphosphate (3 000 Ci/mM in aqueous solution, Amersham), 1 μl DNase (1 μg/ml stock in 5% glycerol) and 2 μl of *E. coli* DNA polymerase 1 (20 units). The reaction mixture was made up to 25 μl with distilled water. Reaction mixtures were kept on ice in sterile microcentrifuge tubes until the enzymes were added. Mixtures were incubated for 2 h at 15°C. DNA was then precipitated by adding 85 μl of isopropanol and 100 μl of 2.5 M ammonium acetate, standing at room temperature for 20 min. After centrifugation for 6 min on 'high' setting (MSE Microcentaur) the supernatant was discarded and the walls of the tube dried. The pellet was redissolved in 200 μl TE.

Binding of DNA to filter

Isolates to be examined were grown on nutrient agar (NA) overnight at 37°C. Growth was then swabbed off the solid medium into 2 ml of sterile PBS. In order to allow comparisons to be made between different isolates, the amount of DNA from each isolate that is bound to the nitrocellulose filter should be the same. Adjustments to the bacterial suspensions were made according to the optical density to provide the same number of cells per 10 μl drop applied to the filter. Nitrocellulose (Schleicher & Schuell: 0.45 μm) was cut into 9 cm × 5 cm sized pieces and marked with isolate identification numbers. Drops (10 μl) of each bacterial suspension were applied slowly to the filters. A hot air dryer was used to keep the spreading distance to a minimum. One filter was prepared for each genomic probe. The cells were lysed *in situ* by passing the filters through a series of solutions as follows:

1. 0.5 M sodium hydroxide, 10 min;
2. 1 M Tris, pH 7.0, 1 min × 3;
3. 1.5 M sodium chloride/1 M Tris, pH 7.0, 10 min;
4. As in 3 with 10 mg/ml pronase, 30 min 25°C;
5. Dried at room temperature for 15 min;
6. 2 × chloroform washes, 1 min;
7. Dried at room temperature for 15 min;
8. Resoaked in 1.5 M sodium chloride/1 M Tris;
9. Baked at 80°C for 2 h *in vacuo*.

Hybridization and washing

Filters were placed in small heat sealable plastic bags with 6 × concentrated sodium chloride/sodium citrate (SSC). After wetting, the SSC was poured off and 12 ml of prehybridization fluid containing 10 μg/ml of freshly boiled and sheared calf thymus DNA (intermittent sonication for 10 min at maximum amplitude, MSE Cabinet Model 100 W) was added to the bags which were sealed. Bags were incubated at 50°C for 1 h. Prehybridization served to block non-specific DNA binding sites on the filter. After 1 h the bags were cut open and the fluid poured off to be replaced by 12 ml of hybridization fluid, with calf thymus DNA (10 μg/ml) and the labelled DNA, both freshly boiled. Formamide is included in the hybridization fluid to increase the 'stringency' of hybridization, that is, to stabilize the mis-matching of DNA strands by providing a polar environment and slowing down the hybridization kinetics. Each increase of 1% in the formamide concentration lowers the DNA melting point (T_m) by 0.7°C (Casey & Davidson 1977). The specific activity of the labelled DNA may be determined by precipitation with trichloroacetic acid (Maniatis *et al.* 1982). Hybridization proceeded statically at 50°C for 15 h. Following incubation, the filters were washed in 2 × sodium chloride, sodium phosphate, EDTA (SSPE) for 20 min; 1 × SSPE + 0.1% SDS for 20 min; 0.1 × SSPE + 0.1% SDS for 20 min. All washings were static at 65°C. After each wash filters were monitored for radioactivity over the spotted areas. Washings were terminated when the level of radioactive background decreased sufficiently, as judged by a mini monitor. Filters were then dried at room temperature on blotting paper.

Autoradiography

After washing, the filters were autoradiographed with Dupont Cronex X-ray film in a Kodak cassette fitted with an intensifying screen. Autoradiography was for 1–3 days depending on how 'hot' the filters were at the start of the autoradiography.

Plasmid profiling

Method

Aeromonas strains were grown until late exponential phase in NB no. 2. Centrifugation of 1.5 ml aliquots in microcentrifuge tubes provided a convenient way to examine many samples at once. The pellet was resuspended in 100 μl of plasmid solution 1 followed by 200 μl of plasmid solution 2 to lyse the cells. Tubes were left on ice for 5 min. Proteins and chromosomal DNA were precipitated by adding 150 μl of potassium acetate solution to neutralize the pH, and chilling again for 5 min. Debris was pelleted by centrifugation. The supernate was decanted into fresh tubes and an equal volume of TE–equilibrated phenol was added. After gentle agitation to mix the phases, they were separated by centrifugation at 'high' setting (MSE Microcentaur). The upper aqueous phase was removed and extracted with diethylether. After centrifugation, the upper, ether phase was discarded and traces of ether were allowed to evaporate from the aqueous layer by warming at 37°C. Plasmid DNA was precipitated by adding 1 ml of ethanol and incubating at room temperature for 20 min. The DNA was pelleted and then dried. Pellets were redissolved in 20 μl of TE. Of this, 10–15 μl were mixed with 3 μl of sample loading buffer and subjected to electrophoresis in 0.8% agarose in 0.5 × Tris–borate–EDTA (TBE) running buffer for 4–6 h at 60 mA. DNA was visualized by staining the gels by the addition of 20 μl of a 10 mg/ml ethidium bromide solution to the running buffer. Fluorescence under u.v. radiation revealed the presence of plasmid bands which were recorded by photography (u.v. and Kodak Wratten 23A filters are required).

For more detailed studies to distinguish between plasmids of the same molecular weight, the restriction enzymes *Bam*HI and *Hin*dIII were used to prepare digests of plasmid preparations. Enzymes were supplied with their appropriate reaction buffers. A 20 μl digest routinely contained 5 μl DNA, 2 μl of reaction buffer, 1 μl spermidine (50 mM), 11 μl distilled water, 1 μl of enzyme. Digests were incubated for 1 h at 37°C. Samples of 20 μl were loaded into slots in a 0.8% agarose gel after the addition of loading buffer. Electrophoresis was performed as previously described. Two dimensional electrophoresis was performed by excising gel lanes containing plasmid bands from gels prepared as above. The agarose segment was u.v.-irradiated for 3 min to 'nick' the DNA molecules. The segment was then fused back into 0.8% agarose at right angles to its previous position before electrophoresis was repeated.

Results

DNA/DNA hybridization

The 'taxonomic spot blot test' provided a semiquantitative method for screening large numbers of *Aeromonas* strains for homology with our reference strains. The degree of relatedness to the reference strain DNA was demonstrated by the degree of 'darkening' observed for the spots after autoradiography and calibrated by densitometry. Two typical autoradiographs are shown in Fig. 1. Autoradiographs giving homology for 21 strains to an *A. hydrophila* probe (Fig. 1a) and an *A. sobria* probe (Fig. 1b) demonstrate the range of homology for the reference DNAs. Spot 17 in Fig. 1a represents the homologous *A. hydrophila* DNA, spots 18, 19 and 20 are *Klebsiella*, *Vibrio* and *Escherichia coli* negative control strains respectively. All other spots gave hybridization with biochemically typed *Aeromonas* isolates. Faint darkening on the control spots indicated that slight relatedness existed between the controls and the *A. hydrophila* genomic DNA. This occurred to a lesser extent with the

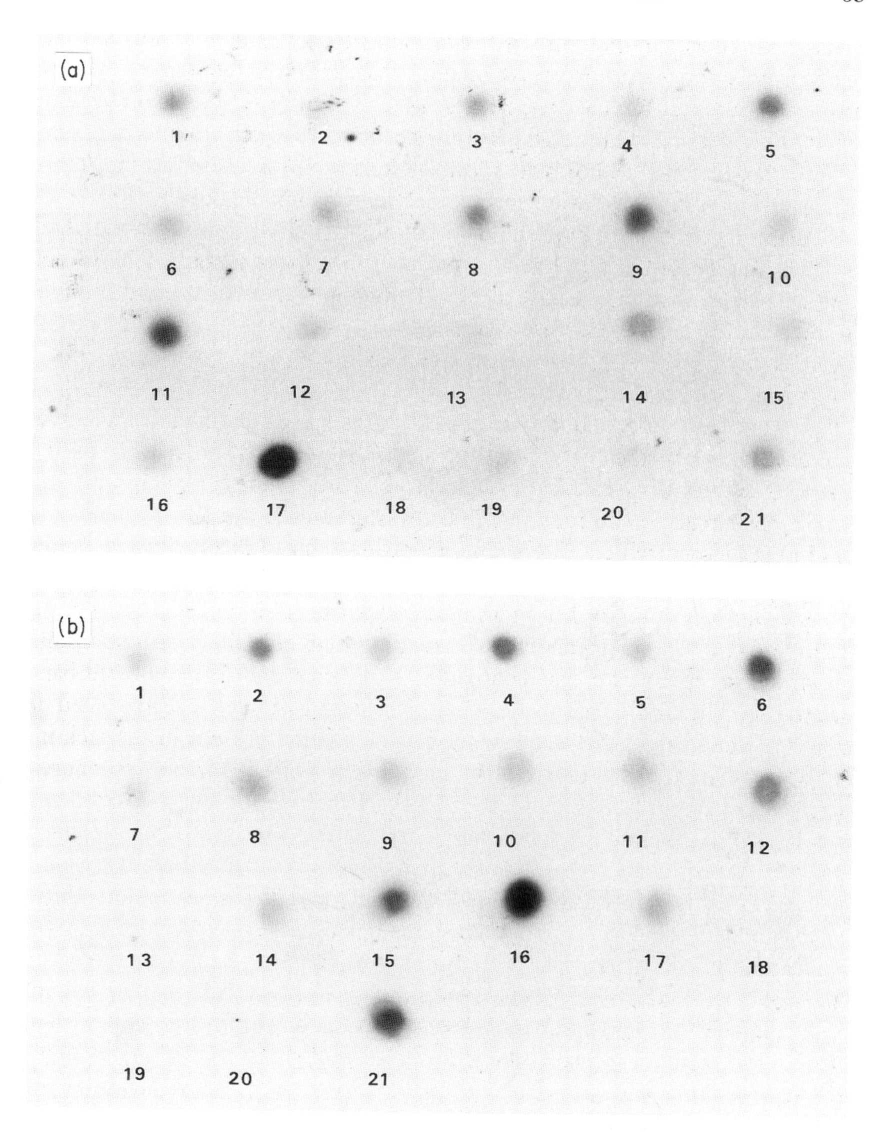

FIG. 1. Autoradiographs of 21 *Aeromonas* strains probed with whole genomic DNA from *A. hydrophila* (a) and *A. sobria* (b). In (a), spot 17 is the homologous strain; and in (b), spot 16 is the homologous strain.

A. sobria DNA. The observed darkening on the autoradiographs was graded to give 6–7 groups. The *A. hydrophila* genomic probe hybridized most efficiently with DNA from isolates biochemically typed as *A. hydrophila*. Those typed as *A. sobria* came next, with *A. caviae* giving the poorest hybridization. The

A. sobria probe hybridized most strongly with other *A. sobria* strains followed by *A. hydrophila*, then *A. caviae*. It was possible to gain a quantitative appraisal of strain relatedness by using a densitometer to scan the spots or by preparing small individual filters for each strain. Labelled DNA can then be measured by radioactive counts. Differential lysis of strains on the filters may introduce some error, as may repeating sequence elements in the DNA. There was evidence that hybridization groups did not always conform to the biochemical typing and that some species delineated by phenotype contained more than one hybridization group.

Plasmid profiling

Plasmids were found in 40% of *Aeromonas* isolates examined from various sources. The molecular size varied from 2.5 kb to large 'F' factor-like plasmids of 70 kb. Recurrent plasmid patterns have emerged. Figure 2a shows plasmid profiles for a range of *Aeromonas* isolates. Lanes 5 and 6 are profiles from clinical isolates which show similarity in their patterns. In Fig. 2b two strains (lanes 1 & 2), isolated from a dairy, exhibit identical profiles. Although plasmids may appear to be of the same size they may be significantly different in their sequence and function. By cutting the plasmids with restriction enzymes, a characteristic fragment ladder can be produced. Usually fragment patterns can be assigned to particular plasmids; identical plasmid patterns provide strong evidence for the same plasmids being present in the isolates.

Interpretation of plasmid profiles may also be impaired by the appearance of different molecular forms of the same plasmid on the gel, i.e. covalently closed circle (supercoiled), open circle and linear. The use of u.v. 'nicking' and two-dimensional electrophoresis allows the number of different plasmids in the same strain to be determined. Plasmids may be transferable within a genus and, in some cases, between genera. Plasmids may be lost as a cell replicates. Thus, before plasmid profiling is applied as a general typing technique, the stability of plasmids constituting characteristic patterns should be examined. For rapid epidemiological surveying, however, the technique can prove very useful, especially where troublesome resistance factors occur in hospitals or other environments. The value of plasmid profiling for *Aeromonas* will become greater as more profile types are found. For example, outbreaks of *Aeromonas*-associated illness such as sepsis or diarrhoea could be traced back to water or foodstuffs.

Acknowledgements

We gratefully acknowledge funding from the Ministry of Agriculture, Fisheries and Food and we thank K.S. Walter for technical assistance.

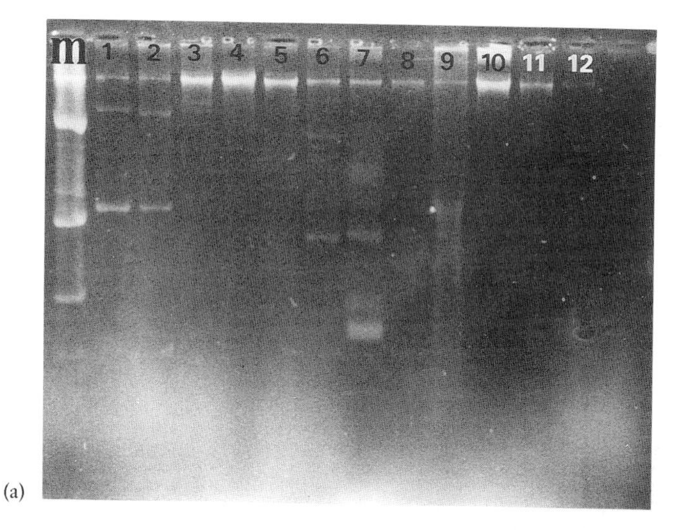

(a)

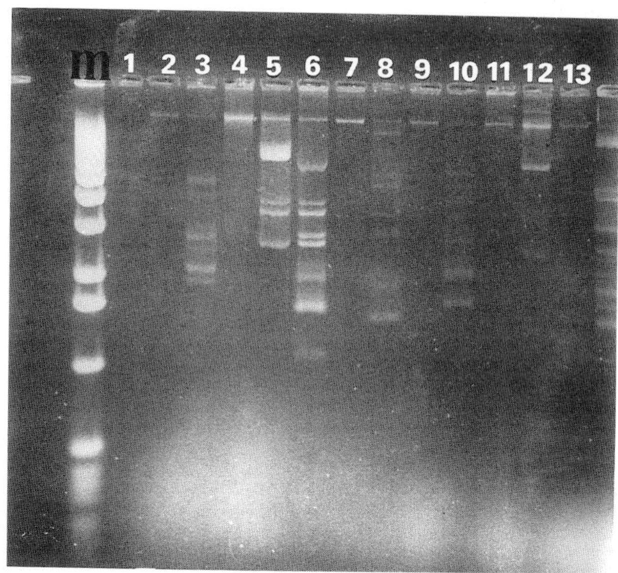

(b)

FIG. 2. Plasmid profiles from clinical isolates (a) and food isolates (b) of *Aeromonas*. In (a) lanes 5 and 6 show some similarities. In (b) lanes 1 and 2 show an identical profile.

References

ABEYTA, C., KAYSNER, C.A., WEKELL, M.M., SULLIVAN, J.J. & STELMA, G.N. 1986. Recovery of *Aeromonas hydrophila* from oysters implicated in an outbreak of foodborne illness. *Journal of Food Protection* **49**, 643–646.

ABEYTA, C. & WEKELL, M.M. 1988. Potential sources of *Aeromonas hydrophila*. *Journal of Food Safety* **9**, 11–22.

ALLEN, A., AUSTIN, B. & COLWELL, R. 1983. *Aeromonas media*, a new species isolated from river water. *International Journal of Systematic Bacteriology* **33**, 599–644.

BEZANSON, G.S., KHAKRIA, R. & PAGNUTTI, D. 1983. Plasmid profiles of value in differentiating *Salmonella muenster* isolates. *Journal of Clinical Microbiology* **17**, 1159–1160.

BRENNER, D.J., FANNING, G.R., RAKE, A.V. & JOHNSON, K.E. 1969. Batch procedure for thermal elution of DNA from hydroxyapatite. *Analytical Biochemistry* **28**, 447–459.

BRENNER, D.J., MCWHORTER, A.C., LEETE-KNUTSON, J.K. & STEIGERWALT, A.G. 1982. *Escherichia vulneris*: a new species of Enterobacteriaceae associated with human wounds. *Journal of Clinical Microbiology* **16**, 319–323.

BRENNER, D.J., STEIGERWALT, A.G., FALCAO, D.P., WEAVER, R.E. & FANNING, G.R. 1976. Characterization of *Yersinia enterocolitica* and *Yersinia pseudotuberculosis* by deoxyribonucleic acid hybridization and by biochemical reactions. *International Journal of Systematic Bacteriology* **26**, 180–194.

BRITTEN, R.J. & KOHNE, D.E. 1966. Nucleotide sequence repetition in DNA. *Carnegie Institute Washington Yearbook* **65**, 78–106.

BURKE, V., GRACEY, M., ROBINSON, J., PECK, D., BEAMAN, J. & BUNDELL, C. 1983. The microbiology of childhood gastritis: *Aeromonas* species and other infective agents. *Journal of Infectious Diseases* **148**, 68–74.

BURKE, V., ROBINSON, J., GRACEY, M., PETERSON, D. & PARTRIDGE, K. 1984. Isolation of *Aeromonas hydrophila* from a metropolitan water supply: seasonal correlation with clinical isolates. *Applied and Environmental Microbiology* **48**, 361–366.

CASEY, J. & DAVIDSON, N. 1977. Rates of formation and thermal stabilities of RNA/DNA and DNA/DNA duplexes at high concentrations of formamide. *Nucleic Acid Research* **4**, 1539–1541.

CHAMPSAUR, H., ANDREMONT, A., MATHIEU, D., ROTTMAN, E. & AUZEPY, P. 1982. Cholera-like illness due to *Aeromonas sobria*. *Journal of Infectious Diseases* **145**, 248–254.

CHANG, B.J. & BOLTON, S.M. 1987. Plasmids and resistance to antimicrobial agents in *Aeromonas sobria* and *Aeromonas hydrophila* clinical isolates. *Antimicrobial Agents and Chemotherapy* **31**, 1281–1282.

CHATTERJEE, B.D. & NEOGY, K.N. 1972. Studies on *Aeromonas* and *Plesiomonas* species isolated from cases of choleraic diarrhoea. *Indian Journal of Medical Research* **60**, 520–524.

DATTA, N. & NUGENT, M.E. 1984. Characterization of plasmids in wild strains of bacteria. In *Advanced Molecular Genetics* ed. Puhler, A. & Timmis, A.M. Ch 1.4, pp. 38–50. Berlin: Springer-Verlag.

DENHARDT, D.T. 1966. A membrane filter technique for the detection of complementary DNA. *Biochemical and Biophysical Research Communications* **23**, 641–646.

FARMER, J.J., HICKMAN-BRENNER, F.W., FANNING, G.R., GORDON, C.M. & BRENNER, D.J. 1988. Characterization of *Vibrio metschnikovii* and *Vibrio gazogenes* by DNA/DNA hybridization and phenotype. *Journal of Clinical Microbiology* **26**, 1993–2000.

FENNELL, C.L., TOTTEN, P.A., QUINN, T.C., PATTON, D.L., HOLMES, K.K. & STAMM, W.E. 1984. Characterization of *Campylobacter*-like organisms isolated from homosexual men. *The Journal of Infectious Diseases* **149**, 58–66.

FLIERMANS, C.B. & HAZEN, T.C. 1979. Immunofluorescence of *Aeromonas hydrophila* as measured by fluorescence photometric microscopy. *Canadian Journal of Microbiology* **26**, 161–168.

FRICKER, C.R. 1987. Serotyping of mesophilic *Aeromonas* spp. on the basis of lipopolysaccharide antigens. *Letters in Applied Microbiology* **4**, 113–116.

GILLESPIE, D. & SPIEGELMAN, S. 1965. A quantitative assay for DNA/DNA hybrids with DNA immobilized on a membrane. *Journal of Molecular Microbiology* **12**, 829–842.

GOTUZZO, E., MORRIS, J.G., BENAVENTE, L., WOOD, P.K., LEVINE, O., BLACK, R.E. & LEVINE,

M.M. 1987. Association between specific plasmids and relapse in typhoid fever. *Journal of Clinical Microbiology* **25**, 1779–1781.

HERMAN, R.L. 1968. Fish furunculosis. *Transactions of the American Fish Society* **97**, 221–230.

HICKMAN-BRENNER, F.W., FANNING, G.R., ARDUINO, M.J., BRENNER, D.J. & FARMER, J.J. 1988. *Aeromonas schubertii*, a new mannitol-negative species found in human clinical specimens. *Journal of Clinical Microbiology* **26**, 1561–1564.

HICKMAN-BRENNER, F.W., MACDONALD, K.L., STEIGERWALT, A.G., FANNING, G.R., BRENNER, D.J. & FARNER, J.J. 1987. *Aeromonas veronii*, a new ornithine decarboxylase-positive species that may cause diarrhea. *Journal of Clinical Microbiology* **25**, 900–906.

HOLMBERG, S.D., SCHELL, W.L., FANNING, G.R., WACHSMUTH, I.K., HICKMAN-BRENNER, F.W., BLAKE, P.A., BRENNER, D.J. & FARMER, J.J. 1986. *Aeromonas* intestinal infections in the United States. *Annals of International Medicine* **105**, 683–689.

KUIJPER, E.J., STEIGERWALT, A.G., SCHOENMAKERS, B.S.C.I.M., PEETERS, M.F., ZANEN, H.C. & BRENNER, D.J. 1989. Phenotypic characterization and DNA relatedness in human faecal isolates of *Aeromonas* spp. *Journal of Clinical Microbiology* **27**, 132–138.

MACINNES, J.I., TRUST, T.J. & CROSA, J.H. 1979. Deoxyribonucleic acid relationships among members of the genus *Aeromonas*. *Canadian Journal of Microbiology* **25**, 579–586

MANIATIS, T., FRITSCH, E.F. & SAMBROOK, J. 1982. *Molecular Cloning*. Cold Spring Harbor, New York: Cold Spring Harbor Laboratory.

MANIATIS, T., JEFFREY, A. & KLIED, D.G. 1975. Nucleotide sequence of the rightward operator of phage lambda. *Proceedings of the National Academy of Sciences* **72**, 1184–1186.

MARKOWITZ, S.M., VEAZY, J.M., MACRINA, F.L., MAYHALL, C.G. & LAMB, V.A. 1980. Sequential outbreaks of infection due to *Klebsiella pneumoniae* in a neonatal intensive care unit:implication of a conjugative R plasmid. *Journal of Infectious Diseases* **142**, 106–112.

MILLERSHIP, S.E. & WANT, S.V. 1989. Typing of *Aeromonas* species by protein fingerprinting: comparison of radiolabelling and silver staining for visualizing proteins. *Journal of Medical Microbiology* **29**, 29–32.

PICARD, B. & GOULLET, P.H. 1987. Epidemiological complexity of hospital *Aeromonas* infections revealed by electrophoretic typing of esterases. *Epidemiology and Infection* **98**, 5–14.

POH, C.L., YAP, E.H., TAY, L. & BERGAN, T. 1988. Plasmid profiles compared with serotyping and pyocin typing for epidemiological surveillance of *Pseudomonas aeruginosa*. *Journal of Medical Microbiology* **25**, 109–114.

POPOFF, M.Y. 1984. *Aeromonas* Kluyer & Van Niel 1936. In *Bergey's Manual of Systematic Bacteriology* Vol 1. ed. KRIEG, N.R. & HOLT, J.G. pp. 545–548. Baltimore: Williams & Wilkins.

POPOFF, M.Y., COYNAULT, C., KIREDJIAN, M. & LEMELIN, M. 1981. Polynucleotide sequence relatedness among motile *Aeromonas* species. *Current Microbiology* **5**, 109–114.

REICHELT, J.L., BAUMANN, P. & BAUMANN, L. 1976. Study of the genetic relationships among marine species of the genera *Beneckea* and *Photobacterium* by means of *in vitro* DNA/DNA hybridization. *Archives of Microbiology* **110**, 101–120.

ROSNER, R. 1964. *Aeromonas hydrophila* as the etiological agent in a case of severe gastroenteritis. *American Journal of Clinical Pathology* **42**, 402–404.

SLADEN, S., MOFFAT, M., CONNERTON, I.F. & FRICKER, C.R. 1988. Studies on the taxonomy of *Aeromonas* spp. *Journal of Applied Bacteriology* **65**, i–xiv.

TENOVER, F.C., WILLIAMS, S., GORDON, K.P., HARRIS, N., NOLAN, C. & PLORDE, J.J. 1984. Utility of plasmid fingerprinting for epidemiological studies of *Campylobacter jejuni* infections. *The Journal of Infectious Diseases* **149**, 279–282.

TOTTEN, P.A., HOLMES, K.K., HANDSFIELD, H.H., KNAPP, J.S., PERINE, P.L. & FALKOW, S. 1983. DNA hybridization technique for the detection of *Neisseria gonorrhoeae* in men with urethritis. *The Journal of Infectious Diseases* **148**, 462–471.

TOTTEN, P.A., FENELL, C.L., TENOVER, F.C., WEZENBERG, J.M., PERINE, P.L., STAMM, W.E. &

HOLMES, K.K. 1985. *Campylobacter cinaedi* (sp. nov.) and *Campylobacter fennelliae* (sp. nov.): two new *Campylobacter* species associated with enteric disease in homosexual men. *The Journal of Infectious Diseases* **151**, 131–139.

TOTTEN, P.A., PATTON, C.M., TENOVER, F.C., BARRETT, T.J., STAMM, W.E., STEIGERWALT, A.G., LIN, J.L., HOLMES, K.K. & BRENNER, D.J. 1987. Prevalence and characterization of hippurate-negative *Campylobacter jejuni* in King County, Washington. *Journal of Clinical Microbiology* **25**, 1747–1752.

Appendix: Reagents

Tris/EDTA (TE)

10 mM Tris, 1 mM EDTA, pH 8.0.

Phosphate buffered saline

Sodium chloride 8.0 g, potassium chloride 0.2 g, potassium dihydrogen phosphate 0.2 g, disodium hydrogen phosphate 1.15 g. Add distilled water to 1 litre.

10 × Nick translation buffer

0.5 M Tris, pH 7.2, 0.1 M magnesium sulphate, 1 mM dithiothreitol, 500 μg/ml bovine serum albumin pentax fraction v. Divide into aliquots and freeze at −20°C.

1 × Sodium chloride/sodium citrate buffer (SSC)

0.15 M sodium chloride, 0.015 M sodium citrate.

10 × 1 M Tris–borate–EDTA buffer (TBE)

108 g Tris base, 55 g borate, 40 ml of 0.5 M EDTA pH 8.0, make up to 1 000 ml with distilled water. Electrophoresis was carried out with 50 mM TBE.

Electrophoresis loading buffer

0.25% bromophenol blue, 40% (w/v) sucrose in water.

15 × Sodium chloride, sodium phosphate, EDTA buffer (SSPE)

Sodium chloride 158 g (0.18 M), sodium dihydrogen phosphate 23.4 g (10 mM), EDTA 5.55 g (1 mM)

TE–equilibrated phenol

Add sufficient Analar grade crystalline phenol to sterile TE buffer to provide 4–5 cm of equilibrated phenol as the lower layer after shaking to dissolve. Do not use if the preparation is brown or becomes so. 8-hydroxyquinoline (0.1% w/v) may be used as an antioxidant.

Potassium acetate solution, 5 M with respect to acetate, pH 4.8

5 M potassium acetate 60 ml, glacial acetic acid 11.5 ml, water 28.5 ml.

DNase-free RNase

Dissolve 10 mg/ml RNase (Sigma) in 0.25 M sodium acetate pH 6.5. Boil for 5 min to inactivate DNase. Cool, divide into batches and freeze at −20°C.

Prehybridization fluid (Totten *et al.*, 1983)

1 × concentration SSPE, 50% (v/v) formamide, 1% (w/v) glycine, 50 mM sodium phosphate (pH 6.5), 5 × concentration of Denhardt's solution (1 × 0.02% (w/v) polyvinyl pyrolidone, 0.02% (w/v) bovine serum albumin, 0.02% (w/v) Ficoll 4 000) made up to 1 litre with distilled water.

Hybridization fluid

50% (v/v) formamide, 1% (w/v) glycine, 20 mM phosphate (pH 6.5), 5 × SSPE, 1 × Denhardt's solution.

Plasmid solution 1

50 mM D-glucose, 25 mM Tris pH 8.0, 10 mM EDTA.

Plasmids solution 2

0.2 M NaOH, 1% (w/v) sodium dodecyl sulphate.

Preparation and Screening of Bacterial Genomic Libraries

J.W. DALE
Molecular Biology Group, Department of Microbiology, University of Surrey, Guildford, Surrey GU2 5XH, UK

A gene library is a collection of recombinant clones, each of which contains a part of the genetic material of the organism you are interested in. If you have enough clones, it should be possible to find any gene you want within the library, given a suitable screening procedure. This means that each time you want to isolate a different gene, you can take an existing library and screen it again with a new probe.

There are two fundamentally different types of gene library: genomic libraries are constructed from the total DNA of the starting organism (RNA in the case of certain viruses), while cDNA libraries are made by producing and cloning complementary DNA with mRNA as a template. A genomic library is therefore representative of the complete genetic information of that organism, while a cDNA library only represents those genes that are actively transcribed under the conditions existing when the mRNA was extracted. cDNA libraries are widely used for eukaryotic organisms since the complexity of the library is reduced, and hence the number of clones to be screened is much smaller. In addition, the absence of introns in the mRNA makes it considerably easier to clone the complete coding sequence of the gene.

With bacteria, these problems do not arise: the genome is usually considerably smaller, and the genes do not contain introns. In addition, there are technical problems that make the construction of a cDNA library more difficult from a bacterial source than from a eukaryote. Genomic libraries are therefore almost always preferred, although on occasions there may be applications for a bacterial cDNA library.

General Strategy

The overall concept of the construction of a bacterial genomic library is illustrated in Fig. 1. The bacterial DNA is first broken into a set of fragments

Genetic Manipulation
0-632-02926-9

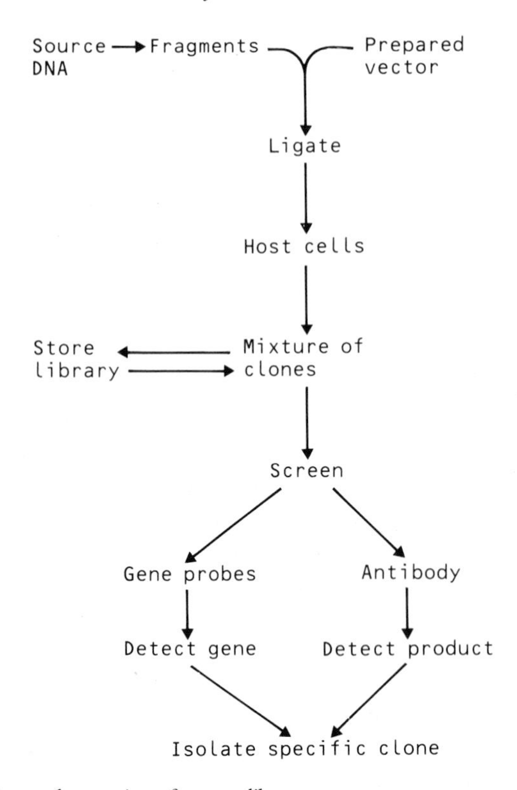

FIG. 1. Construction and screening of a gene library.

of an appropriate size for cloning. This collection of DNA fragments is then mixed with the prepared vector and ligated to produce a pool of recombinant molecules. After the introduction of these into a suitable bacterial host (usually *E. coli*), a mixture of bacterial colonies (or phage plaques) is obtained and this constitutes the gene library. (Strictly, this is an amplified library, since each of the original recombinant DNA molecules is now represented by a large number of bacteria or bacteriophage particles). The gene library can now be stored as a mixture, or screened immediately by transferring the colonies/plaques to a filter that can be tested for the presence of the required gene. The positive clones are then picked from the original plate and purified.

There are three major decisions to be taken in respect of this strategy: the method of fragmentation of the DNA, the choice of vector, and the method of screening. The last is the most important, since it influences the other decisions.

Choice of screening method

There are two principal methods for screening gene libraries:
1 The use of gene probes.
2 Specific antibodies.

In order to generate a gene probe, you have to have some information about the likely sequence of the gene in question. If the corresponding gene from a closely related organism has already been cloned, you can use a piece of DNA from that clone as a probe. If the organisms are very closely related, the homology is likely to be high. Alternatively, if you are able to purify the product of the gene in question, then determination of a short region of the amino acid sequence at the N-terminus would allow you to predict the DNA sequence of the corresponding portion of the gene. There will be some ambiguity owing to the degeneracy of the genetic code, but this can be minimized by taking into account the codon usage of the organism and wherever possible using sequences containing amino acids for which there are only one or two possible codons. Fortunately, it is easy to programme a DNA synthesizer to produce mixed oligonucleotides, and the conditions of hybridization can be adjusted (low stringency) so as to obtain a signal from comparatively low degrees of homology.

A further strategy for obtaining gene probes is to take advantage of the rapidly accumulating amount of sequence information in various databanks. Comparison of the amino acid sequences of proteins with the same function may disclose conserved regions that will probably also be present in your protein. Taking into account the codon usage of your bacterium, synthetic probes can then be generated as above. The redundancy of these probes, and the low stringency of hybridization, may, however, give rise to a high background reaction. The polymerase chain reaction (PCR, see Fig. 2) can be used to generate a better probe. Look for two conserved amino acid sequences and synthesize two oligonucleotides (with a high degree of redundancy) that will act as primers for DNA synthesis, directed towards each other. PCR amplification will generate a product of a predictable size that can be recovered from an agarose gel and used directly as a probe. Increased specificity comes from the use of the two primers, which have to bind to the same DNA fragment: since the product has the same sequence as the gene in question, it can be used as a probe at high stringency.

An alternative to the use of gene probes is to produce antibodies to the protein product. Conventional (polyclonal) antibodies are more easily produced, and tend to have greater sensitivity, but monoclonal antibodies (mAbs) are more specific, and can be produced without the need to purify the product. It can, however, be a major project in its own right to produce mAbs, and would

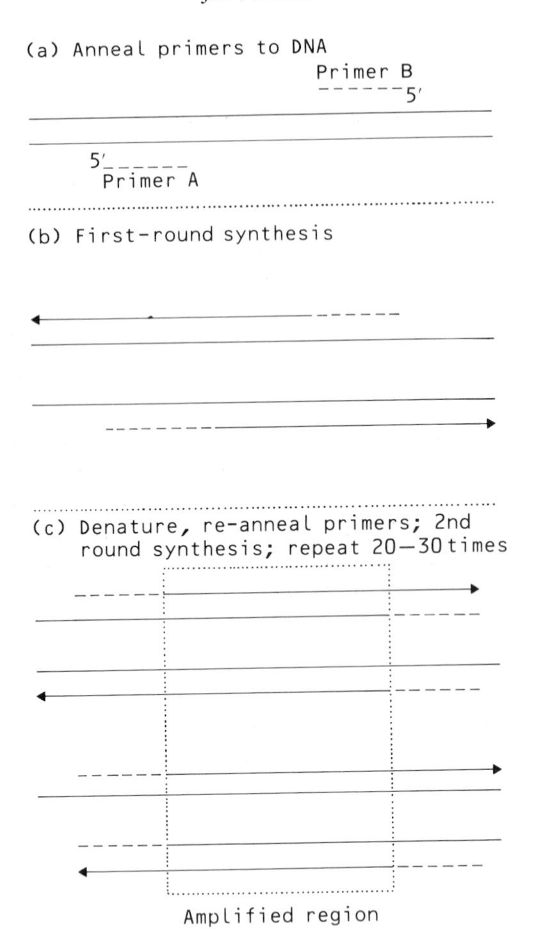

FIG. 2. Use of the polymerase chain reaction to generate a probe for screening a gene library.

not normally be undertaken unless there was some additional use for them.

A third strategy that can be used on occasions is to test the library for its ability to complement specific defects in the host strain, e.g. nutritional requirements. Since this can be used for positive selection, it can be a powerful tool for identifying the presence of the required gene, but only if there is a substantial level of functional expression. The limitations on the application of this strategy are the main reason for not considering it in detail in this chapter.

Choice of vector

The choice of screening procedure has an important influence on the choice of vector. If antibodies are to be used, the gene must be expressed, or at least that part containing the epitope to be detected. If left to its own devices, the gene may not be transcribed in a foreign host, or the gene product may be degraded, especially if only a part of the protein is produced. The first problem is overcome by using an *expression vector*, in which the inserted DNA fragment is placed under the control of a promoter sequence supplied by the vector. There are many plasmid expression vectors available, of which the best known are the pUC series. These contain part of the *lac* operon, in particular the promoter/operator region and part of the β-galactosidase gene. Near the 5′ end (N-terminus) of this gene, a *multiple cloning site* has been constructed, i.e. a sequence containing recognition sites for a number of restriction enzymes. The vector can therefore be cut at any one of these sites to allow the insertion of foreign DNA fragments. The resulting insertional inactivation of the β-galactosidase gene allows recombinant plasmids to be distinguished from the vector. More importantly, the cloned gene can now be expressed from the *lac* promoter.

The resulting product may still be subject to proteolytic degradation, however, due to incorrect folding. This can be overcome by the use of a different vector (e.g. λgt11), in which the cloning site is near the 3′ end (C-terminus) of the β-galactosidase gene. If a fragment of DNA is inserted at this site (in the correct orientation and maintaining the reading frame of the β-galactosidase gene), it will give rise to a fusion protein in which the required peptide is added to the C-terminus of the β-galactosidase. This fusion protein is normally considerably more stable than the peptide product would be on its own. Although originally designed for cDNA libraries, considerable success has also been obtained with bacterial genomic libraries in λgt11.

It has to be borne in mind that expression of a foreign product may be deleterious to the host cell, and that such a gene may therefore be under-represented in a gene library. Vectors such as pUC and λgt11 attempt to overcome this by the use of an inducible promoter, so that expression of the gene only occurs if an inducing agent such as IPTG is added. The use of phage vectors rather than plasmids also helps in this respect, since continued viability of the host cell is not essential. Additional reasons for preferring bacteriophage vectors are that the storage of mixed phage suspensions is much more reliable and that the screening of phage plaques can be done at much higher density. The use of bacteriophage vectors does, however, require some understanding of the basic biology of λ phages. See Dale (1987) for more information.

If the library is to be screened with a gene probe rather than with antibody, expression of the product is not required and alternative phage vectors that do not encourage expression of the cloned gene should be used. One such vector, to be considered in more detail below, is another lambda phage, λEMBL4.

Fragmentation procedure

Having chosen a screening procedure and a vector, it is then necessary to determine the procedure for fragmenting the DNA. The easiest method is to use restriction enzymes. However, an enzyme like *Eco*RI, with a 6-base recognition site, will cut the DNA into fragments that are, on average, several kilobases long, and there is a reasonable chance that either the required gene will have a recognition site within it (and hence will not be cloned intact) or, even worse, will be surrounded by a region of DNA that is unusually devoid of *Eco*RI recognition sites and hence be present on a fragment that is so large that it will be cloned with reduced efficiency or not at all. The randomness of the library can be improved by using a 4-base enzyme (such as *Mbo*I or *Sau*3A); if digestion is allowed to go to completion, the fragments will be too small to be useful, but a partial digestion with such an enzyme will yield a pseudo-random library that is suitable for screening with a gene probe.

A vector link λgt11 demands an even greater degree of randomness however, since production of the appropriate fusion proteins (with the requirement for maintaining the reading frame) requires that, ideally, there should be at least one clone in the library where the inserted DNA starts from any specified base in the bacterial DNA: both orientations are required. The only way of generating such a completely random library is by mechanical shearing, e.g. by repeated passage of the DNA through a syringe needle, or by sonication.

Number of clones required

It is useful to have some idea of the number of independent recombinant clones that are required in the library in order to be reasonably confident of finding the gene of interest. This can be calculated from the formula:

$$N = \ln(1 - P)/\ln(1 - f)$$

where
N = number of clones needed;
P = probability of a specific gene being present;
f = fraction of the genome represented by a single clone.
The parameter f is obtained by dividing the average size of insert in the recombinant clones by the total size of the genome.

Thus, for a genomic library of *E. coli* (4×10^6 bp), if the average insert size is 4 kb (= 4×10^3 bp), the number of clones needed for a 95% probability ($P = 0.95$) is about 3 000. If larger inserts are used, the number of clones needed is correspondingly smaller. It should be appreciated that this calculation assumes a random distribution of clones, with no sequences being under or over-represented, which may not be the case. In particular, if the library is subjected to repeated amplification it may degenerate; any cloned fragment that results in even slightly reduced growth will be outgrown by other recombinants at each stage and will thus be reduced in frequency or even disappear altogether.

For a library prepared with λgt11, the calculation is different: in order to obtain a truly complete λgt11 library with all possible fusions being present, inserts in both directions starting at each base in the genome are needed. The complete total (for an *E. coli* library) is then in excess of 1×10^7 clones. Fortunately, such a complete library is usually not essential since many different fusions can be detected by the same antibody. It is usually advisable however to work with λgt11 libraries of at least 10^6 clones.

Construction of a Genomic Library by using λEMBL4

The vector

Cloning with bacteriophage lambda vectors is constrained by the phenomenon of *packaging limits*; viable phage particles can only be produced if the DNA is between 37 and 52 kb in length. (Note that this does not depend on the nature of the genes contained in that DNA. The phenomenon is not caused by a requirement for specific genes but by a physical requirement for that amount of DNA in order to maintain the physical integrity of the phage particle.) 'Wild type' λDNA is 48.5 kb long, so would have a maximum *cloning capacity* of about 3 kb. The cloning capacity can be increased by deleting non-essential regions, but if more than 11 kb is deleted, the vector DNA is too small for stable phage particles to be produced. A lambda vector into which foreign DNA is inserted (an *insertion vector*) therefore has a limited cloning capacity. The vector λgt11 for example has a maximum capacity of about 6 kb.

Replacement vectors such as λEMBL4 (Fig. 3) have a pair of restriction sites on either side of a non-essential region, which generate three fragments: left and right *arms*, and the non-essential *stuffer fragment*. The principal role of the stuffer fragment is to increase the size of the vector above that required for packaging, thus enabling a considerable increase in cloning capacity. With λEMBL4, for example, the cloning capacity is between 9 and 23 kb. Note that there is a positive selection for inserts of at least 9 kb; smaller inserts, or no

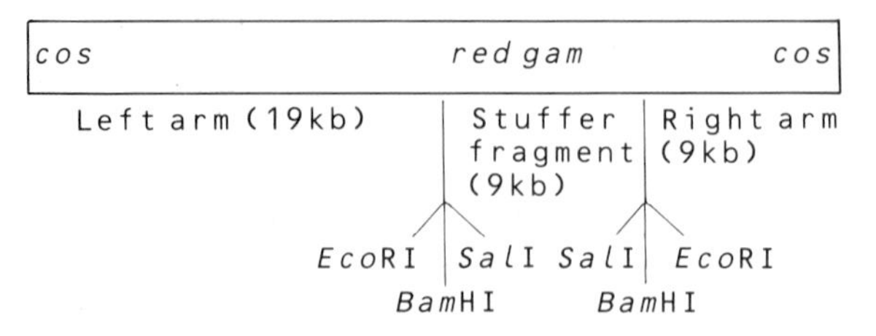

FIG. 3. Structure of lambda replacement vector EMBL4.

insert at all, will not produce viable phage. In the case of λ EMBL4, the stuffer fragment plays an additional role: the presence of the *red* and *gam* genes makes the phage unable to grow in a P2 lysogen (*spi*, sensitive to *P2 i*nhibition). Replacement of the stuffer by an insert fragment makes the phage *spi*$^-$; these recombinant phage can then be selectively grown in a P2 lysogen.

Preparation of the vector

Prepared 'arms' of λEMBL4, after digestion with *Bam*HI, can be purchased (e.g. from Amersham International), and this course of action is strongly recommended, especially to newcomers to the field. A cheaper, but very labour intensive, alternative is to prepare the vector yourself. For details of the method, see Maniatis *et al.* (1982). The steps involved are:

1 Growing the phage on a suitable *E. coli* host strain.
2 Extracting and purifying the phage DNA.
3 Digesting the phage DNA with *Bam*HI and separating the arms from the stuffer fragment.

With λEMBL4, it is useful to carry out a second digestion with *Sal*I. This cuts the ends of the stuffer fragment (see Fig. 3) yielding two very short *Sal*I–*Bam*HI DNA fragments that can easily be removed by selective precipitation with isopropanol. The stuffer fragment now contains *Sal*I sticky ends which cannot be religated to the *Bam*HI sticky ends of the vector arms.

Preparation of bacterial DNA fragments

Bacterial DNA can be isolated and purified by standard procedures, although with some organisms severe problems can be encountered in digesting with restriction endonucleases. This may be due to the presence of modified DNA sequences, or to the presence of contaminating material that is not easily removed. Firmly adhering proteins are sometimes not extracted by treatment with phenol/chloroform, and require the use of proteolytic enzymes (such as proteinase-K). Other materials such as polysaccharides can be even more difficult to remove. In our experience, some of these problems can be overcome by using DNA extracted from relatively young (mid-log-phase) cultures, rather than allowing the organisms to grow into stationary phase.

The bacterial DNA is then partially digested with a restriction endonuclease such as *Sau*3A or *Mbo*I. It is advisable to carry out a series of trial digests to establish the appropriate conditions. A very small amount of enzyme is sufficient, with digestion proceeding for a very short time, e.g. about 5 min when using 1 unit of *Sau*3A per microgram of DNA.

The digested DNA is then size-fractionated to ensure that the majority of DNA fragments are of the required size. Since the vector is selective for fragments between 9 and 23 kb, it might be considered superfluous to remove other fragments. However, there are two further considerations: the presence of smaller fragments will give rise to abortive ligation products that do not result in viable recombinant phage, and hence will reduce the efficiency of the cloning process. Secondly, it is possible to get multiple inserts, i.e. two (or more) of the smaller fragments may be recombined with the vector arms to produce a DNA molecule of the required size. While this may not prevent detection of a specific DNA sequence, it will result in considerable confusion when characterizing the cloned fragment and its relationship to adjacent regions of the bacterial DNA.

The latter problem is especially serious with this strategy, since, once a clone has been isolated, there is no easy way of determining whether it arises from a single contiguous DNA region. There will be a *Sau*3A site at the junction of the two fragments, but since the original digest was only partial, there will be a lot of *Sau*3A sites within single pieces of DNA. An additional procedure is therefore adopted to reduce the likelihood of multiple inserts, which is to treat the partially digested DNA with alkaline phosphatase (calf intestinal phosphatase, CIP). This removes the phosphate groups from the 5′ ends of the DNA fragments. Since ligation requires the presence of 5′ phosphates, the dephosphorylated fragments cannot be ligated to one another. The vector arms are *not* dephosphorylated, and can therefore be joined to an insert fragment.

Ligation

A number of conditions affect the efficiency of ligation, the most important of which are the total DNA concentration and the ratio of vector to insert DNA. The overall DNA concentration should be high in order to favour intermolecular (rather than intramolecular) recombination. With lambda vectors this also has the advantage of favouring the production of multiple length molecules, which are good substrates for *in vitro* packaging. The use of high DNA concentrations does demand that the DNA is of high purity. In particular, careful ethanol precipitation and washing with cold 70% ethanol is essential to remove traces of salt that would adversely affect ligation.

The ratio of vector to insert DNA is normally expressed in molar terms, not in weight, which means that you have to take into account the size of the molecules concerned. So for a plasmid vector of 5 kb and an insert of 1 kb, a 3:1 vector:insert ratio (molar) would be 15:1 in weight terms, i.e., for every one molecule of insert (1 kb) there would be three molecules of vector (total $3 \times 5 = 15$ kb).

With λ vectors, the calculation is complicated by the fact that the 'vector' consists not of a single molecule as with a plasmid vector, but of two arms, and, furthermore, by the requirement for the production of multiple length recombinant molecules of the form (left arm–insert–right arm)$_n$. An extensive theoretical treatment is provided by Maniatis *et al.* (1982), the upshot of which is that for a λ vector with arms totalling 31 kb and an insert size of 20 kb, the optimal ligation conditions would consist of 43 μg/ml of inserts and 135 μg/ml of vectors arms. In practice, several ligations with different ratios of DNA are recommended.

In vitro *packaging*

Although λDNA can be used in transformation (transfection) in much the same way as plasmid DNA, much greater efficiency is obtained by *in vitro* packaging. This involves the use of extracts from cells infected with defective bacteriophages so that the resulting mixture contains all the components necessary for the packaging and assembly of mature phage particles, but without any contaminating λDNA. The DNA sequence between two *cos* sites is packaged into the pre-existing empty phage heads to produce mature phage particles which can then be used to infect a sensitive indicator bacterium.

Buffers and other materials

TE buffer:
  10 mM Tris–HCl, 1 mM EDTA, pH 8; NA45 membranes (Schleicher & Schuell).

NA45 elution buffer:
  0.05 M arginine, 1 M NaCl.

5 × ligation mix:
  0.25 M Tris–HCl pH 7.6, 50 mM $MgCl_2$, 5 mM ATP, 5 mM dithiothreitol, 25% (w/v) polyethylene glycol-8 000.

SM (phage) buffer:
  50 mM Tris–HCl, 200 mM NaCl, 1 mM $MgSO_4$, 0.01% gelatin, pH 7.5.

L-broth:

15 g tryptone, 5 g yeast extract, 5 g NaCl, water to 1 litre.

Top agar:

Add 0.7 g of agar to 100 ml of L-broth. Boil to dissolve the agar, distribute in 3 ml volumes and autoclave. For screening libraries, and for obtaining phage lysate for DNA preparation, agarose is often used instead of agar. It is less likely to stick to the filter, and does not contain the impurities that may contaminate your DNA.

E. coli strains: LE392 F^-, *hsdR*514, *supE*44, *supF*58, *lac*Y1

P2392 P2 lysogen of LE392.

Method

1 *Partial digestion with Sau3A.* Assume that trial digests have shown that with 1 unit of enzyme per μg of DNA, the time required to obtain a suitable range of restriction fragments is about 5 min.

Take 5–10 μg of bacterial DNA, add 0.1 volume of restriction buffer (10 ×), warm the tube to 37°C and start the reaction by adding the restriction enzyme. Incubate in a 37°C water bath. After 3 min, remove ⅓ of the reaction volume to a tube containing an equal volume of phenol/chloroform and mix vigorously to stop the reaction. Continue incubating the rest of the digest, and remove further samples in a similar manner at 5 and 7 min.

Spin the tubes to separate the layers and remove the aqueous layers to fresh tubes.

2 *Size selection.* To each tube, add 0.1 volume 3 M sodium acetate and 2.5 volumes of ethanol. Place on ice for 30 min and centrifuge for 15 min. Carefully remove all of the ethanol and resuspend the DNA in 10 μl of TE buffer. Separate the fragments on a 0.7% agarose gel, using the whole of each sample, and with a set of standard markers to calibrate the gel. After electrophoresis, stain and photograph the gel in the usual way, but keeping exposure to u.v. to the minimum, as it will be necessary to recover DNA from the gel.

Determine which track has given the best results: the major part of the DNA fragments occur in the 9–23 kb region. With luck, all three will be usable. The suitably sized fragments are then recovered from the gel. This can be done by using NA45 membranes (Young *et al.* 1985). Make a slit in the gel, across the width of the selected track and in a position corresponding to fragment sizes of about 9 kb. Put a piece of NA45 membrane in the slit. Resume electrophoresis until all of the required fragments have migrated as far as the slit, and have therefore bound to the membrane.

It is usual to put a second piece of NA45 membrane at the top of the size range to intercept any larger fragments. For this procedure, however, there will be very few such fragments and they will not interfere significantly. Furthermore, it is difficult to separate 23 kb fragments from larger ones, as they will all run very close together.

Remove the membrane from the gel and put it into 400 μl of elution buffer. Incubate at 70°C for 2 h, shaking it from time to time. Remove the membrane from the tube, and cool the eluate to room temperature. Add 400 μl of isoamylalcohol (water saturated), mix and allow to stand for a few min. Remove the top (alcohol) phase. To the aqueous phase, add 400 μl of isopropanol to precipitate the DNA. Recover the DNA precipitate as usual and redissolve in TE buffer.

An alternative procedure for size selection is that of selective precipitation. Add 0.15 volume of 0.3 M sodium acetate pH 6, followed by 0.3 volume of isopropanol. After standing on ice for 15 min, precipitate the DNA by centrifugation for 10 min. Then wash the pellet and redissolve in TE buffer. This procedure is simpler and can give better recovery of larger fragments (> 15 kb) which can be difficult to elute from the NA45 membrane. However, the NA45 procedure is easier to adapt for a wide range of sizes. Figure 4 shows a comparison of results obtained by the two procedures.

3 *Dephosphorylation.* This can be achieved by adding calf intestinal phosphatase (CIP); although TE buffer is not optimal for this enzyme, there will be sufficient activity to achieve the desired result. Incubate for 30 min at 37°C, then terminate the reaction and remove all traces of the phosphatase by thorough phenol/chloroform extraction and ethanol precipitation. Wash the DNA pellet with ice-cold 70% ethanol before drying carefully in a vacuum dessicator. Redissolve in a minimal volume (e.g. 10 μl) of TE buffer.

4 *Ligation.* If the average fragment size is assumed to be 14 kb, and the size of the vector arms is 28 kb, then a molar ratio of 2:1 (arms:inserts) is represented by a weight ratio of 4:1 (arms:inserts). If there is a large amount of prepared DNA, use three weight ratios, 2:1, 4:1 and 6:1 (representing molar ratios of 1:1, 2:1 and 3:1. Satisfactory results have been achieved with 2 μg of vector to 1 μg of insert, in a reaction volume of 10 μl. Also set up a control ligation, in the absence of insert fragments, to determine the level of background caused by incomplete removal of the stuffer fragment and incomplete digestion of the vector DNA. The total amount of DNA in each reaction should be 2–5 μg.

Mix the appropriate amounts of vector and insert DNA and carry out an ethanol precipitation, redissolving the pellet in 7 μl of deionized distilled water. (If the DNA is already of high enough concentration and purity, this step can be omitted.) Add 2 μl of 5 × ligation buffer and 1 μl of T4 DNA ligase and incubate the ligation mixture at room temperature (e.g. 20°C) for

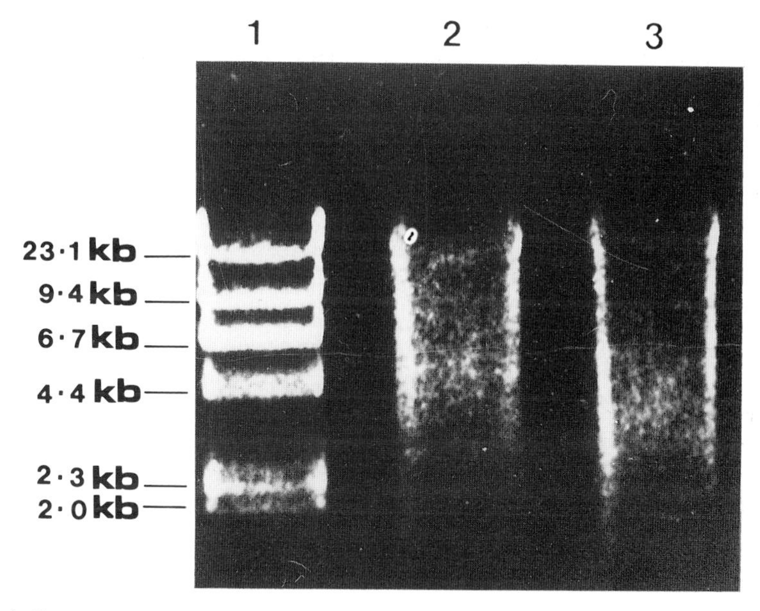

FIG. 4. Size-enrichment of DNA fragments. A sample of DNA from *Mycobacterium tuberculosis* was partially digested with *Sau*3A and divided into two portions. One portion was enriched by selective precipitation, while the second was fractionated by agarose gel electrophoresis and the NA45 procedure. Both were then analysed by agarose gel electrophoresis, in comparison with *Hin*dIII-digested λDNA (track 1). Track 2: enriched by selective precipitation. Track 3: NA45 selection.

4 h. Precipitate the DNA from the ligation reaction with ethanol and redissolve in 3 μl of water.

5 *In vitro packaging.* For each of the ligation mixtures, including the control, carry out an *in vitro* packaging reaction, using a commercially available kit under the conditions recommended by the manufacturer. Include a control packaging reaction with intact λDNA to test the efficacy of the packaging reaction and the sensitivity of the indicator strain. Control DNA is normally provided with the kit.

The product of the *in vitro* packaging reaction constitutes the primary gene library, and can be stored at 4°C after adding a drop of chloroform.

6 *Titration of the library.* Before using the library, it is necessary to determine the number of recombinant phage particles, by titrating the library using a sensitive indicator strain. For a λEMBL4 library, the *E. coli* strain P2 392 is used; this strain is lysogenic for the bacteriophage P2, and will therefore only

allow recombinant phage to grow. The number of non-recombinant phage particles can be determined by using a non-lysogenic host such as LE 392.

Inoculate a single colony of the indicator strain into 10 ml of L-broth containing 10 mM $MgSO_4$ and 0.2% maltose, and incubate, with shaking, overnight at 30°C. Centrifuge the culture and resuspend the cells in 4 ml of 10 mM $MgSO_4$. Keep the cells on ice until required.

Dilute the cells in SM buffer to A_{600nm} of about 0.5. Prepare a 10-fold serial dilution of the packaging mixture (to 10^{-4}), in SM buffer, and add 10 μl of each dilution to 200 μl of prepared indicator cells. Incubate at 37°C for 20 min to allow the phage to adsorb to the cells.

Melt the required number of tubes/bottles of top agar, and allow them to cool to 45°C. Add the infected cells, and pour onto a L-agar plate. It is advisable to do this one at a time, and (at least if you have never done it before) it can be helpful to warm the plates to 37°C before pouring the overlays, to ensure that they do not set before they have been distributed evenly over the surface.

When the overlays have set, incubate the plates at 37°C overnight. (Plaques may be visible at about 6–8 h.) Count the number of plaques on each plate and determine the titre of your libraries.

Screening of a λEMBL4 Library with Gene Probes

Principle of the procedure

At this stage, you will have a primary gene library that should consist of at least 5 000 recombinant phage particles. This number should be sufficient for a bacterial library, but for other organisms with larger genomes, a much more extensive library will be required.

The first step is to amplify the library, to enable you to carry out a number of procedures with it, as well as distributing samples to all your friends! This is done by infecting a bacterial culture and allowing the phage to replicate, after which the progeny is harvested. It is important to be aware of the fact that, in the primary library, each phage arises from a single recombinant DNA molecule and is therefore an independent individual, while in an amplified library there will be many hundreds of identical phages that arise from one such individual. When screening an amplified library, therefore, the positive clones identified are quite likely to be identical re-isolates of the same recombinant. Furthermore, some recombinant phages will grow better than others and will therefore be present in relatively higher numbers in the amplified library. Repeated amplifications will give rise to degeneration of the library.

Identification of the required clones with gene probes relies on the ability of single-stranded complementary DNA molecules to reassociate (hybridize). Material from phage plaques on an agar plate is transferred to a membrane, the DNA is released and bound to the filter in a single-stranded form. The filter is then incubated with the labelled probe, washed to remove unbound probe, and the position of the hybridized probe is correlated with the position of the phage plaques on the original plate. These plaques can then be picked off and the phage purified. The procedure is summarized in Fig. 5.

Buffers

20 × SSC: 3 M NaCl, 0.3 M sodium citrate, pH 7.
Denaturing solution: 0.5 M NaOH, 2.5 M NaCl.
Neutralizing solution: 3 M sodium acetate pH 5.5.

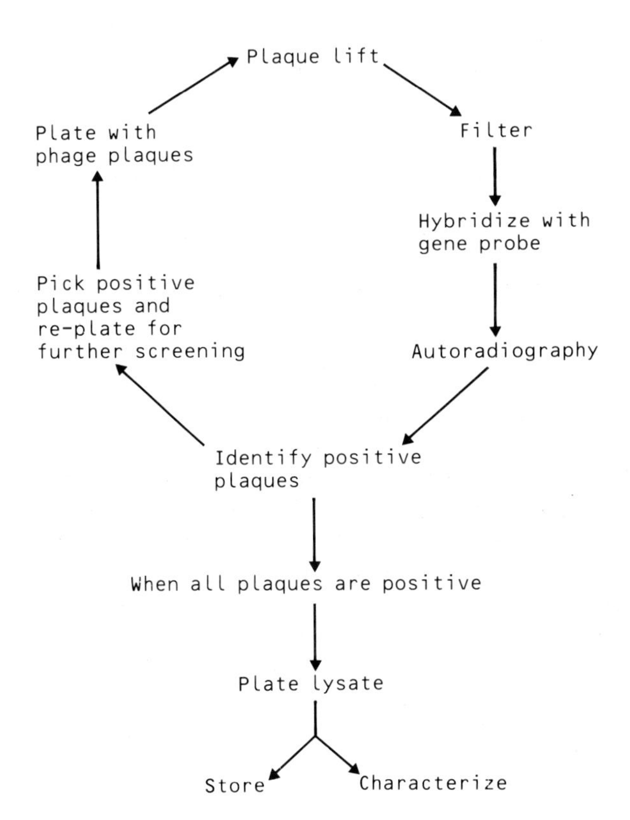

FIG. 5. Screening gene libraries.

Method

1 *Amplification.* This can be done by a plate lysate technique, which is essentially similar to the procedure described above for titrating the library. Add 100–500 μl of the stock library to 3 ml of plating cells and allow the phage to absorb for 30 min at 37°C. Add this suspension to 30 ml of top agar (at 45°C) and pour this over a large L-agar plate (230 × 230 mm). After overnight incubation at 37°C, flood the plate with 15 ml of SM buffer. Shake the plate gently at 4°C for at least 6 h (or overnight), collect the buffer and wash the plate with a further 5 ml of buffer. Pool the two extracts, and add 1 ml of chloroform, to lyse any bacterial cells present. After shaking for 15 min, clarify the suspension by centrifugation at 4000 rpm for 10 min. Recover the supernatant and add chloroform (0.3% v/v). Keep a sample at 4°C as your working stock and store the remainder at −20°C or −70°C after adding DMSO to 7% (v/v).

2 *Plaque lifts.* Plate out an aliquot of the amplified library by the technique described above, using an L-agar overlay. Large plates are required (e.g. 230 × 230 mm) to obtain enough plaques for screening. Aim to get as high a density of plaques as possible, while still retaining the ability to just distinguish separate plaques. Do not incubate the plates for too long (8 h should be enough) and carry out the plaque lifts straight away. Do not underestimate the ability of phage particles to continue diffusing through the overlay, even after plaques have apparently stopped 'growing'.

Cut a piece of nylon membrane (Hybond-N) slightly smaller than the size of the plate and place it on the agar surface. Mark it in several places (e.g. by piercing the membrane and agar with a sterile needle) to facilitate subsequent orientation. After 1 min, remove the membrane and place it on a filter paper pad soaked in denaturing solution. After 7 min, transfer it to a second pad soaked in neutralizing solution and leave it for a further 7 min. Then allow it to dry in air, on a dry filter paper pad, and bake it at 80°C for 1 h. As an alternative to baking, the DNA can be covalently bound to the filter by wrapping it in Saran Wrap and placing it on an u.v. transilluminator for 4 min. While the filter is being processed and hybridized, store the original plate at 4°C.

3 *Hybridization.* The probe can be labelled, with ^{32}P, ^{35}S or biotin, by nick translation or (preferably) by random primer extension. Synthetic oligonucleotides can be labelled by the use of polynucleotide kinase. The probes must be boiled immediately before use to denature them, i.e. to separate the two DNA strands. The hybridization conditions need to be optimized for different types of probe and different degrees of homology. The procedure is therefore

described only in general terms. See Arrand (1985) and Mason & Williams (1985) for a detailed description of labelling and hybridization procedures.

The filter is prehybridized with a solution containing salmon sperm DNA (sonicated, denatured) to saturate the non-specific binding sites. This buffer is then replaced with a similar buffer containing the denatured labelled probe and hybridization is allowed to proceed overnight. After removal of the hybridization buffer, the filter is washed with solutions of decreasing ionic strength. The stringency of the washing procedure can be adjusted to take account of the predicted strength of binding of the probe to the target DNA. If it is known to be homologous to the target, a high stringency wash (e.g. 0.1 × SSC/0.1% SDS at 68°C) will give low background.

However, if a probe derived from the corresponding gene from another organism is to be used the degree of homology may be considerably lower, and a lower degree of stringency is indicated. This can be achieved by using higher ionic strength (e.g. 1 × SSC or even 3 × SSC) or a lower temperature.

If a ^{32}P labelled probe is used, the signal is detected by autoradiography. Remove the filter from the bag, place it on a glass plate and cover it with Saran wrap. Put a piece of X-ray film on the filter, followed by an intensification screen. Mark the film and the filter to aid orientation. Put the whole lot into a cassette and leave at −70°C for a period of time between several hours and several days, depending on the strength of the signal.

4 *Location and purification of required clones.* After autoradiography, any positive clones will be detected as black spots on the X-ray film. Align the film with the original plate to identify the corresponding plaques on the plate. Pick these plaques with a Pasteur pipette and resuspend in 100 μl of SM buffer. If the plaque density is high, it may not be possible to identify the specific plaque that corresponds to that spot; in this case, recover the corresponding area of the overlay, add up to 1 ml of SM buffer and vortex to disrupt the agarose. Remove the agarose by centrifugation and use the supernatant to inoculate a fresh plate at a lower density for rescreening by the same procedure.

Several rounds of plaque purification and retesting may be necessary before a stage is reached when all the plaques on the plate react with the probe. It is commonly found that plaques which give a promising signal on the first plate fail to produce signals on subsequent purification. The original result may be a false positive reaction, or the wrong plaque may have been picked, or the clone may be unstable so that the cloned sequence is lost on subsequent amplification.

Once one or more clones have been plaque-purified to homogeneity, produce an amplified suspension by a plate lysate procedure (see above). Store some of this suspension as described previously, and use the remainder for extracting the DNA for characterization of the insert. Figure 6 a shows an

(a)

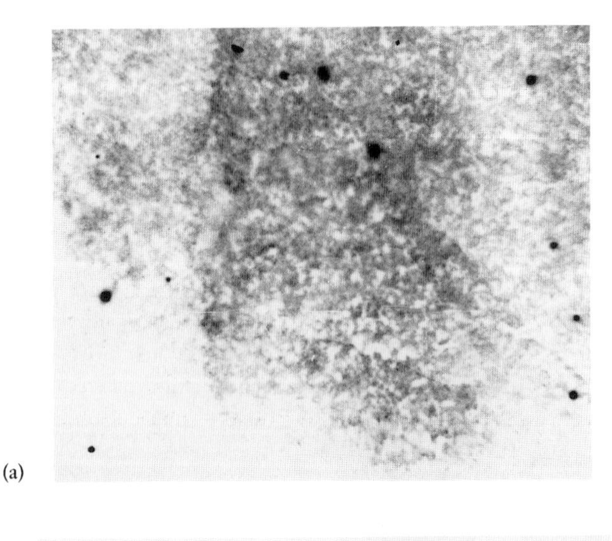

(b)

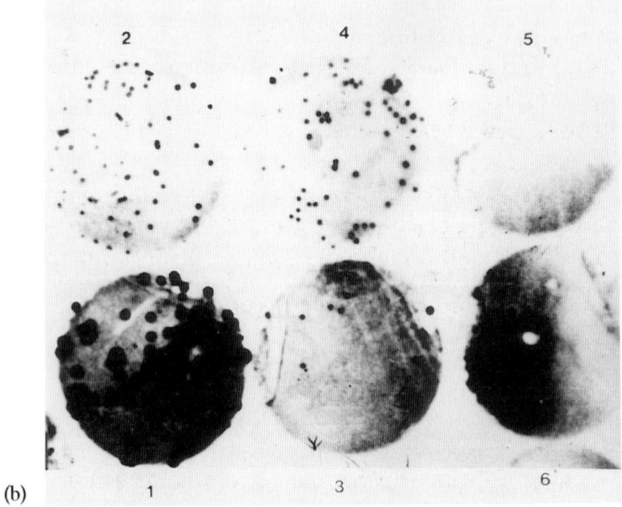

FIG. 6. Screening a gene library with a gene probe: (a) an EMBL4 library of *Mycobacterium tuberculosis* was screened with a ^{32}P-labelled gene probe. Positive signals were detected by autoradiography; (b) six positive plaques were isolated and rescreened at a lower density. Plates 1, 2 and 4 are homogeneously positive; 3 shows a few plaques only, and is still mixed; 5 and 6 are negative, and were either false positive originally or the positive phages have been lost in the process. Problems: in 1, the autoradiograph is over-exposed and could be repeated with a shorter exposure. In 6, the dark area is probably due to insufficient prehybridization of this membrane. See Table 1.

example of the results obtained on screening a library with a gene probe, and Figure 6 b shows the effect of rescreening plaque-purified clones.

Construction of a λgt11 Library and Screening with Antibodies

Much of the procedure for construction of a library in λgt11 is similar to that for a λEMBL4 library and will therefore not be described in detail. In addition, the cloning and screening procedures with λgt11 have been described fully by Huynh *et al.* (1985) and Young *et al.* (1985). The major points of difference from λEMBL4 arise from the design of λgt11 as an insertion vector (with a cloning capacity up to 6 kb) with the ability to express cloned genes from the inducible *lac* promoter giving rise to β-galactosidase fusion proteins (see Fig. 7). It is therefore primarily useful for constructing libraries that are to be screened with antibodies.

Buffers and other materials required

IPTG: 10 mM in water; sterilize by filtration.
X-gal (5-bromo-4-chloro-3-indolyl-β-D-galactopyranoside): 40 mg/ml in dimethylformamide.
Overlays: use 4 μl of IPTG and 40 μl of X-gal in 3 ml of top agar.
TBS: 50 mM Tris–HCl, 150 mM NaCl, pH 7.5.
TBST: TBS containing 0.05% Tween 20.
Peroxidase detection reagent: dissolve 30 mg 4-chloro-1-naphthol in 10 ml AnalaR methanol. Add 50 ml TBS followed by 30 μl of H_2O_2 (30%).
E. coli *strains*: Y1088 (ATCC 37195): *E. coli* Δ *lac*U169 *sup*E *sup*F *hsd*R^- *hsd*M^+ *metB trpR tonA*21 *proC*::Tn5 (pMC9)
Y1089 (ATCC 37196): *E. coli* Δ *lac*U169 *pro*A^+ Δ *lon araD*139 *strA hflA*150 (chr::Tn10) (pMC9)
Y1090 (ATCC 37197): *E. coli* Δ *lac*U169 *pro*A^+ Δ *lon araD*139 *strA sup*F [*trp*C22::Tn10] (pMC9)
NB: pMC9 is pBR322-*lac*I^q. Each strain should be maintained and cultured in the presence of ampicillin (50 μg/ml) to ensure retention of this plasmid.

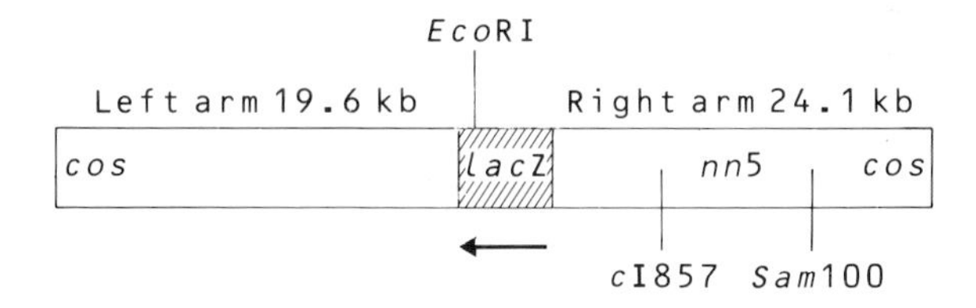

FIG. 7. Structure of the vector λgt11.

Method

1 *Fragmentation of DNA to be cloned.* In order to obtain the full range of in-frame fusions, it is necessary to use a genuinely random fragmentation procedure, such as repeated passage through a 21G syringe needle. This needs to be done as forcefully as possible, and it may be necessary to dilute the DNA to obtain efficient shearing. After at least 50 passages, remove a sample and check the size distribution by agarose gel electrophoresis. Repeat the shearing procedure, checking the size distribution periodically, until the major part of the DNA is less than 6 kb in length. This may require several hundred passages.

Fill in the ragged ends of the DNA fragments, using T4 DNA polymerase, and use *Eco*RI methylase to protect any internal *Eco*RI sites. Remove these enzymes by phenol/chloroform extraction and ethanol precipitation before ligating *Eco*RI linkers to the ends of the fragments. Inactivate the ligase by heating to 70°C for 15 min. Adjust the buffer conditions (carrying out a further ethanol precipitation if necessary) and digest with *Eco*RI endonuclease to expose the *Eco*RI sticky ends.

Size-fractionate the DNA on a 0.8% agarose gel, selecting fragments between 2 and 6 kb. Smaller fragments can be included if desired, but fragments larger than 6 kb will give rise to abortive products that are too large to be packaged into phage particles. The DNA fragments can be recovered by the NA45 procedure described above.

The DNA fragments recovered from the gel may be contaminated by traces of the large excess of linker molecules that are present. This problem can be overcome by ethanol precipitation prior to gel electrophoresis, using a short centrifugation time (e.g. 5 min).

2 *Ligation.* Since λgt11 does not provide positive selection for recombinants, it is advisable to use dephosphorylated vector arms to reduce the level of non-recombinant phage particles. Prepared vector arms (*Eco*RI cut and dephosphorylated) are commercially available (Stratagene, obtainable from Northumbria Biologicals). Do *not* dephosphorylate the insert as well!

The same considerations as before apply to the amount of vector and insert DNA to be used. Theoretical calculations indicate higher vector:insert ratios (such as 10:1 or even 20:1, by weight), but satisfactory results have been achieved with 1 μg of vector to 250 ng of insert. After ligation, the primary library is produced by *in vitro* packaging.

3 *Titration of the library.* The proportion of recombinant clones in the library can be easily estimated since insertion of foreign DNA will (usually) cause inactivation of the β-galactosidase. In the presence of IPTG and X-gal,

plaques arising from parental vector phage will be blue while recombinant phage particles will give rise to colourless plaques. Two additional features of λgt11 must be remembered:

1 It contains an amber mutation (*S*100) and will therefore only produce plaques on amber suppressor (supF) hosts.

2 It is a *c*I$_{857}$ phage, which results in a temperature-sensitive repressor; it will therefore be repressed at 30°C and can be induced by shifting the culture to 42°C.

Assay a sample of the library, with Y1088 as the host strain, using L-agar plates containing ampicillin (50 μg/ml) and overlays containing IPTG and X-gal. Incubate the plates at 42°C. Determine the total number of phage in the library, and the percentage of recombinants (colourless plaques). If this is satisfactory, a portion of the library can be amplified by the same procedure but omitting the IPTG and X-gal. Avoiding induction of the β-galactosidase gene reduces the likelihood of degeneration of the library due to deleterious effects arising from the fusion proteins.

4 *Antibody screening.* The host strain used for this purpose is Y1090. After infection with the library, overlay a plate of L agar containing ampicillin (50 μg/ml); do *not* include IPTG or X-gal. Incubate the plate at 42°C for 3.5 h to induce phage multiplication. One available screening procedure, summarized in Fig. 8, uses a biotinylated secondary antibody, followed by streptavidin–biotinylated peroxidase complex. This has been found to be a sensitive procedure, despite the drawback of frequent detection of false-positive clones that express biotinylated products (Collins *et al.* 1987).

Soak a nitrocellulose filter in IPTG (10 mM) and allow it to dry in air. Lay this filter on the agar surface and incubate for a further 3.5 h at 37°C; this induces expression of the β-galactosidase gene. Mark the filter and the plate to aid subsequent orientation. Remove the filter from the plate and wash it in TBST before incubating overnight in TBST containing 3% BSA to block non-specific protein binding sites on the filter.

Replace the blocking solution with TBST + BSA + antibody. After 1 h, wash the filter twice with TBST and then incubate for 2 h in TBST + BSA + biotinylated second antibody. For example, if the primary antibody is raised in mice, the second antibody can be biotinylated rabbit anti-mouse. Wash the filter in TBST and then incubate for 1 h in TBST containing streptavidin–horseradish peroxidase complex (Amersham). After washing in TBS, develop the filter with the detection reagent. Positive plaques exhibit a purple halo within 15 min. These plaques can then be picked and purified as described above.

Note that if a conventional antiserum (polyclonal) is being used as the primary antibody, it will usually be necessary to adsorb out any *E. coli* anti-

(i) React filter with primary antibody

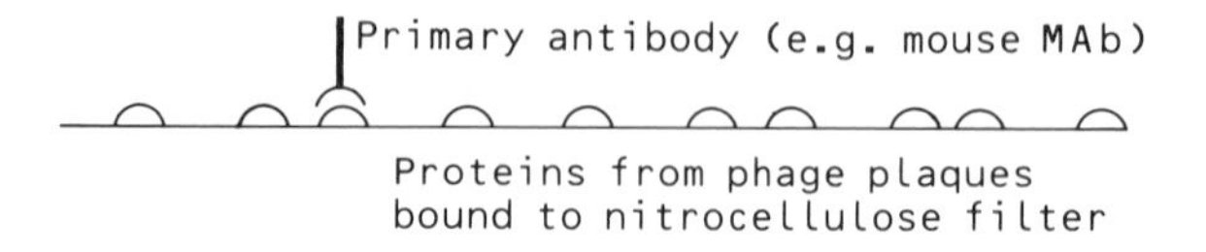

(ii) Bind secondary antibody (biotinylated)

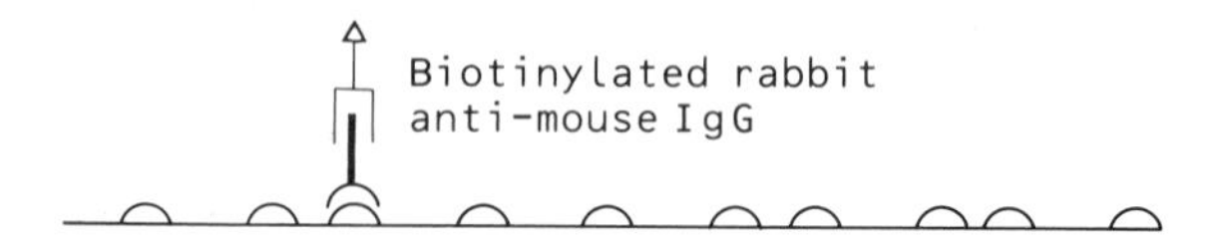

(iii) React with streptavidin–biotinylated peroxidase complex

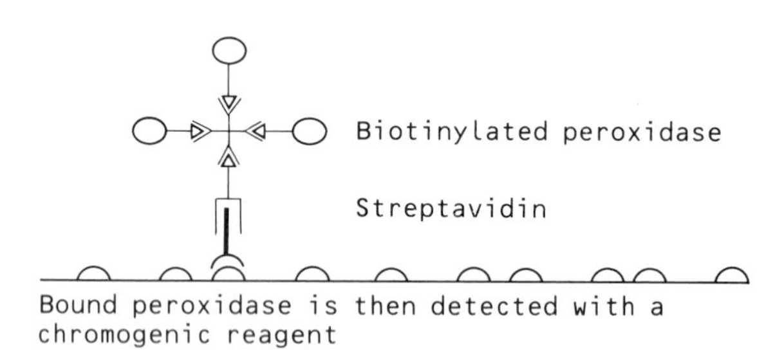

Bound peroxidase is then detected with a chromogenic reagent

FIG. 8. Antibody screening with biotin–streptavidin.

bodies. Rabbit serum in particular often contains considerable levels of *E. coli* antibody that will cause extensive non-specific reactions. Adsorption of antiserum can be conveniently carried out by using nitrocellulose discs. Prepare a sonicated extract of *E. coli* in PBS, place a number of nitrocellulose discs in this extract, and shake for about 5 min. Remove unbound material by two brief rinses in TBST. Allow these discs to dry and store them at 4°C. To adsorb a serum sample, place a prepared disc in suitably diluted serum and leave it for 30 min. Remove the disc and discard it, and repeat the procedure with a fresh disc. The serum is then ready for use, or can be stored at −20°C or −70°C until needed.

5 *Preparation of lysogens.* In order to characterize the antigen produced by these clones, it is necessary to produce lysogens. The procedure, using the

host strain Y1089, is described by Huynh *et al.* (1985), but we have found it necessary to use a considerably higher input ratio, of about 100 phage per bacterium. After infection, the cells are plated out to obtain single colonies, and the plates are incubated at 30°C. Lysogens are detected by testing these colonies for inability to grow at 42°C, although this test is not completely reliable.

Single colonies that grow at 30°C but not at 42°C are then purified and used to inoculate 30 ml of L-broth which is incubated at 30°C for 3–4 h, to $A_{600} = 0.5$. Transfer the flask to a shaking water bath at 42°C for 20 min, add IPTG (10 mM) and transfer to 37°C for 15–30 min. Do not incubate too long at this stage or the cells will lyse.

Pellet the cells and prepare extracts for SDS PAGE electrophoresis by boiling in sample buffer. After electrophoresis, transfer the bands to a nitrocellulose filter by electroblotting and develop with antibodies as described above.

TABLE 1. *Troubleshooting*

Causes	Action
1 *No plaques*	
Wrong (or contaminated) indicator strain	Purify indicator strain from original stock. Check sensitivity by plating a known phage (if available), or by transfection (transformation) with vector DNA or ligated phage arms
Failure of ligation	Carry out a control ligation using vector arms (+ stuffer fragment if appropriate). Examine ligation products by gel electrophoresis. Repurify insert DNA by phenol/chloroform extraction and ethanol precipitation. Use different vector:insert ratios/DNA concentration
Failure of *in vitro* packaging	This should not happen with commercial kits, unless your DNA is unsuitable. Test with control λ DNA. If it still does not work, and you have established that your indicator strain is correct, complain to the suppliers
2 *No positive clones detected with gene probe*	
Stringency too high	Repeat with a lower temperature for washing and/or higher ionic strength
Lack of sensitivity	Check labelling of probe. Use longer period of autoradiography

TABLE 1. (contd.)

Causes	Action
Inadequate hybridization	Ensure that hybridization solution covers the filter completely. If multiple filters are used, ensure they are separated
Not enough plaques screened	Repeat with more plaques!
Gene not present in the library	Make another library, with a different strategy/vector if necessary
Unsuitable synthetic probe	Check the design of your probe. Try a different one. Check the labelling of the probe
Cloned probe not homologous	Devise another strategy or give up!
3 *Background too high (using a gene probe): all plaques positive, extensive black areas on autoradiograph, or large numbers of non-specific spots (see Fig. 6b)*	
Stringency too low	Wash the filter at higher stringency and repeat autoradiography
Inadequate prehybridization	Ensure that prehybridization buffer covers the filter completely. If multiple filters are used, ensure they are separated
Autoradiography problems	Check cassette for radioactive contamination. Is the cassette light-proof? Film may have been exposed before use
4 *No positive clones detected with antibody*	
Wrong concentration of antibody	Repeat with different antibody dilutions
Failure of detection system	Check that your second antibody is raised against the species that your primary antibody comes from. Include a positive control: a small spot of cell extract from the original source of your gene
Not a protein antigen	This usually applies to mAbs, which may recognize other types of antigen, or non-linear epitopes. Use a different mAb
Lack of sensitivity	mAbs often show less sensitivity than polyclonal serum. Use a different mAb, or a polyclonal serum
(See also the problems under (2) above)	
5 *High background with antibody screen*	
E. coli antibodies present	Adsorb serum with *E. coli* antigen
(see also problems under (3) above)	

TABLE 1. (contd.)

Causes	Action
6 *Non-specific clones detected with biotinylated second antibody*	
Biotinylated proteins detected	Test all filters in duplicate without primary antibody; saturate biotin with streptavidin then saturate streptavidin with biotin; *or*, use a different detection procedure

Acknowledgements

I am indebted to past and present workers in my laboratory who have adapted and tested the procedures that I have described; I would like in particular to acknowledge the contribution made by Mick Moss, Zainul Zainuddin, Margaret Collins and Sue Wall.

References

ARRAND, J.E. 1985. Preparation of nucleic acid probes. In *Nucleic Acid Hybridisation: a Practical Approach*, eds Hames, B.D. & Higgins, S.J. Ch. 2, pp. 17–45. Oxford: IRL Press.

COLLINS, M.E., MOSS, M.T., WALL, S. & DALE, J.W. 1987. Expression of biotinylated proteins in mycobacterial gene libraries. *FEMS Microbiology Letters* **43**, 53–56.

DALE, J.W. 1987. Cloning in bacteriophage λ. In *Techniques in Molecular Biology*, eds Walker, J.M. & Gaastra, W. Ch. 9, pp. 159–177. London: Croom Helm.

HUYNH, T.H., YOUNG, R.A. & DAVIS, R.W. 1985. Constructing and screening cDNA libraries in lambda gt10 and lambda gt11. In *DNA Cloning*, ed. Glover, D.N. Vol. 1, pp. 49–78. Oxford: IRL Press.

MANIATIS, T., FRITSCH, E.F. & SAMBROOK, J. 1982. *Molecular Cloning: a Laboratory Manual.* Cold Spring Harbor, New York: Cold Spring Harbor Laboratory.

MASON, P.J. & WILLIAMS, J.G. 1985. Hybridisation in the analysis of recombinant DNA. In *Nucleic Acid Hybridisation: a Practical Approach*, eds Hames, B.D. & Higgins, S.J. Ch. 5, pp. 113–137. Oxford: IRL Press.

YOUNG, R.A., BLOOM, B.R., GROSSKINSKY, C.M., IVANYI, J., THOMAS, D. & DAVIS, R.W. 1985. Dissection of *Mycobacterium tuberculosis* antigens using recombinant DNA. *Proceedings of the National Academy of Sciences of the United States of America* **82**, 2583–2587.

DNA Probes for Detection and Identification of Bacteria

J.J. McFadden and A. Knight
Molecular Biology Group, Department of Microbiology, University of Surrey, Guildford, Surrey GU2 5XH, UK

Principle of Nucleic Acid Hybridization

DNA and RNA hybridization follow essentially the same principles and only DNA will therefore be referred to. The RNA hybridizations follow similar principles.

DNA denaturation

When an aqueous solution of double-stranded DNA is heated, the u.v. absorbance, due to the bases, increases as a result of local reversible denaturation, or looping out, of short stretches of DNA. At low temperatures, this effect is reversible – the DNA will spontaneously reanneal on cooling and the absorbance will return to the initial value. If, however, the temperature increases beyond the melting temperature (T_m), the denaturation becomes irreversible, the strands separate and, on cooling, the absorbance remains high. The T_m of dsDNA is a function of the thermal stability of DNA, which in turn is a function of the number of base pairing interactions per unit length. In perfectly homologous DNA this will be a function of the base ratio (G+C content) of the DNA, since GC pairing involves 3 hydrogen bonds; whereas AT pairing involves only 2. T_m is also affected by the ionic strength of the medium (being raised with increasing ionic strength), the presence of ions (e.g. Mg^{+}), polysaccharide, protein and organic solvents (e.g. formamide). Under standard conditions, measurement of T_m is used to estimate the G+C content of DNA and has been widely used to estimate base ratios of bacterial DNA.

DNA reassociation

The reassociation of denatured double-stranded DNA, at temperatures below the T_m, is a spontaneous reaction. The rate-limiting step is the collision of the

Genetic Manipulation
0–632–02926–9

complementary DNA strands capable of forming a stable hybrid. The reaction therefore follows approximately second-order kinetics with the reaction rate being proportional to the product of the concentration of complementary DNA. An obvious, though important, point is that each single-stranded reactant will hybridize to only a molar equivalent of homologous DNA. Measurement of rates of DNA reassociation may be used to estimate the complexity of DNA and to detect repeated sequences in DNA (Britten & Kohne 1968).

DNA hybridization

Most applications of DNA probes involve strand annealing between heterologous DNA, a reaction known as DNA hybridization. The formation of stable DNA hybrids between heterologous DNA strands is possible only at temperatures below the T_m of the hybrids. As described above, the T_m of heterologous DNA is largely a function of the number of base-pairing interactions and, therefore, in this case, the DNA homology between the strands forming the hybrid.

Since the T_m is affected by all of the factors mentioned above (ionic strength, organic solvents), DNA hybridization will be similarly affected. The term *stringency* is an important concept in this context, but is a loosely defined term encompassing all the factors affecting hybrid stability: temperature, salt concentration, presence of organic solvents. Conditions of low stringency (e.g. high salt, low temperature) allow DNA hybrids to form between DNA species with low levels of DNA homology; whereas high stringency conditions (e.g. low salt, high temperature, presence of formamide) allow hybridization only between DNA species with high levels of DNA homology. This simple principle, coupled with methods for detection of the hybrid, is the basis of most DNA probe applications.

Conditions are usually chosen that allow hybrids to form only between highly homologous DNA, and so, for most DNA probe applications, hybridizations are carried out at temperatures 20–30°C below the T_m of homologous DNA. This will allow hybrids to form between DNA strands with approximately 15–25% mis-match of total base-pairing interactions (Bonner *et al.* 1973). However, this will vary depending on the G+C content, length of probe and distribution of homologous sequences.

Hybrid detection

In some of the earliest studies hybridization reactions were monitored spectrophotometrically, utilizing the hyperchromicity of DNA to measure the conversion of ssDNA to dsDNA. The advantages of this method are its relative simplicity and non-reliance on radioactive labelling. However, the requirement for large amounts of DNA, a spectrophotometer with thermostatically-controlled cuvette chambers and the inability to monitor more than one hybridization at a time, makes this an impractical method for most DNA probe applications.

Most other methods require one of the DNA species involved in the hybridization to be labelled (usually radiolabelled) and the sequestering of the labelled DNA into hybrid is monitored. Many hybridization experiments are performed with both reactants in solution and numerous techniques may be used to assess the amount of double-stranded hybrid formed. Hydroxyapatite will, under appropriate conditions, only bind dsDNA and can therefore be used to determine the fraction of labelled probe present in hybrid. S1 nuclease can also be used since, again under appropriate conditions, it will specifically degrade ssDNA.

An alternative strategy is to perform hybridizations with the target DNA bound to a solid support. The labelled DNA, in solution, is allowed to hybridize with the DNA bound to a solid support. The amount of hybrid formed can simply be determined by washing unbound probe from the solid support and measuring the bound radioactivity by scintillation counting, autoradiography, etc. In Southern (Southern 1975) and Northern hybridizations, the solid support is a membrane (usually nitrocellulose or nylon based) that has been blotted with DNA (or RNA) transferred from a gel in which the target DNA (or RNA) samples have been separated by electrophoresis. Target DNA may also be applied directly to a membrane by filtration, replica plating, or direct application. In addition, other solid supports such as activated cellulose, sephacryl, polystyrene or magnetic beads may be used.

Application of DNA Hybridization

DNA homology measurements

The rate of hybridization between two species of genomic DNA will be proportional to the concentration of homologous DNA and therefore measurement of hybridization kinetics has been widely used to estimate DNA homology between bacterial species.

In DNA hybridization measurements, however, DNA samples differing by less than approximately 5% in homology can not be accurately resolved; the technique cannot therefore be used to differentiate between very closely related strains or species. It should always be remembered that two DNA samples that are indistinguishable by homology studies are not necessarily identical. DNA homology measurements are also not useful for measuring the relationships between diverse species or genera with very low levels of homology.

In this case, sequencing of ribosomal RNA (Woese 1987) has provided the most useful tool.

In order to detect small differences in DNA sequence between closely related strains and species, more sensitive methods are required. DNA sequencing is the most sensitive method available to detect small sequence changes, but it is not yet a technique that can rapidly be applied to a large number of samples. A more readily applied technique requiring little expertise is the detection of restriction fragment length polymorphisms (RFLPs).

Restriction fragment length polymorphism detection

If two dsDNA fragments are identical at all but one base, and if that base difference lies within the recognition sequence of a restriction endonuclease (RE), digestion of both samples with that restriction endonuclease will produce products of differing size (Fig. 1). The products may be resolved by electrophoresis through, for example, 1% agarose, so that banding differences are obtained. These banding differences are referred to as restriction fragment length polymorphisms (RFLPs).

Most REs recognize either a 4 or 6 bp sequence; therefore in any random sequence the frequency of sites will be either approximately every 256 bp (4^4) or 4096 bp (4^6) respectively; the exact value depending on the relative base composition of restriction target site and DNA sample. For small genomes (e.g. viruses) electrophoresis and staining of restriction fragments produced by digestion allows easy detection of RFLPs. As the genome size increases, however, the number of fragments produced increases proportionally, so that for a typical bacterial genome of approximately 5 million base-pairs, somewhere in the region of 1000 fragments would be produced by a typical 6 bp RE. The highly complex banding patterns obtained are then only poorly resolved by electrophoresis. Nevertheless with good quality DNA and careful electrophoresis on long agarose gels, complex banding patterns may be seen, particularly in the high molecular weight region. These banding patterns, sometimes referred to as genomic fingerprints, may be used to differentiate between closely related species and even within species (Marshall *et al.* 1981; Kaper *et al.* 1982).

The direct examination of RE digests of genomic DNA, although useful and relatively simple to perform, is limited by the resolving power of the gels. The use of pulsed field gel electrophoresis (Carle *et al.* 1986) to resolve the limited number of very large DNA fragments generated by REs that recognize rare sites in bacterial DNA has recently been used to produce simpler and more easily interpreted genomic fingerprints. Though this technique will certainly be highly useful in large-scale mapping of bacterial genomes, the requirement for very high molecular weight DNA, costly apparatus and con-

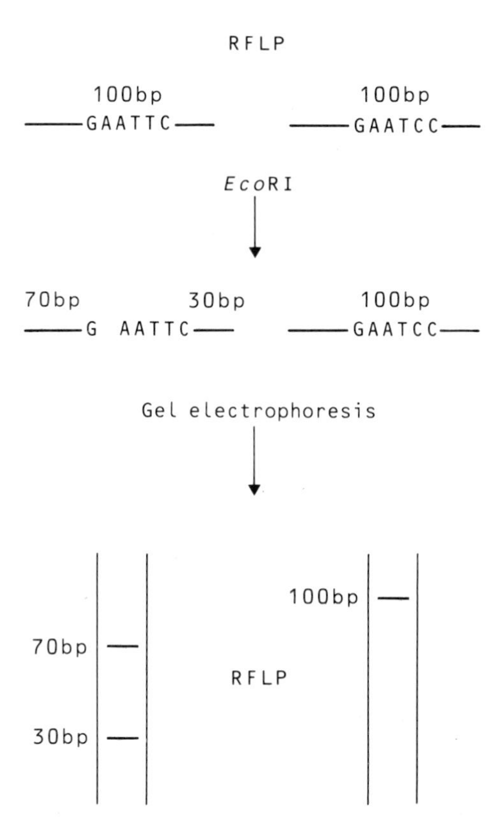

FIG. 1. Illustration of how two DNA molecules, each of 100 bp, but differing in a single-base substitution, may be differentiated by RFLP analysis. The DNA molecules are digested with a restriction endonuclease that recognizes a site containing the base substitution, in this case *Eco*RI. When this site is present (left), the DNA is cut into two fragments of 70 bp and 30 bp; whereas in the other DNA molecule the site is altered such that the enzyme does not cut the DNA, leaving a full-length fragment of 100 bp.

siderable expertise in performing the electrophoresis, presently inhibits its widescale use.

An alternative method of simplifying complex genomic fingerprints is to use a specific DNA probe to examine only those DNA fragments containing sequences homologous to the probe. DNA samples are digested with a RE and electrophoresed as above through 1% agarose. However, after electrophoresis, the DNA in the gel is denatured (in 0.5 M NaOH, 2 × 30 min washes in 5 volumes), neutralized (2 × 30 min washes in 0.5 M Tris–HCl pH 7) and Southern blotted to a nylon membrane (e.g. Hybond, Amersham, UK) by means of transfer buffer (25 mM sodium phosphate pH 7). The membrane is then prehybridized in hybridization buffer [3 × SSC (1 × SSC is 0.15 M sodium chloride, 0.015 M sodium citrate pH 7), 0.5% SDS, 5 × Denhardt's reagent (100 × Denhardt's reagent is 0.1% Ficoll, 0.1% polyvinylpyrrolidine, 0.1% bovine serum albumin), 10 μg/ml of single-stranded sheared DNA, 10 mM sodium phosphate pH 7] for 4 h at 65°C and then labelled probe is added and the membrane is hybridized for 18 h at 65°C. After hybridization, the membrane is washed in 1 × SSC, 5 × 5 min washes at 65°C and then one wash for 30 min at 65°C. Visualization of the bound DNA probe (by autoradiography for radiolabelled probes) reveals simple banding patterns that can be used to distinguish between strains and species (see Kaper *et al.* 1982; Tomkins *et al.* 1986; McFadden *et al.* 1987a for a few of many examples of the use of this technique). Figure 2 shows a number of *Mycobacterium avium* complex strains, probed with a DNA probe that produces clearly distinguishable banding patterns for most of the strains. The use of DNA probes to visualize RFLPs has a number of advantages over direct visualization of genomic fingerprints. Firstly, RFLPs can be clearly visualized

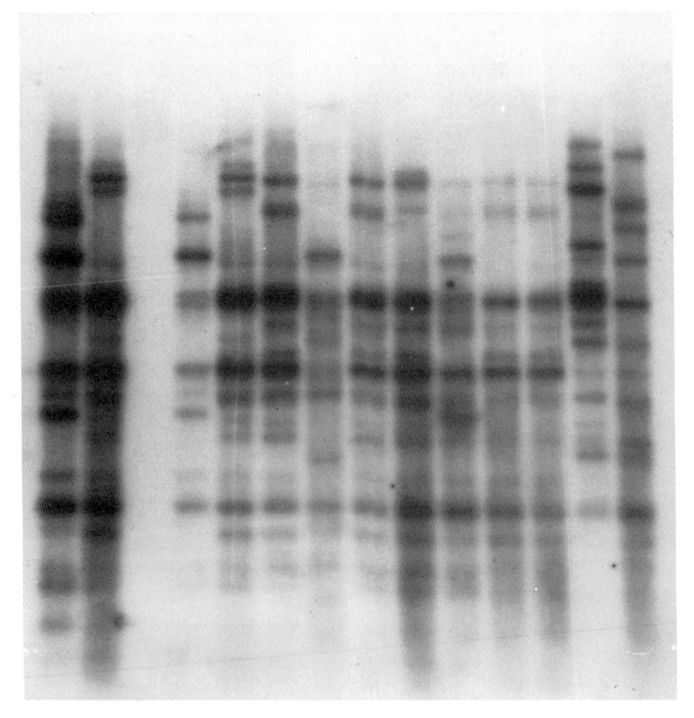

FIG. 2. RFLP typing of *Mycobacterium avium* complex strains. DNA (1 μg) from each of 13 *M. avium* complex strains was digested with *Pvu*II, electrophoresed through 1% agarose, blotted onto nylon membrane and hybridized to ^{32}P-labelled probe pMB20 and autoradiographed.

(e.g. Fig. 2). Secondly, once the information obtained with one DNA probe has been recorded, the probe may be removed and the membrane hybridized with further DNA probes. Essentially each (independent) DNA probe may be used to examine a different portion of the bacterial genome for RFLPs. This technique enables strains, such as the *M. avium* complex strains in Fig. 2, that are indistinguishable by biochemical analysis, to be clearly differentiated.

Most RFLP typing on bacteria has been performed with radiolabelled DNA probes, as in Fig. 2. Handling and disposal problems coupled with short shelf-life has inhibited the adoptation of this technique beyond research laboratories. However, techniques presently available for non-radioactive labelling of DNA are simple to perform and similar in sensitivity to radiolabelling with, for example, ^{32}P. Figure 3 shows DNA from *Neisseria meningitidis* blotted and probed with a conventional radiolabelled (with ^{32}P) DNA probe (Fig. 3 part 3) and detected by autoradiography. In Fig. 3 part 1 (a and b), a similar filter was probed with a DNA probe that was labelled with peroxidase and detected with a substrate giving a product that can be visualized by enhanced chemiluminescent (ECL labelling system, Amersham, UK) exposure of a photographic film. Banding patterns may be seen with less than an hour of exposure of the photographic film and, after exposure, the probe may be removed to allow the membrane to be reprobed; as with radiolabelled probes. Fig. 3 part 2 shows a photograph of a similar membrane, probed with the same probe that has been labelled by incorporation of digoxygenin-modified nucleotides (DIG-labelling system, Boehringer Mannheim, W. Germany) and then detected with alkaline phosphatase conjugated antibody to digoxygenin and visualized with a substrate giving visible blue bands. This system is perhaps a little simpler to perform than the ECL system but, at present, the membranes cannot be re-used. It can be seen that both non-radioactive systems give results comparable to those obtained with radiolabelled probes.

An additional advantage of the use of DNA probes to detect RFLPs is that the data generated can be used to estimate levels of base substitution between closely related DNA species (Upholt 1977). The number of conserved (homologous) bands in common between two DNA samples is expressed as a fraction (F) of the total number of hybridizing bands:

$$F = \frac{\text{number of bands common to both samples}}{\text{total number of bands}}. \tag{1}$$

F will, of course, be a function of the level of base substitution between the two DNA samples. The value of F is a function of the probability of base substitutions occurring within the RE sites located within the restriction enzyme target sequence(s). Essentially each probe and RE reacts with a number of DNA bases along the target sequence. The probability of these being the same or different between two DNA samples should (if base

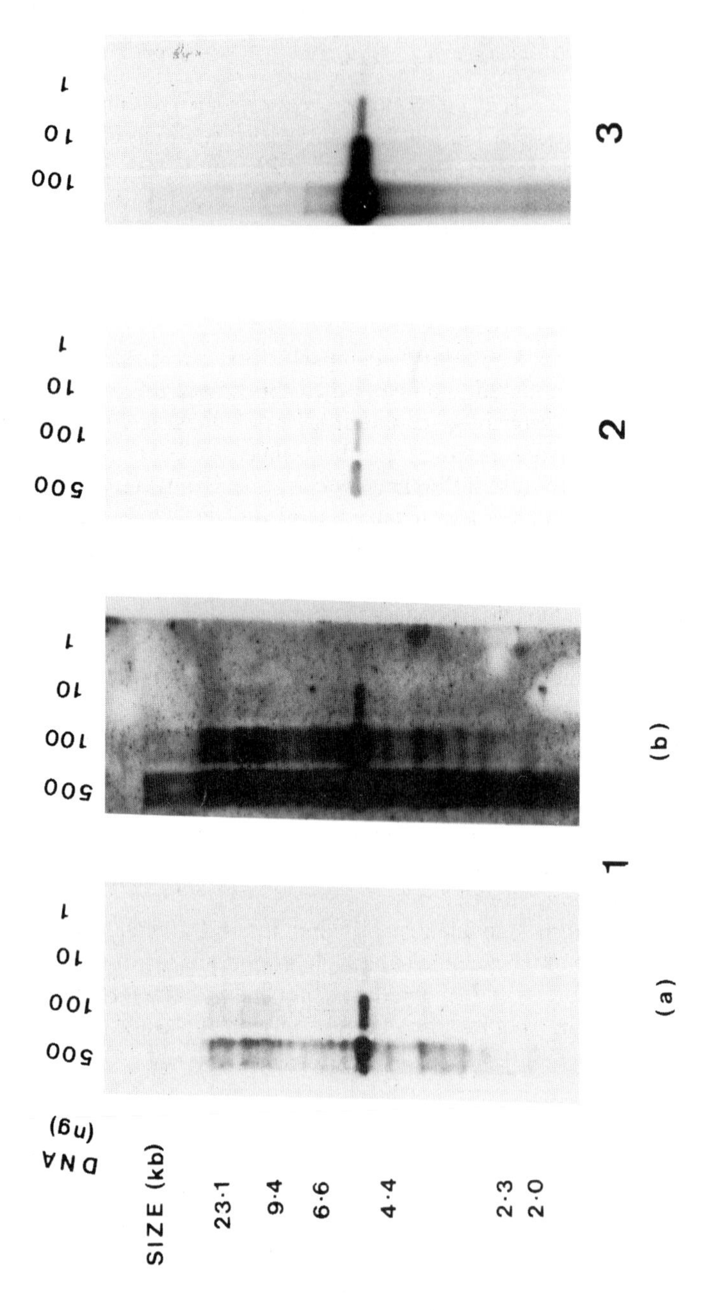

FIG. 3. Detection of *Hin*dIII fragments of *N. meningitidis* DNA. DNA samples digested with *Hin*dIII, electrophoresed through 1% agarose, blotted onto nylon membrane. Membranes were probed with probe labelled by enhanced chemiluminescence (1) with 1 min exposure of film (1a) and 30 min exposure of film (1b); digoxygenin labelling system (2), and conventional ^{32}P-labelled probe (3) detected by autoradiography.

substitutions are randomly distributed) follow binomial distribution laws such that the level of base substitution (P) may be expressed as a function of F:

$$P \simeq 1 - [-F + (F^2 + 8F)^{1/2}/2]^{1/n}$$

where n is the number of bases in the restriction enzyme site (usually 4 or 6).

This equation is only valid if their are no gross DNA rearrangements and the probability of multiple substitutions within the RE site is low. However, the equation can be used to estimate the base substitution between DNA samples having greater than approximately 80% DNA homology. The standard error σ of estimating P may be determined using the equation:

$$\sigma = [P(1 - P)/n]^{1/2}$$

Clearly the greater number of bands compared, the lower the standard error. It should be remembered, however, that the choice of probe may significantly influence the probability of detecting base changes. Thus DNA probes from conserved genes (such as the heat-shock genes or ribosomal RNA genes) will show few band differences between strains, whereas probes homologous to transposons or integrated bacteriophages may over-represent the average genetic differences between strains. For an accurate estimate of base substitution, the best probes to use are randomly-derived clones homologous to single copy DNA with no evidence of rearrangements. The data obtained may be used to construct phylogenetic trees between closely related strains or species (McFadden *et al.* 1987b).

Strain identities as determined by RFLP analysis have been found not necessarily to correlate well with serotyping, phage typing or other phenotypic methods (e.g. McFadden *et al.* 1987b; Hampson *et al.* 1989; Harshman & Riley 1980; Tomkins *et al.* 1986). This may be expected since RFLP typing is dependent only on the genotype of a strain, whereas other measurements examine the phenotype and may therefore be influenced by either very small genetic changes or by cultural differenes.

The use of DNA probes to identify RFLPs that differentiate between strains and species for the determination of bacterial relationships is essentially a variation of numerical taxonomy. Examination of equation (1) reveals that F is equivalent to the similarity coefficient S_j used in numerical taxonomy:

$$S_j = \frac{\text{number of characters positive in both strains}}{\text{total number of positive characters measured}}.$$

In numerical taxonomy, the presence or absence of a positive result with a phenotypic test is used to compare strains; whereas, in the RFLP typing, the presence or absence of a homologous DNA band is used. Both methods show whether a genetic character is the same or different between two strains; RFLPs, however, do not require phenotypic expression of that character. In addition, RFLP typing does not require viable organisms, and DNA extracted from infected tissue may also be used (McFadden *et al.* 1987b). RFLP typing may be therefore particularly useful for non-cultivable bacteria such as *Mycobacterium leprae* (Clark-Curtiss & Docherty 1989).

Examination of some pathogens by RFLP typing has revealed that the population structure appears to be clonal, in that very few, if any, RFLPs are found within some bacterial populations (McFadden *et al.* 1987a; Clark-Curtiss & Docherty 1989). To detect clonality, or, more generally, to estimate the confidence limits on the identity of two strains for which no RFLPs were detected, it is necessary to calculate the maximum fraction of base substitution (P_m) compatible with identical banding patterns. The binomial test statistic Z may be used to estimate this value:

$$P_m = Z^2/(N + Z^2)$$

where N is the total number of independent bases examined, which is equal to the number of fragments examined times the number of bases (n) in the recognition sequence of the restriction enzyme(s) producing the fragment pattern. The value of Z is obtained from tables of area under the normal curve: for a level of significance of 95%, Z is 1.64, and therefore the equation simplifies to:

$$P_m = 2.7/(N + 2.7).$$

Detection

DNA probes have a number of advantages over conventional diagnostic techniques, including speed, specificity, stability of reagents and sensitivity. The application of DNA probe technology to diagnostic bacteriology has, however, been slow, chiefly because of the success of the competing technology. More than one hundred years of refinement has rendered the simplicity, sensitivity and economy of conventional diagnostic microbiology a hard act to follow. Armed with little more than petri dishes, media, a bunsen burner and a few simple biochemical tests, any microbiologist worth their salt may detect and identify most clinically important bacteria present in reasonable concentration in a biological sample, in less than 48 h. The development of these techniques by Koch, Pasteur and their followers was surely one of the key advances in medical science, allowing control and, in some cases, virtual eradication of diseases such as cholera, diphtheria and tuberculosis from developed countries. What do DNA probes have to offer therefore, that can not be provided more simply and economically by standard microbiological techniques?

The most obvious advantage is one of speed. Most DNA probe applications for diagnosis take less 4 h. Therefore in those cases where rapid diagnosis may be critical (e.g. meningitis) DNA probes may be useful, particularly with 'slow-growing' or 'difficult-to-grow' bacteria. Culture of many bacteria such as mycoplasmas, mycobacteria, spirochetes, legionellas, chlamydia and rickettsias may be slow, difficult, require tissue culture or not be possible at present. It may be expected therefore that the use of DNA probes for bacterial detection would be most advantageous with these organisms. Accordingly, it is not surprising that some of the first DNA probes to become available have been for these 'difficult-to-grow' bacteria.

Another important advantage of DNA probes is specificity. DNA probes can potentially differentiate between bacterial strains that have only a single-base difference in their whole genome (providing the base difference is known). It is unlikely that any other technique would be capable of matching this specificity. The probe may also be designed to obtain specific information, such as possession of a drug-resistance or virulence plasmid or gene. With the development of the polymerase chain reaction (discussed below), DNA probes may be even more sensitive than culture. A final advantage is the stability of DNA probes and their ease of manufacture.

DNA probes have now been developed for detection of most bacterial pathogens and have already proved valuable in a wide variety of uses. The sensitivity of direct detection of bacterial DNA or RNA is, however, limited by quantity of target nucleic acid in the sample, which can only hybridize to, and form hybrids with, a molar equivalent of the probe. Detection of the limited amount of probe DNA bound to the target in hybrids therefore sets a lower limit on sensitivity. Use of probes homologous to sequences present in high copy number in the bacteria cell may increase the sensitivity. The company Gen-Probe presently market a number of probe tests based on hybridization to ribosomal RNA (see, e.g. Kiehn & Edwards 1987; Musial *et al.* 1988), which is present at several thousand copies per cell. Repetitive genomic DNA sequences may similarly achieve increased sensitivity (e.g. Eisenach *et al.* 1988). Most DNA probes, however, are only capable of directly detecting upwards of approximately 10^4-10^5 bacteria (Shoemaker *et al.* 1985; Butcher *et al.* 1988). This is clearly less than what may be achieved by sensitive culture; therefore most presently available DNA probe tests require prior culture of the organisms from infected tissue, which clearly limits their value.

It seems unlikely that advances in DNA probe labelling techniques will achieve the required several orders of magnitude increase in specific activity that would allow direct detection of bacteria at levels similar to those that can be achieved by sensitive culture techniques, i.e. 1–10 bacteria. Therefore, to detect lower numbers of bacteria, either the target nucleic acid or the hybridized probe must be amplified.

The Polymerase Chain Reaction

The recent development of the polymerase chain reaction (PCR) at Cetus Corporation, California, USA (Saiki *et al.* 1985, 1988) looks set to revolutionize DNA probe technology. The technique provides exquisite sensitivity, at least equal and usually greater than that provided by culture, combined with the speed and specificity of DNA probe assays. In addition, the PCR is remarkably simple to perform. It provides an analogous amplification to that by culture, but instead of the bacterium amplifying itself by replication, a fragment of the bacterial DNA is rapidly amplified by successive rounds of DNA replication by means of the enzyme DNA polymerase (Fig. 4).

The first step in the PCR is to ensure that bacterial DNA is available for PCR amplification: the bacterial cells must first be lysed. Several procedures utilize denaturing agents such as alkali for this purpose, although we have found that, for bacteria we have so far examined (*E. coli*, mycobacteria), boiling in water releases sufficient DNA for PCR. This has the additional advantage of simultaneously denaturing the DNA, ready for PCR amplification. DNA oligonucleotides, usually 15–25 bases homologous to sequences 50–2000 bases apart on the target DNA, on opposite strands, are then added and allowed to anneal to the target DNA. This step, essentially a DNA hybridization governed by all the factors discussed above, provides the specificity of the reaction. For mycobacterial DNA we anneal at 68°C for 30 s. The annealed oligonucleotides are then allowed to prime DNA synthesis by DNA polymerase, synthesizing a copy of each of the strands from the primers. DNA between the primers is effectively duplicated during this step. For a 300 bp amplified

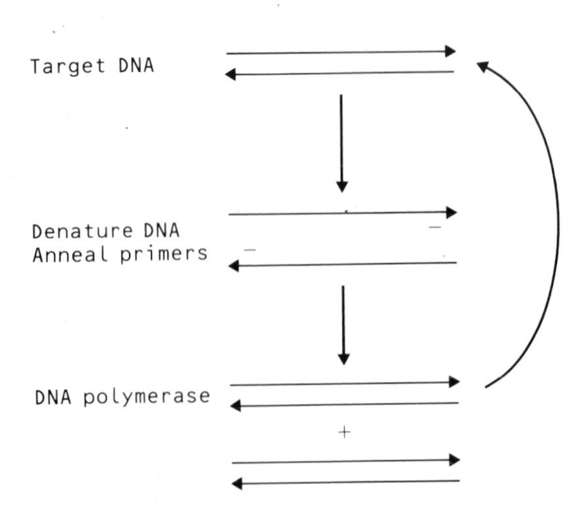

FIG. 4. Polymerase chain reaction.

mycobacterial DNA product, a 1 min incubation at 72°C in this extension step is used. The DNA is then denatured by heating (95°C, 30 s), the primers are allowed to anneal again and the cycle is repeated 20–35 times. It can be seen that on each cycle the amount of DNA between the primers is doubled, allowing an exponential increase in the quantity of this DNA. Simple calculations predict that 25 cycles should provide a (2^{25}) 33 million-fold amplification. However, saturation of enzyme, annealing of template and other factors may limit amplification. Nevertheless, 1 million-fold amplification of a single copy of a dsDNA molecule of size 300 base-pairs will yield approximately 0.3 pg of DNA, a quantity that is detectable by DNA hybridization. With the introduction of thermostable DNA polymerase from *Thermus aquaticus* (Saiki *et al.* 1988), the reaction is easily automated, may be performed and analysed in a few h and therefore provides a rapid and powerful amplification step for DNA detection and analysis.

The PCR product may be detected directly by gel electrophoresis and ethidium bromide staining. We are able to detect PCR product from approximately 100–1000 bacterial cells by direct examination of gels. The reaction may be performed in crude solutions—we are able to obtain good yield of PCR product merely by boiling a bacterial colony in water and submitting a small fraction of the supernatant (1%) directly to a PCR reaction. Specific identification of bacteria may therefore be achieved in a few h. Greater sensitivity may be achieved by Southern or dot-blotting the PCR product and hybridization with a labelled probe. Alternatively, a second 'nested' PCR may be used to detect small quantities of PCR product. Specific detection of single bacterial cells may be achieved by PCR (Hartskeerl *et al.* 1989; Hance *et al.* 1989; Vary *et al.* 1990), by these techniques.

The PCR is clearly capable of detecting and identifying mycobacteria in clinical samples in a matter of a few h from receipt of sample. It should, however, be remembered that, in common with most direct applications of DNA probes, present PCR systems would almost certainly detect dead bacteria as well as live organisms. The technique may not therefore be suitable for applications where the presence of dead bacilli may be important, such as monitoring drug therapy. The extremely high sensitivity of the PCR provides additional problems as great care must be taken in sample preparation to avoid false positives due to contamination, particularly with PCR product. However, microbiologists are well used to working with the possibility of contamination in mind and the sterile procedures that are required in microbiological work can easily be modified to handle the PCR.

References

BONNER, T.E., BRENNER, D.J., NEUFELD, B.R. & BRITTEN, R.J. 1973. Reduction in the rate of DNA reassociation by sequence divergence, *Journal of Molecular Biology* **81**, 123–128.

BRITTEN, R.M. & KOHNE, D.E. 1968. Repeated sequences in DNA. *Science* **161**, 529–540.

BUTCHER, P.D., MCFADDEN, J.J. & HERMON-TAYLOR, J. 1988. Investigation of mycobacteria in Crohn's disease tissue by Southern blotting and DNA hybridisation with cloned mycobacterial genomic DNA probes from a Crohn's disease isolated mycobacteria. *Gut* **29**, 1222–1228.

CARLE, G.F., FRANK, M. & OLSON, M.V. 1986. Electrophoretic separation of large DNA molecules by periodic inversion of the electric field. *Science* **234**, 65–68.

CLARK-CURTISS, J.E. & DOCHERTY, M.A. 1989. A species-specific repetitive sequence in *Mycobacterium leprae* DNA. *Journal of Infectious Diseases* **159**, 7–15.

EISENACH, K.D., CRAWFORD, J.T. & BATES, J.H. 1988. Repetitive DNA sequences are probes for *Mycobacterium tuberculosis*. *Journal of Clinical Microbiology* **26**, 2240–2245.

HAMPSON, S.J., PORTAELS, F., THOMPSON, J., GREEN, E.P., MOSS, M.T., HERMON-TAYLOR J. & MCFADDEN, J.J. 1989. DNA probes demonstrate a single highly conserved strain of *Mycobacterium avium* infecting AIDS patients. *Lancet* **1**, 65–68.

HANCE, A., GRANDCHAMP, B., LEVY-FREBAULT V., LECOSSIER, D., RAUZIER, D. & GICQUEL, B. 1989. Detection and identification of mycobacteria by amplification of mycobacterial DNA. *Molecular Microbiology* **37**, 843–849.

HARSHMAN, L. & RILEY, M. 1980. Conservation and variation of nucleotide sequences in *Escherichia coli* strains isolated from nature. *Journal of Bacteriology* **144**, 560–568.

HARTSKEERL, R., DE WIT, M.Y.L. & KLATSER, P. 1989. Polymerase chain reaction for the detection of *Mycobacterium leprae*. *Journal of General Microbiology* **155**, 2357–2364.

KAPER, J.B., BRADFORD, H.B., ROBERTS, N.C. & FALKOW, S. 1982. Molecular epidemiology of *Vibrio cholera* in the US Gulf coast. *Journal of Clinical Microbiology* **16**, 129–134.

KIEHN, T.E. & EDWARDS, F.F. 1987. Rapid identification using a specific DNA probe of *Mycobacterium avium* complex from patients with acquired immunodeficiency syndrome. *Journal of Clinical Microbiology* **25**, 1551–1552.

MARSHALL, R.B., WILTON, B.E. & ROBINSON, A.J. 1981. Identification of leptospiria serovars by restriction endonuclease analysis. *Journal of Medical Microbiology* **14**, 163–166.

MCFADDEN, J.J., BUTCHER, P.D., THOMPSON, J., CHIODINI, R. & HERMON-TAYLOR, J. 1987a. The use of DNA probes identifying restriction-fragment-length polymorphisms to examine the *Mycobacterium avium* complex. *Molecular Microbiology* **1**, 283–291.

MCFADDEN, J.J., BUTCHER, P.D., CHIODINI, R. & HERMON-TAYLOR, J. 1987b. Crohn's disease-isolated mycobacteria are identical to *Mycobacterium paratuberculosis*, as determined by DNA probes that distinguish between mycobacterial species. *Journal of Clinical Microbiology* **25**, 796–801.

MUSIAL, C.E., TICE, L.S., STOCKMAN, L. & ROBERTS, G.D. 1988. Identification of mycobacteria from culture by using the Gen-Probe Rapid Diagnostic System for *Mycobacterium avium* complex and *Mycobacterium tuberculosis* complex. *Journal of Clinical Microbiology* **26**, 2120–2123.

SAIKI, R.K., GELFAND, G.H., STOFFEL, S., SCHARF, S., HIGUCHI, S.J., MULLIS, K., HORN, G. & ERLICH, H.A. 1988. Primer-directed enzymic amplification of DNA with a thermostable DNA polymerase. *Science* **239**, 487–491.

SAIKI, R.K., SCHARF, S., FALOONA, F., MULLIS, K., HORN, G., ERLICH, H.A. & ARNHEIM, N. 1985. Enzymic amplification of β-globin genomic sequences and restriction site analysis for diagnosis of sickle cell anemia. *Science* **230**, 1350–1354.

SHOEMAKER, S.A., FISHER, J.H. & SCOGGIN, C.H. 1985. Techniques of DNA hybridization detect small numbers of mycobacteria with no cross-hybridization with non-mycobacterial respiratory organisms. *American Review of Respiratory Diseases* **131**, 760–763.

SOUTHERN, E. 1975. Detection of specific DNA fragments separated by gel electrophoresis. *Journal of Molecular Biology* **98**, 503–517.

TOMKINS, L.C., TROUP, N., LABIGNE-ROUSSEL, A. & COHEN, M.L. 1986. Cloned random

chromosomal sequences as probes to identify *Salmonella* species. *Journal of Infectious Diseases* **154**, 156–162.
Upholt, W.B. 1977. Estimation of DNA sequence divergence from comparisons of restriction endonuclease digests. *Nucleic Acids Research* **4**, 1257–1265.
Vary, P., Andersen, P., Green, E., Hermon-Taylor, J. & McFadden, J. 1990. Use of highly specific DNA probes and the polymerase chain reaction to detect *Mycobacterium paratuberculosis* in Johne's disease. *Journal of Clinical Microbiology* **28**, 933–937.
Woese, C.R. 1987. Bacterial evolution. *Microbiological Reviews* **51**, 221–271.

Extraction and Purification of Eukaryotic mRNA

P.G. SANDERS, T. JACKSON, JANE NEWCOMBE, SARAH BELL, G.E. SCOPES AND A. KNIGHT
Molecular Biology Group, Department of Microbiology, University of Surrey, Guildford, Surrey GU2 5XH, UK

The purification of good quality undegraded RNA is a necessary prelude to many of the procedures used to analyse gene expression and for cDNA cloning. Purified RNA can be used in Northern blot analyses to determine whether a particular gene is expressed in the cells or tissue under study, S1 mapping and primer extension can be used to locate the 5′ ends of RNAs to determine the precise point(s) at which transcription starts. Subsequently, cDNA cloning of RNA species (see the following chapter) allows the nucleotide sequence of the RNA to be quickly determined. Such studies have enabled rapid progress to be made in the analysis of gene organization and expression and have played a major role in the discovery of introns and gene splicing, the manipulation and analysis of those viruses which only have RNA genomes, and the expression of commercially or medically valuable products in a variety of host species.

The various protocols described in this chapter have been used to extract RNA and mRNA for S1 mapping, Northern analysis and cDNA cloning. Although our original intention was to focus on the extraction of mRNA primarily for cDNA cloning of viral and mammalian cell mRNAs we have expanded the protocols to include a method we have used successfully with a filamentous fungus, *Fusarium sporotrichioides*. From this range of protocols, newcomers to this field may therefore find one to suit their own particular eukaryotic organism.

The protocols described are often modifications of the developments of others and we have tried to acknowledge this. Our apologies to those whose contributions we may inadvertently have overlooked. For additional protocols describing relevant techniques which are not discussed here, such as gel electrophoresis of RNA, see Maniatis *et al.* (1982) and Berger & Kimmel

0–632–02926–9

(1987). Both of these contain excellent practical techniques and background information.

Basic Precautions Prior to RNA Extraction

Removal of RNases

The major preoccupation of those who work with RNA is the removal of all potential sources of RNase contamination. RNases are particularly stable enzymes and retain their activity after boiling and, even after exposure to strong denaturing agents, can rapidly renature and regain activity. The experimenter is a potent source of RNases and gloves must be worn at all times when purifying RNA. In addition, disposable 'plasticware' should be used if possible for all RNA work as this is normally found to be free of RNases.

If glassware has to be used for some steps, it must be treated to remove any RNases contamination. This can be achieved by soaking in 0.1% diethylpyrocarbonate (DEPC) for at least 12 h followed by thorough washing in DEPC-treated water and autoclaving. Autoclaving breaks down the DEPC to alcohol and carbon dioxide. DEPC should be handled carefully as it is an irritant to the skin, mucous membranes, eyes and respiratory tract and may be a carcinogen. Until inactivated, solutions should only be used in a fume hood.

Water and reagent solutions, except those containing Tris buffers in which DEPC rapidly breaks down, can be treated with 0.1–0.5% DEPC followed by autoclaving. A common problem, however, is that it is difficult to remove the DEPC completely and a lingering DEPC odour is often present even after autoclaving. For stocks of deionized or distilled water, residual DEPC can be removed by boiling in a fume hood until the smell has disappeared. If removal of DEPC is incomplete, its potent carboxymethylating activity can modify the RNA that is being extracted. In these situations, it is probably better to dispense with DEPC treatment and undertake initial experiments without it. One of the authors has avoided DEPC in all work up to the isolation of mRNA from total RNA with no obvious detrimental effects; his colleagues, however, tend to treat all solutions and glassware. An alternative strategy is to bake all glassware at 180°C for at least 8 h or 250°C for 4 h.

Reagents

It is sensible to keep a separate stock of reagents for RNA work and to buy only the purest grade. No material removed from stock bottles should ever be returned and if spatulas are used these must be baked as described above. In laboratories where work involving RNases is ongoing try not to keep

RNases, or samples containing these, with your RNA or reagents for RNA work. This does happen and samples do get degraded.

Safety

Many of the reagents used in these protocols are potentially harmful and gloves, safety goggles and a fume hood should be used when making up and if possible, when using solutions. Strong solutions of phenol, for example, can be lethal if an area only the size of the hand is affected by a spill.

RNA Extraction Protocols

The following protocols have been used to prepare RNA and mRNA from fungi and mammalian cells in culture, including virally infected mammalian cells where the viral sequences were the ones of interest. Other protocols for the extraction of RNA from viruses are described in the chapter by Gibson *et al.*, this volume. For a method involving lithium chloride precipitation of RNA which is not included in these protocols, see Birnboim (1988).

The first two of the three protocols do not allow separation of cellular compartments, and extract total cellular RNA. The third protocol involves separation of nuclei and cytoplasmic fractions, allowing the RNA in these compartments to be extracted separately.

The protocols also involve the use of similar methods at certain stages, particularly with regard to phenol or phenol/chloroform extraction and ethanol precipitation of nucleic acids. These are outlined now rather than repeating them for every protocol.

Phenol and phenol/chloroform extractions

Many older protocols for extraction of nucleic acids suggest redistilling phenol in a nitrogen atmosphere prior to use, which we have done for RNA work. With high quality phenol which is not discoloured by oxidation products, this may not be necessary. Some suppliers of molecular biology reagents also sell 'molecular biology' grade phenol but this is relatively expensive.

Phenol is prepared by melting high quality crystalline phenol at 65°C in a water bath and saturating this with deionized water containing 0.3% NaCl. This can be stored at 4°C in tightly stoppered dark bottles. 8-Hydroxyquinoline can be added if required and is essential for removing vanadyl complexes.

Before use, the phenol is buffered and for routine use the buffer is 1.0 M Tris–HCL pH $\leqslant 8$ (do not go above pH 8).

When extracting RNA, the use of phenol at pH 4–6 selectively partitions the DNA into the interphase and organic phase, leaving the RNA in the aqueous phase. These conditions are used in protocol 1. We have, however, obtained good RNA yields with phenol/chloroform at pH 8.0 by use of protocol 3. Phenol alone should not be used for RNA extractions below pH 7.6 as poly(A) mRNA is lost to the organic phase under these conditions (Perry *et al.* 1972; see also discussion in Wallace 1987).

For phenol/chloroform extractions, buffered phenol is mixed with an equal volume of chloroform. It has been found that omiting isoamylalcohol which is normally used with chloroform to reduce foaming (ratio 24:1, chloroform:isoamylalcohol) does not appear to affect our yields of RNA or DNA.

Ethanol precipitation

To precipitate RNA from aqueous solutions add 1/10th volume 3.0 M sodium acetate pH 5.2, or NaCl to 0.1 M followed by 2.5 volumes of absolute ethanol. Leave at −20°C for 1 h or, preferably, overnight. For dilute solutions −70°C is preferable. Spin at 12 000 ***g*** for 10 min, remove the supernatant, wash the pellet with 80% ethanol and re-spin. Remove the supernatant and vacuum dry the pellet before redissolving in deionized water.

Protocol 1: extraction of RNA from filamentous fungi

This protocol is based on that of Paietta *et al.* (1987). The fungal mycelium grown in liquid culture is rapidly washed, air dried and then frozen in liquid nitrogen to prevent RNase activity. The frozen mycelium is then disrupted before removal of RNases and other proteins by SDS and phenol/chloroform extraction. The phenol in this protocol is initially saturated with water and then buffered with extraction buffer. At a pH of 5.0 the DNA is partitioned into the interphase. RNA prepared in this way has been used for Northern blot analyses.

Reagents

Extraction buffer; 100 mM sodium acetate, 1 mM EDTA, 1% SDS, pH 5.0.
Phenol/chloroform, buffered with extraction buffer.
Chloroform.
Ethanol.
Liquid nitrogen.

Method

1 A 50 ml liquid culture of *F. sporotrichioides* is quickly washed with deionized water and suction dried using a Buchner funnel. This provides about 3 g of mycelium (wet weight).

2 The mycelium is then rapidly frozen in liquid nitrogen and rapidly ground to a powder in a coffee grinder, taking care not to transfer liquid nitrogen to the coffee grinder along with the mycelium. The ground mycelium is collected in DEPC-treated 50 ml centrifuge tubes on ice (the authors use polyallomer 'Oak Ridge' tubes with screw caps).

3 Add 10 ml extraction buffer and disrupt the mycelium further using a Polytron homogenizer for 10 s at setting 4, followed by 30 s on ice. This is repeated until microscopical examination of the slurry shows complete lysis.

4 Perform a 2 min clearing spin at 5 000 ***g*** and 4°C to remove debris.

5 The supernatant is transferred to a 50 ml 'Oak Ridge' tube and extracted with an equal volume of phenol/chloroform. The phases are mixed by rotation for at least 10 min.

6 The phases are separated by centrifugation for 15 min at 5 000 ***g*** at room temperature.

7 The top aqueous phase is removed and re-extracted until clear. This may take up to three extractions depending on the amount of material extracted.

8 The aqueous phase is extracted with an equal volume of chloroform and the phases are separated by centrifugation for 10 min at 5 000 ***g*** at room temperature. The aqueous phase is removed and re-extracted with chloroform.

9 After centrifugation the RNA is ethanol precipitated from the supernatant (see *Ethanol precipitation*, above).

10 Purified RNA can be kept at −70°C in distilled water as described on p. 119.

Protocol 2: extraction of RNA from virally infected mammalian cells with guanidinium thiocyanate

Guanidinium salts are strong protein denaturants and there are a variety of protocols based on their use for RNA extraction. The protocol described below is based on the methods of Chirgwin *et al.* (1979) and MacDonald *et al.* (1987) and involves the differential precipitation of DNA and RNA from guanidinium thiocyanate and guanidinium hydrochloride. It has been used to extract RNA from rubella virus-infected Vero cells and RSV-infected Hep2 cells and is recommended for cells with high RNase levels. Degradation of intact RNA is minimized by lysis or homogenization in 4 M guanidinium thiocyanate to denature cellular proteins including RNases. The addition of DTT or 2-mercaptoethanol also aids in the breaking of intramolecular disulphide bonds.

Variations of guanidinium-based protocols are described by MacDonald *et al.* (1987) and by Maniatis *et al.* (1982). A commonly used alternative protocol involves pelleting the RNA through caesium chloride after guanidinium lysis and this may be better for use with small samples. Extraction of RNA with guanidinium salts is not very efficient with dilute RNA solutions.

Reagents

Phosphate buffered saline solution A (PBSA):
- 2.68 mM potassium chloride;
- 1.47 mM potassium dihydrogen phosphate.
- 0.137 M sodium chloride;
- 8.06 mM disodium hydrogen phosphate.

Guanidinium thiocyanate solution:
- 4 M guanidinium thiocyanate (50 g);
- 0.5% sodium lauryl sarcosine (0.5 g);
- 25 mM sodium citrate pH 7.0 (2.5 ml 1 M);
- 0.1 M 2-mercaptoethanol (0.7 ml).

The figures in parentheses indicate amounts of reagent for 100 ml of solution. Make up the given quantities to 100 ml with deionized water, filter through Whatman number 1 filter paper and adjust the pH to 7 with NaOH.

Guanidium hydrochloride solution:
- 7.5 M guanidinium hydrochloride buffered with 0.025 volume of 1 M sodium citrate pH 7.0, 5 mM DTT;
- 1 M acetic acid;
- Ethanol.

Method

Removal of infected cells from tissue culture flasks:
The cells can be removed either by scraping or by trypsinization. Scraping off the cells is faster but good yields of RNA have been obtained following trypsinization of Chinese hamster cells even when the trypsinization process has been slow and the cells have taken up to 30 minutes to come off the surface. Good yields of RNA from rubella-infected Vero cells have been obtained following trypsinization. Obviously the faster the cells are removed and lysed the better.

Mechanical removal of cells:
1 Pour the medium from the cells (if the infected cells produce large amounts of extracellular virus the supernatant can be stored at −70°C for virus purification and later extraction of virion RNA), and wash with ice-cold PBS. Scrape off the infected cells in 10 ml of ice-cold PBS per 150 cm^2 flask, 25 ml of PBS per roller bottle (10^8 cells) or 4 ml of PBS per 90 mm cell culture dish. A rubber policeman or a scraper designed for use in roller bottles should be used. Centrifuge at 2000 ***g*** for 2 min at 4°C to pellet the

cells. To obtain a good pellet use disposable universals with conical bottoms.

2 Resuspend the pellet from four 150 cm^2 tissue culture flasks in 3.0 ml of 4 M guanidinium thiocyanate solution. Adjust this appropriately for larger volumes of cells.

For certain cell lines and for using this method with tissues the cells can be distrupted by homogenization once they are in the guanidinium solution. If homogenization is to be used, do not add the sarkosyl to the guanidinium solution but add it immediately after homogenization and mix well.

An alternate method described by Macdonald *et al.* (1987) is to apply the denaturing solution directly to the tissue culture cells.

3 Transfer the solution to a 10 ml centrifuge tube and add a further 3 ml guanidinium thiocyanate, mix well by vortexing. Shear the DNA by passage through a 19–25 gauge needle to reduce the viscosity (this can be omitted if the cells are homogenized in step 2).

4 Centrifuge at 10 000 ***g*** at 10°C for 10 min to pellet any insoluble material.

5 Transfer the supernatant to a fresh centrifuge tube.

6 Add 0.025 volumes of 1 M acetic acid to reduce the pH to 5.

7 Add 0.5 volumes of ethanol and precipitate at −20°C overnight. To prevent freezing at −20°C, 0.75 volumes of ethanol can be added but this increases DNA contamination of the pellet.

8 Centrifuge at 10 000 ***g*** for 10 min to pellet the RNA.

9 Remove the supernatant and resuspend the pellets in a total of 3 ml of guanidine hydrochloride. Briefly heating to 68°C may aid dispersion of the pellet.

10 Repeat steps 6 and 7, leaving the RNA to precipitate for at least 3 h. Only add 0.5 volumes of ethanol at this stage.

11 Repeat step 8.

12 Remove the supernatant. For some cell lines a second precipitation from guanidinium hydrochloride may be necessary when the amount of guanidinium hydrochloride can be halved.

13 After one or two precipitations, resuspend the pellet in 0.5 ml absolute ethanol to remove the last traces of guanidium hydrochloride. Because of the small volumes being handled at this stage, transfer the suspension to a microcentrifuge tube. Wash out the original tube with 0.5 ml ethanol and add to the RNA suspension. Centrifuge for 10 min to pellet the RNA.

14 Decant the ethanol and dry briefly under a vacuum to remove the last traces of the ethanol.

15 Resuspend the pellet in 200 μl deionized water (or go to step 17) and store at −70°C.

16 The RNA can be analysed as described below (p. 122).

At this stage mRNA is usually purified from the total RNA by using either oligo(dT) cellulose or mAP paper. If analysis suggests that further purification of the total RNA is required, rather than the selection of poly(A) mRNA, then the following steps taken from MacDonald *et al.* (1987) can be performed.

17 Add EDTA to the RNA solution to 10 mM pH 7.0.

18 Extract the RNA solution with 3 volumes of chloroform:*n*-butanol (4:1, v:v) vortexing if necessary to mix the phases.

19 Spin for a few seconds to separate the phases, remove the supernatant and reextract any interface and organic phase with 100 μl EDTA (pH 7.0). Centrifuge and remove the supernatant. A third extraction may be required.

20 Precipitate the RNA by adding 0.1 volumes of 3 M sodium acetate and 2.5 volumes of ethanol. Store at −20°C overnight and centrifuge (10 000 ***g*** in microcentrifuge) for 10 min to pellet the RNA.

21 Wash the pellet with 80% ethanol, respin for 5 min, decant the ethanol and dry under vacuum.

22 Resuspend the pellet in deionized water, analyse and store at −70°C.

Protocol 3: extraction of cyptoplasmic RNA from mammalian cells in culture

This relatively quick and versatile procedure is based on that of Preston (1977) and was originally used for the extraction of RNA from herpes virus-infected cells. The protocol has been used to extract cytoplasmic RNA from Chinese hamster ovary, hybridoma cells and from Vero cells uninfected and infected with rubella virus. The extracted RNA has been used in Northern blot studies and for cDNA cloning. mRNA extracted from RNA obtained by this protocol has been stable for over two years at −70°C and then used to make a cDNA library from which the required cDNA was obtained.

Vanadyl ribonucleoside complexes (VRC) (Berger & Berkenmeier 1979) are used to inhibit RNases followed by SDS and phenol/chloroform extraction. High concentrations of Tris are also present to inhibit RNase activity and the cycloheximide present protects the polysomes and hence attached mRNAs. The VRCs are removed by repeated extraction with phenol containing 8-hydroxyquinoline (8HQ) which is black in the presence of the VRC but, when removal is complete, it retains the yellow tint of the 8HQ.

This is the only protocol described which allows for the separation of nuclei and cytoplasm. This means that the RNA present in these compartments can be obtained separately, although the cytoplasmic fraction will also contain mitochondrial RNA species. If nuclear RNA is required, it can be obtained from the nuclear pellet as noted in step 5. Alternatively, DNA can be extracted from the nuclei by following appropriate protocols (see Maniatis *et al.* 1982) and the note in step 5, although the vanadyl complex does appear at times to inhibit the activity of protease-XI (K) used in many DNA extraction protocols. This requires further investigation.

Reagents

Keep all reagents on ice.
PBS: (p. 118).
Lysis buffer:
0.2 M Tris–HCl pH 7.5;
0.14 M NaCl;
10 μg/ml cycloheximide;
2 mM VRC;
0.5% NP40 (detergent).
Make up in 19 ml amounts without the VRC and store at −20°C. Add 1 ml VRC just before use as it is inactivated by Tris.
TSE:
10 mM Tris–HCl pH 8.5;
5 mM EDTA;
0.5% SDS.
Phenol/chloroform containing 0.1% 8-hydroxyquinoline.

Method

1 Trypsinization or scraping can be used to remove the cells from the tissue culture flask etc., as in the previous protocol. The cells must be kept as cold as possible. Precool tubes, centrifuge buckets and buffers used in all steps up to the addition of TSE and phenol/chloroform.

After the cells have been pelleted and the PBS has been removed:

2 Resuspend each pellet (either from 1 roller bottle (10^8 cells) or combined cells to this quantity from flasks or dishes) in 5 ml of lysis buffer. Adjust proportionally for other quantities of cells. As two tubes are needed for centrifugation, splitting one sample into two may ease the manipulations.

3 Pipette the pellet gently up and down at least 4 times through a *wide bore* 10 or 5 ml pipette to resuspend the cells and to allow the cytoplasmic membrane to lyse. Violent pipetting and use of a narrow tipped pipette will cause the nuclei to rupture, increasing DNA and nuclear RNA contamination.

4 Spin at 2000 ***g*** for 3 min at 4°C.

5 Decant the supernatant into a 100 ml conical flask or 50 ml centrifuge tube containing 5 ml TSE, 10 ml water and 15 ml phenol/chloroform/8HQ at room temperature. Conical flasks provide a larger surface area, allowing more rapid mixing and faster inactivation of RNases. Take care to minimize contamination by the nuclear pellet. Mix rapidly by swirling or shaking.

Leave the supernatant in the phenol/chloroform for 5–10 min with occasional mixing at room temperature. NB. If RNA or DNA are wanted from the nuclear pellet, the appropriate lysis and extraction can be carried out at the same time (for RNA), or later for the DNA. (For DNA, the lysed nuclear pellet can be stored at −70°C until required and then extracted by this protocol with RNase and proteinase-X1 (K) treatments between the lysis step and the phenol/chloroform extraction.) Lyse the nuclei in 1% SDS and 50 mM/EDTA. To extract nuclear RNA, first shear the DNA by repeated passage of the lysate through a 19-gauge syringe needle and then follow this protocol from step 5.

6 Transfer the extraction mixture to a 50 ml tube and spin at 2 500 rpm for 10 min at room temperature.

7 Remove the aqueous (top) layer with a *wide bore* 10 ml pipette and transfer to another 100 ml flask or tube.

8 Add 10 ml phenol/chloroform and mix.

9 After 10 min, spin as in step 6 and collect the aqueous phase. If one roller bottle was used, a third extraction with phenol/chloroform may be necessary, but if five, or less, 90 mm dishes were used, two extractions should suffice. (If complete removal of the VRC is required at this stage, repeat the phenol/chloroform extractions until the 8HQ stays yellow. Alternatively, if mRNA is to be extracted by oligo-dt cellulose chromatography (p. 124), then the VRC can be left in to protect the RNA. It will be removed when the oligo-dT column is washed prior to mRNA elution.).

10 Remove the aqueous supernatant and mix with an equal volume of chloroform. Mix gently, then centrifuge as in step 6 to separate the phases.

11 Repeat steps 9 and 10 with fresh chloroform.

12 After two chloroform extractions, place the final aqueous phase into a centrifuge tube and precipitate with ethanol.

13 When drying the RNA pellet after ethanol precipitation in a vacuum dessicator, place parafilm over the top of the tube and pierce with a few pin holes. Although we have not had problems of RNase contaminations at this stage, try not to use the dessicator also used by colleagues who treat their DNA samples with RNase!

14 Add 50 μl of water per 90 mm dish or 250 μl of water per roller bottle. Stand the tubes on ice while the RNA dissolves, gently swirling occasionally.

15 Quantify a sample of the RNA as in the following section.

One roller bottle of CHO cells (10^8 cells) normally provides about 2 mg of RNA by this method.

Analysis of Purified RNA

A spectrophotometric analysis can be made to determine the quantity of RNA obtained and to provide an estimate of its purity. DNA and VRC contamination will also affect the apparent nucleic acid concentration and, to determine the amount of DNA present, gel electrophoresis is recommended. A matched set of quartz cuvettes should be set aside for RNA work, soaked in DEPC and thoroughly washed before use.

The absorbtion of an aliquot of the RNA at 230, 260 and 280 nm should be determined and pure RNA preparations should have the ratios:
230/260 of 2.3
280/260 of 2.0
An OD of 1 at 260 nm is equivalent to 40 μg/ml for RNA.

The quickest way to analyse visually the quality of RNA extracted is to run a sample on a 2% agarose gel in Tris–acetate electrophoresis buffer. After electrophoresis, staining with ethidium bromide and illuminating with u.v., two strong bands representing the 18S and 28S eukaryotic ribosomal RNA should be present. If these are not sharp, then it suggests that RNase activity had degraded the material. Be careful not to run this gel in apparatus used by colleagues for running samples containing RNase. The gel apparatus can be DEPC-treated. (NB: appropriate shielding of the face etc. should be worn when using u.v.)

Alternatively, the RNA can be electrophoresed through a formaldehyde-containing gel and visualized or used to prepare a Northern blot with a common gene as a probe (e.g. actin). The quality of the RNA can also be assessed by translation of a sample in a rabbit reticulocyte lysate cell-free system. The production of high molecular weight proteins as visualized by PAGE and autoradiography is a sign that the preparation contains undegraded RNA.

Isolation of Polyadenylated mRNA from Total RNA

About 1–5% of the RNA extracted is messenger RNA, most species of which have poly-adenylic acid tails allowing for purification by affinity chromatography. There are two matrices routinely used for polyA mRNA purification by affinity chromatography, either oligo-dt cellulose (Edmonds *et al.* 1971; Aviv & Leder 1972) or polyU sepharose.

The latter matrix is better for mRNAs with shorter polyA tails but elution conditions are more exacting as they can require the use of formamide. We present a protocol employing oligo-dt cellulose together with a more recent method which uses message-affinity paper (mAP) (Werner *et al.* 1984). mAP paper has polyU sequences bound to it, which in turn reversibly bind polyA tailed mRNAs. The binding of the paper depends on the batch but is typically >25 μg polyA RNA per cm^2 with 60–80% of this being released by the elution protocol. The return of polyA mRNA from either of these protocols is usually 1–2% of the total RNA added.

Oligo(dT) chromatography for the isolation of poly(A) mRNA

Reagents

Loading buffer:
20 mM Tris–HCl pH 7.6;
0.5 M NaCl;
0.1% SDS*;
1 mM EDTA.

Washing buffer:
20 mM Tris–HCl pH 7.6;
0.1 M NaCl;
1 mM EDTA.

Eluting buffer:
10 mM Tris–HCl pH 7.6;
1 mM EDTA.

NB. *SDS is often found in all three buffers but, as it can precipitate with the mRNA and interefere with subsequent manipulations, it should not be placed in the washing or elution buffers. One author would also omit it from the loading buffer but, if the column is washed well, residual SDS should not be a problem.

Method

1 Hydration of oligo(dT) cellulose: the binding capacity of the oligo(dT) cellulose should be checked for the particular batch used. 0.2 g oligo (dT_{12-18}) cellulose (Pharmacia) is resuspended in 10 ml loading buffer. This can be used immediately or left overnight. The gel slurry is loaded into a disposable Poly-Prep column (Bio-Rad) to give a bed volume of 0.5 ml.

2 Equilibriation of the column: this is done by eluting sequentially with the following: 4 ml water, 4 ml 0.1 M NaOH, 4 ml 5 mM EDTA, 4 ml water and, finally, with 8 ml of loading buffer (at least 5 column volumes). In many protocols, washing the column with loading buffer alone is considered adequate.

3 Sample preparation and loading: the RNA sample (up to 2.8 mg) is diluted with an equal volume of 2 × loading buffer at room temperature. The sample is heated to 65°C for 5 min, cooled on ice to room temperature and loaded onto the column.

4 2 ml loading buffer is then run through the column, the eluate collected, heated as above and reloaded.

5 Washing the column: 4 ml loading buffer is passed through the column, followed by a 4 ml washing buffer.

6 Elution of mRNA: the RNA is eluted with 3 ml elution buffer. Fractions of 500 μl are collected and the absorbance of each measured at 260 and 280 nm in cuvettes reserved for RNA work.

7 Fractions containing poly(A) RNA are pooled and chromatography repeated (step 8) or the mRNA is precipitated with ethanol. Poly(A) RNA normally elutes either in fractions 1 and 2, or 2 and 3.

8 After re-equilibriating the column and restoring the NaCl level of the mRNA to 0.5 M, a second cycle of chromatography can be performed.

Although two cycles are often recommended, the loss of material may outweigh the gain in purity. After one cycle the mRNA is approximately 30% poly(A). If two cycles are performed the fractions containing the mRNA are identified and precipitated as in steps 6 and 7.

9 Poly(A) mRNA is resuspended in deionized DEPC-treated water, divided into 5 μg amounts and stored at −70°C.

The quality of the mRNA can be checked (p. 122). Translation of a small amount of mRNA and/or Northern blot analysis using a common probe (e.g. actin) is the most economical way of checking for degradation.

Isolation of polyadenylated mRNA using Hybond mAP Paper

Reagents

Hybond mAP paper (Amersham);
5.0 M NaCl;
0.5 M NaCl;
Ethanol.

Method

1 Dissolve 1 mg of RNA in 450 μl deionized water in a 1.5 ml microcentrifuge tube.

2 Heat at 65°C for 5 min and then place on ice. Spin for a few seconds to collect all of the liquid.

3 Add 50 μl 5 M NaCl to give a final concentration of 0.5 M NaCl. A ribonuclease inhibitor such as human placental ribonuclease inhibitor can be added at this stage (see the following chapter and Roth 1958).

4 Preparation of mAP paper: the binding capacity of 1 cm^2 is >25 μg of poly(A) mRNA. 1 cm^2 of paper is normally used for 1 mg of RNA. If 2 cm^2 are used adjust the solution volumes accordingly. Some people may find it more convenient to cut the paper into two or three strips with sterile scissors to make sure it is all covered by the RNA solution. Shake the paper in 10 ml 0.5 M NaCl for 10 min.

5 Add the paper to the RNA and shake at room temperature for 20–25 min.

6 Remove the paper with heated, flamed forceps and wash in 2 × 10 ml 0.5 M NaCl for 10 min and 1 × 70% ethanol for 2–10 min.
Air dry the paper on a sterile plastic petri dish.

7 Elution of the RNA: place the paper into a microcentrifuge tube and add 700 μl of deionized water. Heat at 70°C for 5 min.

8 Use flamed forceps to pull out the paper from the water and trap the paper in the lid of the microcentrifuge tube. Spin briefly to remove the liquid from the paper.

9 To quantify the amount of mRNA eluted, measure the OD of the solution at 260 and 280 nm. A 1% return of poly(A) mRNA is normally obtained by this method.

10 Recover the 700 μl and ethanol precipitate.
Resuspend the pellet after washing in distilled water at 1 μl/μg in 5 μg amounts.

Storage of RNA Samples

Purified RNA should kept at −70°C either in 70% ethanol or distilled water. The latter is preferable as the RNA can be used directly after being thawed. It is also sensible to divide the purified mRNA into small amounts so that the same sample does not have to be repeatedly frozen and thawed every time a sample is needed. A concentration of 1 μg/μl of mRNA is useful for cloning in quantities of 5 μg. When mRNA has been extracted from total RNA the vials must be treated with DEPC and thoroughly washed before being used to store the mRNA.

Acknowledgements

We wish to thank our colleagues who have helped with work involving these protocols, especially Karen Dawe for the work on *Fusarium sporotrichiodes*. Work in the group has been funded by the MRC, Wellcome Trust, Action Research for the Crippled Child and the SERC.

References

Aviv, H. & Leder, P. 1972. Purification of biologically active globin messenger RNA by chromatography on oligo thymidylic acid-cellulose. *Proceedings of the National Academy of Sciences of the United States of America* **69**, 1408–1412.

Berger, S.L. & Berkenmeier, C.S. 1979. Inhibition of intractable nucleases with ribonucleoside-vanadyl complexes: isolation of messenger ribonucleic acid from resting lymphocytes. *Biochemistry* **18**, 5143–5149.

BERGER, S.L. & KIMMEL, A.R. eds 1987. *Methods in Enzymology: Guide to Molecular Cloning Techniques* Vol. 152. New York & London: Academic Press.

BIRNBOIM, H.C. 1988. Rapid extraction of high molecular weight RNA from cultured cells and granulocytes for Northern analysis. *Nucleic Acids Research* **25**, 1487–1497.

CHIRGWIN, J.M., PRZYBYLA, A.E., MACDONALD, R.J. & RUTTER, W.J. 1979. Isolation of biologically active ribonucleic acid from sources enriched in ribonuclease. *Biochemistry* **18**, 5294–5299.

EDMONDS, M., VAUGHN, M.H. & NAKAZOTO, H. 1971. Polyadenylic acid sequences in the heterogeneous nuclear RNA and rapidly labelled polyribosomal RNA of HeLa cells: Possible evidence for a precursor relationship. *Proceedings of the National Academy of Sciences of the United States of America* **68**, 1336–1340.

MACDONALD, R.J., SWIFT, G.H., PRZYBYLA, A.E. & CHIRGWIN, J.M. 1987. Isolation of RNA using guanidium salts. In *Guide to Molecular Cloning Techniques*. eds Berger, S.L. & Kimmel, A.R. New York & London: Academic Press.

MANIATIS, T., FRITSCH, E.F. & SAMBROOK, J. 1982. *Molecular Cloning. A Laboratory Manual.* Cold Spring Harbor, New York: Cold Spring Harbor Laboratory.

PAIETTA, J.V., ATKINS, R.A., LAMBOWICZ, A.M. & MARZLUF, G.A. 1987. Molecular cloning and characterisation of the *cys*-3 regulatory gene of *Neurospora crassa. Molecular and Cellular Biology* **7**, 2506–2511.

PERRY, R.P., LATORREE, J., KELLY, D.E. & GREENBERG, J.R. 1972. On the lability of poly(A) sequences during extraction of messenger RNA from polyribosomes. *Biochimica et Biophysica Acta* **262**, 220–226.

PRESTON, C.M. 1977. The cell free synthesis of herpesvirus-induced proteins. *Virology* **78**, 349–353.

ROTH, J.S. 1958. Ribonuclease VII. Partial purification and characterisation of a ribonuclease inhibitor in rat liver supernatant fraction. *Journal of Biological Chemistry* **231**, 1085–1095.

WERNER, D., CHEMLA, Y. & HERZBERG, M. 1984. Isolation of poly(A)+ mRNA by paper affinity chromatography. *Analytical Biochemistry* **141**, 329–336.

WALLACE, D.M. 1987. Large and small scale phenol extraction. In *Methods in Enzymology: Guide to Molecular Cloning* Vol. 152. eds Berger, S.L. & Kimmel, A.R. New York: Academic Press.

cDNA Cloning

P.G. Sanders[1] and A.J. Easton[2]

[1] *Molecular Biology Group, Department of Microbiology, University of Surrey, Guildford, Surrey GU2 5XH, UK; and* [2]*Department of Biological Sciences, University of Warwick, Coventry CV4 7AL, UK*

The considerable improvements in nucleic acid cloning techniques in recent years mean that any DNA or RNA sequence can be cloned and isolated if a suitable selection procedure is available. For many, cDNA cloning is the first step towards expression of a particular protein either for identification of the required clone or, ultimately, for some medical or biotechnological purpose. From the nucleic acid sequence of the cDNA the amino acid sequence can be deduced. This is much simpler than sequencing the actual protein, which may not in fact be easily purified, and this has led to interesting discoveries regarding protein evolution. cDNAs can also be used as probes to isolate genomic sequences and the comparisons of cDNAs to their respective genomic clones allows the positions of introns, exons and overlapping genes to be defined. Currently the potential for protein engineering is being explored and the manipulation of cDNA sequences to produce mutant (engineered) proteins is of major importance to this work.

The advantage of cDNA cloning over the production of a genomic library (see chapter by Dale, this volume) is that the cDNA library is enriched for those sequences which are expressed in the cell or tissue under investigation. There are fewer 'non-coding' sequences such as introns present and the size of the library is therefore smaller than a genomic library containing all of the DNA of the cell.

How Many Clones are Required?

The number of clones required will depend on the abundance of the particular RNA of interest and has been well reviewed by Williams (1981). The probability of finding a particular clone in a library can be determined as follows:

$$N = \ln(1 - P)/\ln\left(1 - \frac{1}{n}\right)$$

Genetic Manipulation

0-632-02926-9

where P = the probability of obtaining the required sequence;
n = the minimum number of clones required; and
N = the total number of clones to be analysed (Clarke & Carbon 1976).

Assuming a low abundance mRNA is the required target, it has been estimated that there are 10 670 of these present at a copy number of less than 14 each in a typical cell population of about 36 790 mRNAs. To have a 99% probability of having the desired clone present in a library, 1.69×10^5 clones will be required. Conversely only 7 200 clones would be required for a 99% probability of finding a clone for a moderately or highly abundant message.

Cloning Strategies

Cloning strategies are continually evolving in order to produce large libraries of clones with the minimum number of manipulative steps. Allowing for the fact that it is the nature of the probe which determines the cloning strategy, those attempting cDNA cloning for the first time should keep to a simple strategy and make as large a library as possible in an easy-to-use vector. This should ensure that the desired sequence is at least present and has been amplified. The library can then be probed with DNA, cDNA or RNA probes which are either radiolabelled or modified for detection by newer non-radioactive methods. If antibodies recognizing the protein product of the desired cDNA are available then expression cloning becomes a possibility. Strategies for the isolation of clones by expression of cDNAs in mammalian cells have been developed but these will not be discussed here (Tsai *et al.* 1989 and references cited therein).

The strategies that we describe have been used for the production of cDNA libraries from virally infected mammalian cells and uninfected myeloma cells in both bacteriophage lambda and pUC plasmids (Vieira & Messing 1982). The libraries produced have facilitated the isolation of cDNA clones for pneumonia virus of mice, turkey rhinotracheitis virus, respiratory syncytial virus, human astrovirus and rubella virus. From the myeloma cells we have obtained cDNAs for specific immunoglobulin heavy and light chains.

The authors routinely produce libraries from mRNA primed with oligo (dT), random hexanucleotides and specific primers in pUC13 but for maximum cloning efficiency where the mRNA is of low abundance we use λgt10 and identify the required clones using radiolabelled probes (Young & Davis 1983; Huynh *et al.* 1985) or λgt11, when we have antibodies available to detect expressed fusion products. In our hands λgt10 is a much easier phage to work with than λgt11.

After obtaining purified, undegraded mRNA there are a number of protocols available for the production of cDNA and subsequent cloning.

Those described in detail are shown in Fig. 1. Figure 2 shows the relationships between the protocols so that those using different strategies can see how to proceed.

The priming of first strand-synthesis is normally undertaken by hybridizing oligonucleotides composed of 12–18 thymidylic acid residues to the poly-adenylic acid (poly(A)) tail of the mRNA. This has some drawbacks as a few species of mRNA do not have poly(A) tails and these are also absent from the genomes of some RNA viruses. In these situations, or where the RNA is not an mRNA, either random priming using synthetic hexanucleotides or, if nucleic acid or amino acid sequence data are available from the same or a related gene, specific oligonucleotide primers (perhaps exhibiting a degree of

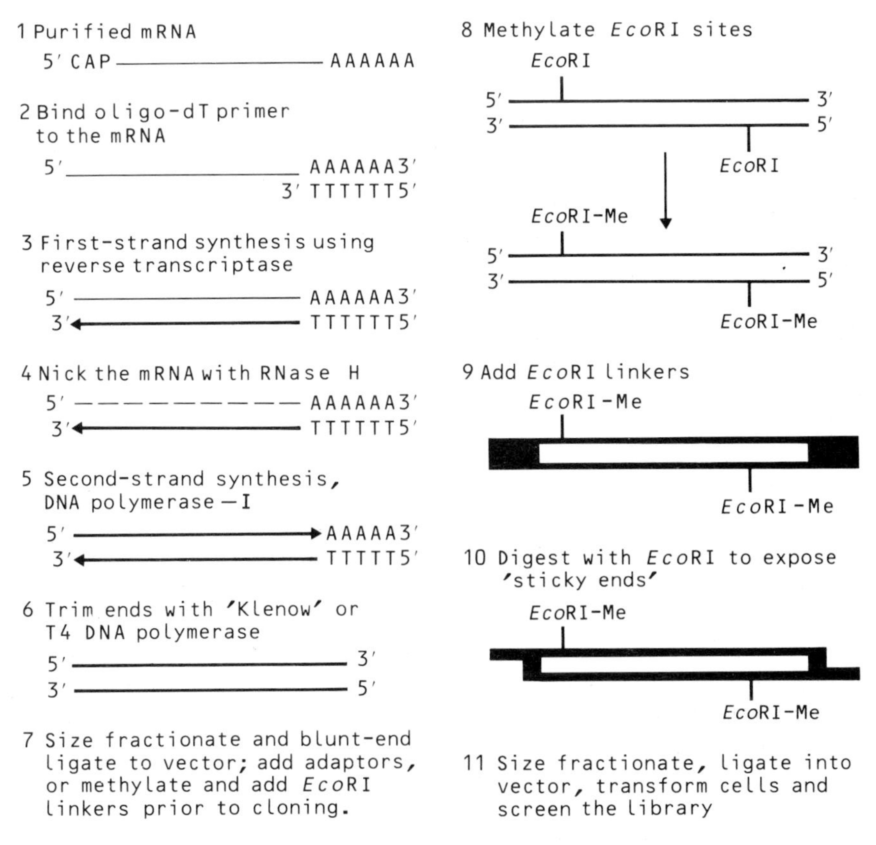

FIG. 1. Diagrams of the basic steps of cDNA synthesis and cloning as presented in this chapter. For the preparation of mRNA for cDNA synthesis see chapter by P.G. Sanders *et al.*, this volume. Although describing oligo-dT priming, the strategies are also relevant for random primer and specific oligonucleotide priming of cDNA synthesis.

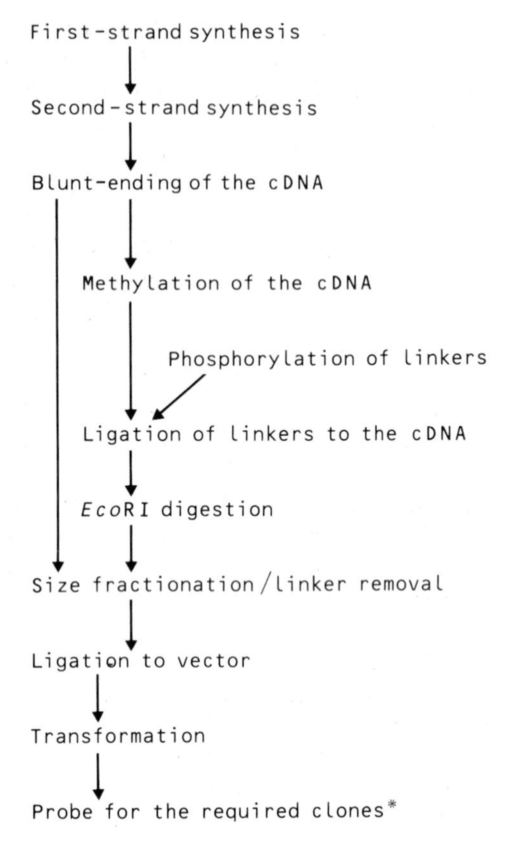

FIG. 2. A diagram showing the inter-relationships between the steps in cDNA synthesis and cloning as described in this chapter.* This step is not covered in detail.

codon redundancy) can be made. The use of random hexanucleotides is considered valuable for obtaining the 5′ ends of mRNAs which may not be reached by priming with oligo(dT) at the poly(A) tail. Although random primers can often lead to primarily short cDNAs we have used them to generate cDNAs of up to 3 kb. Another approach would be to add a poly(A) tail to the 3′ terminus of poly(A) minus RNAs with dATP and terminal transferase which could then be used for oligo(dT) priming.

A number of reverse transcriptase enzymes are now available. We use the avian myeloblastosis virus (AMV) enzyme but have also used the RAV enzyme from Amersham to make first-strand cDNAs of over one kb from relatively crude RNA preparations, by using specific primers, prior to amplification by the polymerase chain reaction. The other commonly used enzyme is from mouse mammary tumour virus.

The most usual strategy at present for second-strand synthesis is based on the use of RNase H to nick and partially degrade the mRNA after first-strand cDNA synthesis (see Gubler 1987). The RNA oligonucleotides left bound to the first cDNA strand then act as primers (Okazaki-like fragments) for second-strand cDNA synthesis with *E. coli* DNA polymerase-I. This enzyme also removes the RNA primers during second-strand synthesis. This technique dispenses with the need to use S1 nuclease to remove a hairpin which forms at the 3′ end of the first strand in the older techniques. S1 nuclease may also degrade or nick the DNA in regions where the double-strand can 'breath' and S1-based methods may not produce full length cDNAs. The population of ds (double-stranded) cDNAs that is produced after second-strand synthesis will have many molecules with ragged ends. For efficient cloning into the vector as many of these as possible must be trimmed and made to have blunt ends. Molecules with recessed 3′ termini can be filled in using DNA polymerase-I or, more often, the 'Klenow' fragment of DNA polymerase-I. A more efficient enzyme, however, is T4 DNA polymerase which can fill in 3′ recesses and trim back 3′ overhangs (Toole *et al.* 1984). Excessive exonuclease activity is repressed by having an excess of deoxynucleotide triphosphates present for the polymerization reaction.

The dscDNA can then either be blunt-end-ligated into the vector or synthetic oligonucleotide linkers may be ligated onto the blunt-ends. Linkers are short double-stranded oligonucleotides which can be digested to expose one or more restriction sites. Blunt-ended ligation is not efficient but such a vast excess of linkers is added that their ligation to the cDNA is favoured. The linkers must then be digested to expose the restriction site that they contain, usually *Eco*RI. To prevent digestion of the cDNA by the same enzyme, any internal restriction sites are protected by methylation before the addition of linkers. There is no obligation to use *Eco*RI linkers and methylases are now available to protect other restriction sites. The cDNA is then separated from unattached and digested linkers and can be size-fractionated at the same time so that only molecules over a certain length, say 500 base-pairs, are cloned. Alternatively, adaptors can be used (Bahl *et al.* 1978). These are variations on linkers whereby two oligonucleotides are joined to the end of the cDNA to provide a 'ready cut' restriction site for cloning. This removes the need to methylate the cDNA to protect any internal restriction sites (see chapter by Carter & Hayle, this volume). The blunt-ended or linkered/adapted cDNAs are then ligated into the vector. For rare cDNAs from libraries prepared by means of oligo(dT) or random primers, λ is the vector of choice as the subsequent packaging reaction provides a most efficient way of transfecting the host bacterium. Recent improvements in *E. coli* transformation protocols and strain development mean that for highly and moderately abundant mRNAs, plasmid vectors offer a viable alternative and the protocols presented have been used mainly for cloning in plasmid vectors. Plasmid vectors for cDNA cloning have also been developed which allow for the expression cloning of mRNAs whose products may be lethal to the host (Stanley and Luzio, 1984).

For a fuller discussion of the different techniques than is possible here see Berger & Kimmel (1987), Watson & Jackson (1985), Huynh *et al.* (1985) and, for an older but useful review, Efstratiadis & Villa-Komaroff (1979).

Many of the commercial firms producing reagents for molecular biology have started to market kits for cDNA synthesis and the subsequent cloning. They are not all listed here but those produced by Amersham have been used with good results. For those with suitable budgets who only require a one-off cloning it may well be worth considering the purchase of a ready tried and tested system from a commercial organization. If repeated clonings are contemplated then this can rapidly become very expensive.

The following protocols are based around the cloning of viral RNAs from infected mammalian cells using AMV reverse transcriptase for the first-strand synthesis, RNase H and 'Klenow' for second-strand synthesis with filling in of 3′ ends also with the 'Klenow' enzyme. cDNAs are then 'blunt-end' ligated into pUC13. Protocols for blunt-ending with T4 DNA polymerase, methylation and the addition of *Eco*RI linkers are also provided. The protocols end with transformation of *E. coli* by a variation of the calcium chloride procedure. For screening methods for the library the reader is referred to chapters by Carter & Hayle and Dale, this volume. Our apologies go to those whose contributions to the methods described here we have overlooked. We have refrained from recommending suppliers of most reagents as at different times we have obtained satisfactory materials from different manufacturers.

First-strand Synthesis

Analysis of a number of cDNA protocols shows some variation in the basic reagents required for first-strand synthesis with a number of different additions being made by groups in attempts to optimize the reaction. Magnesium ions are required by AMV reverse transcriptase and $MgCl_2$ is normally present at 8 or 10 mM, KCl concentrations are between 40 and 50 mM, although this is not included in all recipes. All reactions take place in 50 mM Tris–HCl pH 8.3 at 42°C, with dNTP concentrations between 500 μM and 1.8 mM and DTT from 0.4 to 10 mM. Those groups who determine the pH of their Tris at room temperature quote a pH of 8.5 for the Tris–HCl. Others add spermidine–HCl (5 mM), sodium pyrophosphate (2 mM), actinomycin D and EDTA (10 mM). DMSO has been used in RT reactions of crude RNA preparations for first-strand synthesis prior to amplification by the polymerase chain reaction (Carman *et al.* 1989). All first-strand synthesis should include an inhibitor of RNase (Roth 1958). This is a protein which is normally extracted from the placenta and known as human placental ribonuclease inhibitor (HPRNase inhibitor). This is sold under a variety of trade names (e.g. RNAsin).

This protocol involves duplicate reactions for both the first and second-strands allowing the use of small amounts of radiolabel to monitor cDNA synthesis. The final reaction volume in each tube is 50 μl containing 2.5–5 μg mRNA, although this can be reduced depending on the amount of mRNA available.

Reaction a

Reagents

10 × first-strand buffer:
- 500 mM Tris–HCl pH 8.3 (at 42°C);
- 500 mM KCl;
- 100 mM $MgCl_2$;
- 500 mM DTT;
- 100 μg/ml oligo dT_{12-18}.

dNTP mixes, one with each dNTP (dATP, dTTP, dCTP and dGTP) at 10 mM. For reaction b, a dNTP mix with dCTP or dGTP is reduced to 500 μM depending on the radiolabel to be added.
AMV reverse transcriptase.
HPRNase inhibitor (RNAsin or equivalent).
If required, deionized water can be DEPC-treated (see previous chapter, this volume).

The final reaction mix is 50 mM Tris–HCl pH 8.3, 50 mM KCl, 10 mM $MgCl_2$, 10 mM DTT and 1 mM of each dNTP.

Method

1 Add 2.5–5.0 μg mRNA in distilled water to an autoclaved 1.5 ml microcentrifuge tube. (Many authors recommend that these should be siliconized and baked or DEPC treated. We do not feel this is necessary.) Add sufficient deionized water so that when all the ingredients have been added the final volume will be 50 μl.

2 Heat the mRNA at 70°C for 5 min to denature secondary structure (if the genome has a very high G+C ratio it may be necessary to use a higher temperature). Treatment with methyl mercuric hydroxide can be used to remove secondary structure of high G+C rich genomes but use of this very toxic compound is not recommended unless absolutely essential (Payvar & Schimke 1979).

3 Cool on ice.
4 Add 5 μl of 100 μg/ml oligo dT_{12-18}. A period for primer annealing appears to be unnecessary, probably because of the high primer concentration.
5 Add 5 μl 10 × RT buffer.
6 Add 5 μl of dNTP mix with each nucleotide at 10 mM.
7 Add 1 μl of 500 mM DTT.
8 Add 50–100 units of HPRNase inhibitor.
9 Add 25–50 units of AMV RT.
10 Mix reagents, spin for a few seconds in a microcentrifuge and place at 42°C for 1 h.

Reaction b: radiolabelled control reaction

In order to monitor first-strand synthesis, a second reaction should be set up but with only one triphosphate (preferably dGTP or dCTP) at 500 μM. To the reaction add 2 μl (20 μCi) of the selected triphosphate (Amersham 3 000 Ci/mmol). Precautions for the use of radio-isotopes must be followed (see Zoon 1987). Amersham also provide information on the safe use of radioisotopes.

After 1 h at 42°C both reactions (a and b) are stopped by the addition of EDTA to 25 mM. Add 2.5 μl of 500 mM Na_2 EDTA pH 8.0.

Analysis of first-strand synthesis

To determine the amount of first-strand cDNA made, the total amount of radioactivity and the amount incorporated have to be determined. This is done by precipitating two samples of the DNA/mRNA hybrids onto separate filter papers and then determining the radioactivity on each disc in a liquid scintillation counter with an appropriate scintillant or by Cerenkov counting. In addition to determining the amount of cDNA produced, gel electrophoresis should be used to estimate the sizes of the first-strand molecules. For details of gel electrophoresis techniques see Maniatis *et al.* (1982) or Rickwood & Hames (1982). 10–20 000 cpm can be removed from the first-strand reaction and analysed by alkaline agarose gel electrophoresis, using a 1.4% gel.

TCA precipitation

Method

1 Label (in pencil) 2 Whatman no. 1 2.5 cm diameter filter paper discs (a) and (b). Spot a 1 μl quantity from the radiolabelled reaction onto each disc.
2 Dry the samples onto the discs under a heat lamp.

3 One disc (b) is washed three times in 50 ml ice cold 10% trichloroacetic acid (TCA). Leave the discs in each wash for 5 min with occasional shaking.
4 Dry the discs by two washes in 95% ethanol.
5 Place the discs into scintillation vials and either count without further treatment on the low energy tritium channel of a scintillation counter (Cerenkov counting) or add a suitable aqueous scintillant and count on the higher energy (^{32}P) channel.
6 Determine the percentage incorporation of radioisotope in the first-strand reaction.

Calculation of the amount of first-strand cDNA produced

The yield of first-strand cDNA is calculated as follows:
In the labelled reaction the limiting amount of the labelled dNTP is 0.5 mM, ignoring any contribution from the radiolabel added. This is equivalent to 0.5 nmol of this dNTP per μl, and therefore in the 50 μl reaction there are 25 nmol.

Assuming the labelled dNTP is dCTP then the molecular weight is circa 330 and 1 nmol contains 330 ng.

Therefore 25 nmol contain 8.25 μg.

There are, however, 4 dNTPs present and therefore the maximum amount of cDNA that could be made is $4 \times 8.25 = 33$ μg with 100% incorporation of radiolabel.
Every 1% incorporation is therefore equivalent to 330 ng of first-strand cDNA.

Purification of the mRNA/cDNA hybrids

The first-strand reactions are now phenol/chloroform extracted and the mRNA–cDNA hybrids recovered by ethanol precipitation. It is important to keep reactions (a) and (b) separate. The 50 μl reactions can be diluted to 100 μl with 50 μl 1 × TE (10 mM Tris–HCl pH 8.0) prior to phenol/chloroform extraction. The phenol/chloroform should also be back-extracted with 50 μl 1 × TE and the 100 and 50 μl samples pooled prior to precipitation. For protocols see the previous chapter, this volume.
Do not pool reactions (a) and (b) at this stage.

Second-strand Synthesis

As with the first-strand synthesis, different groups have used various protocols for second-strand synthesis by RNase H/DNA polymerase-I based reactions. HEPES (100 mM) pH 6.9 or 7.6 has been used instead of Tris–HCl, with

4–10 mM $MgCl_2$, often 70 mM KCl and 40 μM to 1 mM dNTPs. DTT or β-mercaptoethanol are not always included and some reaction protocols include one or more of 10 mM ammonium sulphate, 50 μg/ml BSA and B-NAD. In early protocols, DNA ligase was often added to join any nicks in the cDNA but this is not required. The reaction conditions described here are the same as those of Gubler (1987).

For second-strand synthesis, duplicate reaction mixes are again prepared. This time the unlabelled first-strand reaction products from (a) are radiolabelled to monitor second-strand synthesis.

Reagents

10 × second-strand buffer:
- 200 mM Tris–HCl pH 7.5;
- 50 mM $MgCl_2$;
- 100 mM $(NH_4)_2SO_4$.

The volume of the second-strand reaction is 100 μl.

The final reaction conditions are: 20 mM Tris–HCl pH 7.5, 5 mM $MgCl_2$, 10 mM $(NH_4)_2SO_4$, 100 mM KCl, 50 μg/ml BSA, 40 μM dNTPs.

Method

1 Resuspend the pelleted cDNA/RNA hybrids in sufficient deionized water in order to make the final reaction volume 100 μl.
2 Add 10 μl 10 × second-strand buffer.
3 Add 10 μl 1 M KCl.
4 Add 5 μl 1 mg/ml BSA.
5 Add 1 μl of 4 mM dNTP and mix.
6 To the unlabelled first-strand reaction add 20 μCi ^{32}P-labelled dNTP.
7 Add 30 units of *E. coli* DNA polymerase-I and 1 unit of RNase H.
8 Incubate at 12°C for 1 h.
9 Incubate at 22°C for 1 h.
10 Terminate the reaction by adding EDTA to 20 mM.
11 Take 1 μl quantities, precipitate with TCA and count as above.
12 Take 10–20 000 cpm to analyse the cDNAs by neutral gel electrophoresis on a 1.2% gel and compare to radiolabelled standards. The gels should be electrophoresed in Tris–acetate buffer (Maniatis *et al.* 1982).
13 Combine both reactions, extract with phenol and precipitate with ethanol.
14 Calculate the amount of second-strand cDNA synthesized from the percentage incorporation of radiolabel in step 11 above. The amount of second-strand synthesis can be determined in the same way as for the first-strand synthesis, i.e. a 100 μl reaction containing 40 μM of each dNTP will

contain 4 nmol of each of the four dNTPs and therefore 16 nmol in total. Therefore the maximum amount of second-strand cDNA that can be synthesized is $16 \times 330 = 5280$ ng or 5.28 μg. Every 1% incorporation is therefore equivalent to 52.8 ng of second-strand cDNA. However, because the DNA is now double-stranded for each 1% incorporation, there is a total of 105.6 ng of cDNA.

Preparation of the cDNA for Cloning

Removal of ragged ends from the cDNA to produce blunt-ends

The products of cDNA synthesis will have ragged termini. Two protocols are described for the removal and/or filling in of these ragged ends from the ds cDNA. In the first method the Klenow fragment of DNA polymerase-I is used to fill-in short 3′ tails, although in some protocols total DNA polymerase-I is used. The second strategy uses T4 DNA polymerase to remove overhanging 3′ end and fill in short 3′ regions. The latter method is becoming the method of choice as it can deal with a greater diversity of cDNA molecules. The potential to degrade the cDNA is, however, greater with the T4 polymerase.

Filling in recessed 3′ ends with the Klenow fragment of DNA polymerase-I

Method

The final reaction mixture should contain, in 50 μl:50 mM Tris–HCl pH 8.0, 5 mM $MgCl_2$, 1 mM DTT, 500 mM each dNTP, 1 unit of Klenow per microgram of cDNA.

1 Dissolve the cDNA pellet from step 13 above (second-strand synthesis) in 35 μl deionized water.
2 Add 5 μl 10 × *E. coli* DNA polymerase-1 buffer.
3 Add 5 μl of 5 mM dNTP and mix.
4 Add 5 μl of 10 mM DTT.
5 Add 1 unit of the Klenow fragment of DNA polymerase-I per microgram of cDNA.
6 Incubate at 37°C for 30 min.
7 Stop the reaction by the addition of 0.5 M Na_2EDTA pH 8.0–20 mM.
8 Extract with phenol/chloroform and precipitate with ethanol as above.

Blunt-ending using T4 DNA polymerase

T4 DNA polymerase not only fills in recessed 3′ ends but also removes 3′ overhangs from the dscDNA. The fact that the enzyme has an exonuclease activity means that it could theoretically degrade precious cDNA but the presence of all four dNTPs means that the polymerization reaction takes precedence over the exonuclease activity at the 3′ end of the ds cDNA.

A survey of the literature shows that T4 polymerase prefers buffers containing acetate instead of chloride ions but chloride salts have been used successfully. This obviously simplifies the development of one-tube reactions for cDNA synthesis. The reaction conditions provided are based on those of O'Farrell as described by Maniatis *et al.* (1982).

Reagents

10 × T4 polymerase buffer:
0.33 M Tris–acetate pH 7.9:
0.66 M potassium acetate;
0.1 M Mg acetate;
5 mM DTT;
1 mg/ml BSA;
(Store at −20°C).

The reaction conditions are, in 50 μl: 33 mM Tris–acetate pH 7.9, 66 mM K acetate, 10 mM Mg acetate, 0.5 mM DTT, 100 μg/ml BSA, 200 μM dNTPs.

Method

1 Dissolve the cDNA pellet (see second-strand synthesis above) in 42 μl of deionized water.
2 Add 5 μl of 10 × T4 polymerase buffer.
3 Add 2.5 μl of the 4 mM dNTP mixture.
4 Add 3–5 units of T4 DNA polymerase.
5 Mix and spin for a few seconds in a microfuge.
6 Incubate at 37°C for 30 min.
7 Add 150 μl deionized water, extract in phenol/chloroform and precipitate as before.

Comments

The cDNA prepared so far should have predominantly blunt-termini and can be blunt-end-ligated into a suitable plasmid or phage vector for library construction without further modification. We have successfully inserted blunt ended cDNA, produced by means of oligo(dT), random and specific oligonucleotide primers, into *Sma*I-digested, dephosphorylated pUCI3 (Pharmacia) and produced libraries by transfection into *E. coli* DH5 (BRL) (Hanahan 1983). For cDNA cloning we use 100 μl of each DH5 as recommended

(BRL). For more routine DNA clonings, especially subcloning of fragments, we have found that 20 μl of the high transformation (10^9transformants/μg) efficiency DH5α is sufficient to obtain desired clones at very little cost per transformation. To produce large libraries, however, we have used λgt10 and protocols requiring the addition of *Eco*RI linkers before cloning into the vector. The same strategy can be used to clone into *Eco*RI-cut plasmids.

Methylation of the cDNA

The components of reaction mixtures for methylation again vary with published protocols containing: 50–100 mM Tris–HCl pH 7.5 or 8.0, 1–10 mM EDTA, 0–200 μg/ml BSA, 50–125 μM *S*-adenosyl methionine, 5–20 units *Eco*RI methylase. $MgCl_2$ should not be present as it inhibits *Eco*RI methylase. Some protocols recommend titrating each batch of methylase on suitable DNA fragments containing *Eco*RI sites, before use with precious cDNA.

Reagents

5 × Methylation buffer:
500 mM Tris–HCl pH 8.0;
10 mM EDTA;
50 mM DTT;
1 mg/ml BSA.
Make up 1 ml, allowing for the addition of the equivalent of 12.5 μl 32 mM *S*-adenosyl methionine per ml (this comes with each enzyme from New England Biolab's) before use. Store in 79 μl amounts.
Add 1 μl of *S*-adenosyl methionine to 79 μl of 5 × buffer before use. This gives a final concentration of 80 μM *S*-adenosyl methionine in the reaction vial.

Method

1 Resuspend DNA from stage 7 above in sufficient deionized water to make a final reaction volume of 50 μl.
2 Add 10 μl 5 × methylation buffer containing freshly added *S*-adenosyl methionine.
3 Add 20 units of *Eco*RI methylase.
4 Incubate at 37°C for 20 min.
5 Phenol chloroform extract and precipitate as previously described.

Linker Kinase Reaction

Oligonucleotide linkers may be purchased either unphosphorylated or phosphorylated. If phosphorylated linkers have been purchased then this section can be ignored. If unphosphorlyated *Eco*RI linkers are being used, then in order to attach them to the methylated cDNA they must be phosphorylated at their 5′ termini by means of T4 polynucleotide kinase and ^{32}P-γ ATP. The use of a radiolabel also helps to monitor the efficiency of linker ligation and digestion. The conditions used by different groups for this reaction include variations in Tris–HCl concentrations between 10 and 70 mM, DTT concentrations from 5 to 15 mM and the optional addition of reagents such as spermidine and BSA. The optimal conditions for T4 polynucleotide kinase are, in fact, Tris–HCl pH 7.6, 10 mM Mg^{2+} and 5 mM DTT with 1 mM spermidine which stabilizes the tetrameric form of the enzyme.

With the careful selection of reaction buffer the ligation of the kinased linkers to the cDNA can be performed sequentially in the same tube and this may be preferred by some as it cuts out unnecessary manipulations.

Reagents

10 × kinase buffer:
660 mM Tris–HCl pH 7.6;
100 mM $MgCl_2$;
50 mM DTT;
10 mM spermidine;
2 mg/ml BSA;
10 mM ATP.

The final reaction mix should contain, in 50 μl: 66 mM Tris–HCl pH 7.6, 10 mM $MgCl_2$, 5 mM DTT, 1 mM spermidine, 200 μg/ml BSA, 10–20 units T4 pol, 1 mM ATP, 10 μCi γ ATP, 5 μg *Eco*RI linkers.

Method

The following procedures should be carried out for a 50 μl reaction containing 5 μg of dephosphorylated *Eco*RI linkers:

1 Place 5 μg of linkers (5 μl of 1 mg/ml in deionized water) in a 1.5 ml microcentrifuge tube. Add water to make the final reaction volume 50 μl.
2 Heat at 90°C for 2 min, centrifuge briefly and cool on ice.
3 Add 5 μl 10 × kinase buffer.
4 Add 10 μCi (1 μl) γ ATP (specific activity ≤ 3 000 Ci/mmol)
5 Add 10–20 units T4 polynucleotide kinase.
6 Incubate at 37°C for 1 h.
7 Heat to 70°C for 5 min, centrifuge for a few seconds and allow to cool slowly to room temperature.
8 Remove a sample for analysis of ligation efficiency and store the remaining linkers in 5 μl quantities at −70°C.

Test of Ligation Efficiency

Remove a 1 μl sample of kinased linkers and self-ligate these under the conditions described in the next section.

Split the ligation mix into two and digest one portion with *Eco*RI using the conditions described below but use proportionally less enzyme.

Prepare a 10% polyacrylamide gel and electrophorese a labelled linker control (0.5–1 μl), the ligated linkers and ligated-digested linkers on the gel. Autoradiograph the gel and determine the proportions of ligated and digested products. If satisfactory ligation and subsequent digestion have occurred proceed to the next section.

Ligation of the Phosphorylated Linkers to the cDNA

As described above, the reaction conditions for the phosphorylation of the linkers with T4 polynucleotide kinase are almost identical to those for the ligation of cDNA and linkers with T4 DNA ligase. Reports describing the use of various additives such as Ficoll 400, polyethylene glycol and hexamine cobalt chloride to improve the efficiency of blunt-end ligations have been published (Pheiffer & Zimmerman 1983; Rusche & Howard-Flanders 1985). We have obtained efficient ligation without these.

1 Resuspend the precipitated, methylated DNA (p. 141) in sufficient 1 × TE (10 mM Tris–HCl pH 7.6, 1 mM EDTA) so that the *final* reaction volume will be 30 μl.
2 Add 3 μl of the 1 mg/ml linker solution.
3 Add 3 μl of 10 × T4 kinase buffer (p. 142) as 'ligase buffer'.
4 Remove 1 μl for analysis (see next section), keep in ice.
5 Add 4 units of T4 DNA ligase.
6 Leave at 14°C overnight.
7 Heat inactivate the ligase at 70°C for 10 min.
8 Take 1 μl for analysis (see next section).

*Eco*RI Digestion and Removal of Excess Linkers from the cDNA

The next stages involve (a) digestion of the attached *Eco*RI linkers to expose the *Eco*RI 'sticky ends' and (b) subsequent removal of the vast excess of linkers present in the reaction. If the excess linkers are not removed these will subsequently ligate into the vector producing a large number of 'false positive' recombinant clones.

To the heat inactivated ligation reaction from step 7 in the previous section:

1 Add deionized water so that the final reaction volume will be 300 μl.

2 Add 1 M Tris–HCl pH 7.5 to 50 mM (13.5 µl), 4 M NaCl to 100 mM (7.5 µl), 1 M $MgCl_2$ to 10 mM (27 µl), 10 mg/ml BSA to 100 µg/ml (3 µl).
3 Add 20 µl of *Eco*RI (200 units at 20 units/µl).
4 Incubate at 37°C for 2 h.
5 Take 10 µl for analysis.
6 Analysis of linker ligation and removal.
Analyse the fractions taken in steps 4 and 8 from the previous section and step 5 from this section by electrophoresis with radiolabelled markers through a 4% acrylamide gel (Maniatis *et al.* 1982). Compare the sizes of unligated, ligated and digested fragments to check that ligation and digestion have been successful.
7 Phenol chloroform extract and ethanol precipitate.

Size Fractionation of the Linkered cDNA

Size fractionation at this stage serves two functions. The excess linkers are removed from the cDNA following digestion (previous section) and the cDNA is fractionated to remove small cDNA molecules (< 500 bp).

Without this fractionation, many clones would contain only *Eco*RI linkers and/or small cDNA fragments and screening the library for large clones would be very tedious.

There are a number of matrices that have been used for separation of cDNA and linkers by exclusion chromatography. These include Sepharose CL4B as described here, Sephadex G50, Sephacryl S200, Bio-gel A50 and Ultrogel AcA34.

Method

1 Precipitate the DNA from step 7, previous section and dissolve in 20 µl of 0.3 M NaCl, 10 mM Tris–HCl pH 8.0, 1 mM EDTA (Maniatis *et al.* (1982). The column should be prepared in the same buffer. We have also found that the substitution of 1 × TE (10 mM Tris–HCl pH 8.0, 1 mM EDTA), or even distilled water, both for dissolving the DNA and running the column, gives satisfactory results. Eschenfeldt & Berger (1987), recommend 50 mM NH_4HCO_3 pH 8.0–8.5. This must be removed if T4 polynucleotide kinase is to be used at a later stage.
2 Prepare a 1 ml column of Sepharose CL4B in either a drawn out Pasteur pipette, a 1 ml syringe or a disposable column. Glass wool can be used to plug the Pasteur pipette or syringe and this and any other glass used should be siliconized before use to prevent the nucleic acids binding to it. Although this is more of a problem with single-stranded DNA and RNA, it is, nevertheless, a wise precaution. Equilibrate the column with 0.3 M NaCl, 10 mM Tris-HCl pH 8.0, 1 mM EDTA.

3 Add the cDNA and collect 50 μl fractions.
4 Determine which fractions contain the cDNA by Cerenkov counting of each fraction in a liquid scintillation counter with the low energy tritium channel. Samples (1 μl) should also be taken from each fraction (circa 10 000 cpm) and analysed on a 1.2% agarose gel. Suitable markers are ^{32}P-labelled fragments of *Hae*III-digested ΦX174 and a lambda *Hin*dIII digest. After autoradiography, pool the fractions containing cDNAs over 500 base-pairs in size, precipitate in ethanol (Carrier tRNA, 10 μg/ml final concentration, may be required) and resuspend in 1 × TE. If 1 × TE is used for the column, then precipitation should not be necessary.

Ligating Fragments into pUC Vectors

The blunt-ended or linkered fragments can be cloned directly into *Sma*I or *Eco*RI digested pUC13 respectively. The vectors should be dephosphorylated before use with calf intestinal phosphatase or purchased already digested and dephosphorylated. The selection for recombinants with these vectors involves insertional inactivation of the *lacZ'* gene which prevents suitable mutant host *E. coli* strains from metabolizing the chromogenic substrate 5-bromo-4-chloro-3-indolyl-β-D-galactopyranoside (X-gal). The majority of bacterial recombinants therefore produce 'white' colonies that can be selected from the blue non-recombinant clones. Dephosphorylated vectors are to be preferred as the background of non-recombinants will be reduced and cDNA insertion favoured over vector religation. *rec*A bacterial hosts containing the *lac*I^q mutation should be used for maximum stability of recombinant plasmids. These vectors also allow the expression of clones which insert in the correct reading frame which can be lethal if the protein product is toxic to the cell. This can be overcome by omitting isopropyl-β-D-thiogalactopyranoside (IPTG) from the plating medium, or by switching to lower copy number vectors such as the pEX series (Stanley & Luzio 1984).

To clone blunt-ended cDNAs, the molar ratio of vector to insert should be 3:1 but ratios of 1:1 are also used. The molarity of the cDNA should be calculated from the average size as determined by autoradiography after Sepharose CL4B chromatography. For example, if the average size of the cDNA is 1 000 bp, then the vector to cDNA ratio in micrograms is 7.5:1 as the vector is 2.5 times as large as the average insert.

Ligation reaction for blunt-ended molecules

Reagents

10 × ligase buffer:
500 mM Tris–HCl pH 7.8;

100 mM $MgCl_2$;
4–10 mM ATP;
100 mM DTT.

Method

1 To a 1.5 ml microcentrifuge tube on ice add 10 ng cDNA and 75 ng *Sma*I digested, phosphatased pUC13, with sufficient water to make a 50 μl final reaction volume. The authors use between 50 and 100 ng of DNA in this volume for blunt-end ligations.
2 Add 5 μl of 10 × ligation buffer.
3 Add 1–2 units of T4 DNA ligase.
4 Incubate at 16°C overnight.
5 Set up a control ligation of the vector without the cDNA.
6 The reactions can be stopped by the addition of EDTA to 25 mM before transformation. A five-fold dilution may also increase the number of transformants.

Ligation reaction for 'sticky-ended' molecules

For linkered molecules and other 'sticky-ended' ligations into phosphatased vectors, a ratio of vector:insert of 2:1 can be used. The reaction conditions are as above with DNA concentrations of 5–10 μg/ml and 0.5 u/ml of ligase, 4°C overnight. This should provide $>2 \times 10^6$ transformants of which 90% are recombinant. For protocols for cloning into λ see Huynh *et al.* (1985).

Transformation of Host Bacteria

We use the following protocol but many commercial preparations of bacterial strains such as JM109 and DH5α are available which 'guarantee' extremely high levels of transformation, up to 10^9 transformants per microgram of supercoiled pBR322 or pUC vectors (the number of recombinant molecules will obviously be lower), and these come with their own transformation protocols. For a discussion of bacterial transformation see Hanahan (1985). With the following protocol, which could be refined further, at least 5×10^6 transformants per microgram of supercoiled pUC vectors should be obtained, 10^6 transformants per μg is poor and 5×10^7 per μg can be obtained. The number of recombinants present can vary between 10 and 90% of the transformants.

Method

1 Subculture 1 ml of an overnight culture of JM109 into 75 ml of medium.
2 Shake until the OD_{590} is 0.3–0.4 (usually 90–120 min).
3 Stand the culture on ice for 30 min.
4 Pellet the bacteria from 50 ml at about 5 000 ***g*** for 10 min.
5 Resuspend the pellet in 0.5 times the original 50 ml volume (i.e. 25 ml) of freshly prepared ice cold 0.1 M $MgCl_2$.
6 Immediately pellet as step 4.
7 Resuspend in 0.1 times the previous 25 ml volume (i.e. 2.5 ml) of ice cold and freshly prepared from solid 0.1 M $CaCl_2$.
8 Stand the bacteria on ice for 60 min.
9 Mix 0.2 ml bacteria with a maximum of 20 ng of the ligation mix. (5–20 ng is a suitable range of DNA to use for the transformations.) A five-fold dilution of the ligation reaction before transformation may increase the number of recombinants.
10 Leave on ice for 30 min with occasional mixing.
11 Heat shock at 42°C for 2 min. Then return to ice for 30 min.
12 Add medium to 1 ml, incubate with shaking at 37°C for 1 h.
13 Plate out on selective media containing ampicillin at 100 μg/ml, X-gal at 20–40 μg/ml and IPTG if required at 130 μg/ml. Incubate overnight at 37°C.

Screening for Recombinant Phages and Plasmids

Protocols for cloning into lambda vectors and for screening for recombinants are described in chapters by Dale & Carter and Hayle, this volume, and will not be covered here. These topics are also reviewed by Hames & Higgins (1985) and by Huynh *et al.* (1985).

References

BAHL, C.P., WU, R., BROUSSEAU, R., SOOD, A.K., HSUING, H.M. & NARANG, S.A. 1978. Chemical synthesis of versatile adapters for molecular cloning. *Biochemical and Biophysical Research Communications* **81**, 695–703.

BERGER, L. & KIMMEL, A.R., eds. 1987. *Methods in Enzymology: Guide to Molecular Cloning Techniques*. Vol. **152**. New York & London: Academic Press.

CARMAN, W.F., WILLIAMSON, C., CUNLIFFE, B.A. & KIDD, A.H. 1989. Reverse transcription and subsequent DNA amplification of rubella vira RNA. *Journal of Virological Methods* **25**, 21–30.

CLARKE, L. & CARBON, J. 1976. A colony bank containing synthetic ColE1 hybrid plasmids representative of the entire *E. coli* genome. *Cell* **9**, 91–99.

EFSTRATIADIS, A. & VILLA-KOMAROFF, L. 1979. Cloning of double stranded cDNA. *Genetic Engineering* Vol. 1. eds Setlow, J.S. & Hollaender, A. pp. 15–36. New York: Plenum Press.

ESCHENFELDT, W.H. & BERGER, S.L. 1987. Purification of large double-stranded DNA fragments.

In *Methods in Enzymology: Guide to Molecular Cloning Techniques* Vol. 152, eds Berger, S.L. & Kimmel, A.R. pp. 335–337. New York & London: Academic Press.

GUBLER, U. 1987. Second-strand cDNA synthesis: mRNA fragments as primers. In *Methods in Enzymology: Guide to Molecular Cloning Techniques* Vol. 152. eds Berger, S.L. & Kimmel, A.R. pp. 330–335. New York: Academic Press.

HAMES, B.D. & HIGGINS, S.J. 1985. *Nucleic Acid Hybridisation; A Practical Approach*. Oxford: IRL Press.

HANAHAN, D. 1983. Studies on transformation of *Escherichia coli* with plasmids. *Journal of Molecular Biology* **166**, 557–580.

HANAHAN, D. 1985. Techniques for transformation of *E. coli*. In *DNA Cloning, A Practical Approach* Vol. 1. ed. Glover, D.M. pp. 109–135. Oxford: IRL Press.

HUYNH, T.V., YOUNG, R.A. & DAVIS, R.W. 1985. Construction and screening cDNA libraries in λgt10 and λgt11. In *DNA Cloning, A Practical Approach*. Vol. 1. ed. Glover, D.M. pp. 49–78. Oxford: IRL Press.

MANIATIS, T., FRITSCH, E.F. & SAMBROOK, J. 1982. *Molecular Cloning. A Laboratory Manual*. Cold Spring Harbor, New York: Cold Spring Harbor Laboratory.

PAYVAR, F. & SCHIMKE, R.T. 1979. Methylmercury hydroxide enhancement of translation and transcription of ovalbumin and conalbumin mRNAs. *Journal of Biological Chemistry* **254**, 7636–7642.

PHEIFFER, B.H. & ZIMMERMAN, S.B. 1983. Polymer stimulated ligation: enhanced blunt- or cohesive-end ligation of DNA or deoxyribonucleotides by T4 DNA ligase in polymer solutions. *Nucleic Acids Research* **111**, 7853–7871.

RICKWOOD, D. & HAMES, B.D. eds. 1982. *Gel Electrophoresis of Nucleic Acids: A Practical Approach*. Oxford: IRL Press.

ROTH, J.S. 1958. Ribonuclease VII. Partial purification and characterisation of a ribonuclease inhibitor in rat liver supernatant fraction. *Journal of Biological Chemistry* **231**, 1085–1095.

RUSCHE, J.R. & HOWARD-FLANDERS, P. 1985. Hexamine cobalt chloride promotes intermolecular ligation of blunt end DNA fragments by T4 ligase. *Nucleic Acids Research* **13**, 1997–2008.

STANLEY, K.K. & LUZIO, J.P. 1984. Construction of a new family of high efficiency bacterial expression vectors: identification of cDNA clones coding for human liver proteins. *EMBO Journal* **3**, 1429–1434.

TOOLE, J.J., KNOPF, J.L., WOZNEY, J.M., SULTZMAN, L.A., BUECKER, J.L., PITTMAN, D.D., KAUFMAN, R.J., BROWN, E., SHOEMAKER, C., ORR, E.C., AMPHLETT, G.W., FOSTER, W.B., COE, M.L., KNUTSON, G.J., FASS, D.N. & HEWICK, R.M. 1984. Molecular cloning of a cDNA encoding human antihaemophilic factor. *Nature* **312**, 342–347.

TSAI, S-F., MARTIN, D.I.K., ZON, L.I., D'ANDREA, A.D., WONG, G.C. & ORKIN, S.H. 1989. Cloning of cDNA for the major DNA-binding protein of the erythroid lineage through expression in mammalian cells. *Nature* **339**, 446–451.

VIEIRA, J. & MESSING, J. 1982. The pUC plasmids: an ml3mp7-derived system for insertion mutagenesis and sequencing with synthetic universal primers. *Gene* **19**, 259–268.

WATSON, C.J. & JACKSON, J.F. 1985. An alternative procedure for the synthesis of double-stranded cDNA for cloning in phage and plasmid vectors. In *DNA Cloning, A Practical Approach* Vol. 1. ed. Glover, D.M. pp. 79–88. Oxford: IRL Press.

WILLIAMS, J.G. 1981. The preparation and screening of a cDNA clone bank. In *Genetic Engineering* Vol. 1. ed. Williamson, R. pp. 1–59. New York & London: Academic Press.

YOUNG, R.A. & DAVIS, R.W. 1983. Efficient isolation of genes by using antibody probes. *Proceedings of the National Academy of Sciences of the United States of America* **80**, 1194–1198.

ZOON, R.A. 1987. Safety with ^{32}P- and ^{35}S-labelled compounds. In *Methods of Enzymology* Vol. 152. eds Berger, S.L. & Kimmel, A.R. pp. 25–29. New York & London: Academic Press.

Primer Extension Sequencing of RNA Viruses

C.A. GIBSON,* L.M. DUNSTER AND A.D.T. BARRETT
Department of Microbiology, University of Surrey, Guildford, Surrey GU2 5XH, UK

A number of different techniques are available for sequencing viral, prokaryote and eukaryote derived nucleic acids, of both RNA and DNA origins. These techniques are based on a variety of enzymic degradative and molecular cloning strategies, whilst in recent years there has been a virtual revolution in the field of molecular cloning with development of the polymerase chain reaction (PCR).

In the choice of any technique for nucleic acid sequencing there are inherent advantages and disadvantages which must be assessed in the light of the available materials and the required end result. Thus the majority of molecular cloning techniques have the advantage of requiring only small quantities of DNA or RNA, from which large numbers of nucleotide bases can be sequenced ($\geq 10^7$-fold amplification from single genomic sequences of DNA segments of up to 2 kb in length) (Saiki *et al.* 1988). However, an extremely high stringency of nucleic acid purity is required, particularly in the use of PCR, since the presence of a single copy of contaminating nucleic acid can have disastrous results. Thus the very efficiency of many molecular cloning techniques creates inherent disadvantages, when these techniques are applied to the sequencing of RNA and, in particular, RNA viruses.

The mutational instability of RNA is much higher than that of DNA (error rate of $10^{-3}-10^{-4}$ for RNA compared with $10^{-9}-10^{-10}$ for DNA) resulting in a rapid evolution of RNA viruses and thus producing a genetically heterogeneous mixed population of virions within any given virus preparation. Thus during cloning, it is possible that any minor subpopulation may be

* *Present address: Veterinary Research Institute, 475–485 Mickleham Road, Attwood, Victoria 3049, Australia*

Genetic Manipulation
0–632–02926–9

selected. The validity of any sequence data obtained by molecular cloning must therefore be carefully interpreted in the light of such possible cloning artefacts.

As an alternative, or in combination with molecular cloning, direct sequencing of the total species of template RNA present is possible by any one of three techniques. Two of the techniques examine the RNA molecules directly by cleavage, either by specific ribonucleases or by chemical specific cleavage following chemical modification of the RNA chain. An alternative method for sequencing RNA is dependent upon the production of varying length cDNA copies of the RNA population by means of the enzyme reverse transcriptase (RT) and is known as *primer extension dideoxy chain termination sequencing* (Sanger *et al.* 1977). There are several advantages in the use of this primer extension technique compared with other methods for obtaining RNA sequence data. Firstly, many of the other techniques require multistep protocols prior to analysis of sequence by polyacrylamide gel electrophoresis (PAGE), whereas the primer extension technique can be performed by a simple two-step procedure of primer labelling followed by primer extension. Secondly, with the use of specific synthetic oligonucleotide primers complementary to the RNA sequence, the requirement for RNA preparations absolutely free of any contaminating nucleic acids is reduced.

Primer extension sequencing of RNA viruses is a straightforward technique for rapidly obtaining nucleotide sequence data representative of the majority genotype of a mixed virus population, thereby avoiding cloning artefacts generated by the selection of minor virion subpopulations. Furthermore, primer extension sequencing is a useful quality control procedure for verifying the sequence data obtained by molecular cloning techniques.

We present two modifications of the basic dideoxy chain termination technique of Sanger *et al.* (1977), used for the nucleotide sequencing of negative and positive-strand RNA viruses.

Precautions when Handling RNA

Any manipulations carried out with purified RNA introduce an extreme risk of contamination with adventitious RNase. Its presence is fairly ubiquitous and can be introduced to the RNA sample through contaminated containers, pipettes, tips, reaction buffers, enzymes and from naked skin. Simple precautions can be taken such as the wearing of suitable disposable gloves with frequent changes for all manipulations and handling of equipment and reagents, along with the use of separate equipment dedicated solely for RNA work. Any glassware used should be first treated to remove any RNase present. There is no totally effective way to achieve this, however, pretreatment with diethyl pyrocarbonate (DEPC, Sigma Ltd) is recommended. All glassware should be soaked overnight in a 0.1% v/v solution of DEPC in deionized distilled water, followed by vigorous rinsing in absolute alcohol and baking at 180°C for a minimum of 6 h.

In order to further reduce the risk of introducing RNase all buffers should be prepared with water of the purest source available. We routinely use HPLC-pure water (Aldrich Chemicals Ltd) for the preparation of all buffers and reagents used for sequencing.

Nucleotide Sequencing of Influenza Virus Genes

Influenza viruses are enveloped viruses which possess a single-stranded, negative sense genome composed of eight RNA segments, each segment coding for one or more of the influenza virus-specified polypeptides.

Three major antigenic classes of influenza virus have been identified; namely, Types A, B and C, which are distinguished by distinct serological differences in their nucleoproteins. The type A influenza virus is the major causative agent of pandemics in man and is also implicated in epizootics in seals, horses, swine and fowl.

Virus propagation

Influenza viruses are routinely propagated in fertile domestic hens eggs. Embryonated eggs, 10–12 days old, are selected by candling, followed by the drilling of a small hole in the egg shell and inoculation with influenza virus into the allantoic cavity. Relatively low infectious doses of virus are used, at titres of between 10^3 and 10^5 haemagglutinating units (HAU), in broth saline inoculum for each egg. Low titres are used in order to prevent the generation of defective interfering particles. Following inoculation, the eggs are sealed with molten paraffin wax and incubated at 37°C in a humidified incubator. Incubation times vary for individual influenza types and strains, but are usually between 24 and 72 h.

At the end of the incubation period, the embryos are firstly humanely killed by chilling the eggs at −20°C for 30 min. Care, however, must be taken not to freeze the eggs. Chilling of the eggs also results in constriction of the chick blood vessels, thus reducing the chance of contamination with erythrocytes which would otherwise reduce virus yield by haemagglutination. Harvesting of virus is performed, following chilling, by removing a section of eggshell over the airspace with a commercial egg punch and removing the virus-containing allantoic fluids by aspiration. Care has to be taken at this stage to avoid puncturing the yolk sac, which is generally best displaced to the side with the aid of a pair of blunt-ended forceps. The presence of yolk material in the allantoic harvests complicates subsequent purification of virus.

Clarification of the allantoic fluids is undertaken by centrifugation at 8 000 *g* for 15 min, followed by filtration with a Whatman 114 V coned filter. Filtrate intended for use as purified concentrate can then be stored for up to several days at +4°C or until required, following the addition of a suitable bacteriocide such as 0.1% w/v sodium azide. However, azide is highly toxic and should be handled with care; it must be avoided if any of the allantoic fluids are required for use in subsequent infection.

Before purification, virus is concentrated by ultracentrifugation at 36 000 *g* for 90 min (Beckman T19 rotor). The resulting virus pellets are then resuspended in about 2 ml of PBS. Virus suspensions are further homogenized by subsequent passaging through needles of sequentially narrower gauge, followed by ultrasonication for 30 s. Purification of virus suspensions is carried out by ultracentrifugation in 10–40% (w/v) sucrose/PBS density gradients (35 ml gradient per 2 ml suspension) in a Beckman ultracentrifuge and SW28 rotor at 22 000 rpm for 60 min (65 000 *g*). The virus band is carefully removed from the gradient with a Pasteur pipette and pelleted through a PBS wash in a Beckman T50 rotor at 35 000 rpm (110 000 *g*) for 90 min. The resulting pellets are swollen in 200 μl Tris–saline buffer at 4°C overnight.

RNA extraction

RNA is extracted from purified virus preparations by a cold phenol–SDS method, in 1.8 ml microcentrifuge (MCC) tubes:

virus in Tris–saline	250 μl
1 M sodium acetate (pH 5.0)	20 μl
0.1 M EDTA (pH 7.0)	20 μl
10% (w/v) SDS	40 μl

The contents of the tube should be rapidly vortex mixed in order to lyse the virus. It should be remembered that during the extraction procedure there is a risk of RNase activity: all manipulations must therefore be carried out as rapidly as possible and disposable gloves should be worn. The RNA is separated from the lipid, protein and other material by a minimum of 3 consecutive extractions with 300 μl of aqueous saturated phenol. Washes should be continued until a good phase separation with a clear interface is obtained. At each step, the upper aqueous phase should be carefully removed to a fresh MCC tube by means of an adjustable automatic pipettor. The final aqueous phase is then given two ether washes with an equal volume of ether.

Nucleic acid is precipitated from the final aqueous layer by the addition of 20 μl 2 M NaCl and 2.5 × volume of cold absolute alcohol (ethanol stored at −20°C). When large quantities of nucleic acid are present a blue-white precipitate may form at the solvent interface. Maximum precipitation is achieved by thorough vortex mixing and storage of the samples for a minimum period of 2 h at −70°C, 15 min on dry ice, or overnight at −20°C.

Pelleting of the precipitated nucleic acids is carried out by centrifugation at 13 000 rpm for 10 min (MSE Microcentaur). The aqueous supernatant is carefully decanted, and any remaining liquid in the tube is collected at the base by a brief centrifuge pulse and is then removed by means of a small volume automatic pipettor, the pellet can then be either air dried or vacuum desiccated for several min.

For convenience, the nucleic acid pellets can be resuspended in 50 μl HPLC pure water, from which 5 μl is removed for spectrophotometric analysis, the remainder being stored at −70°C until required.

Primer Extension Sequencing of Influenza A Viruses

All reactions are carried out in 1.8 and 0.55 ml plastic microcentrifuge tubes (Sarstedt Ltd).

5′-end-labelling procedure

TABLE 1. *The preparation of reaction mix in a 0.55 ml MCC tube*

Reagent	Volume (μl)
Primer (1.0 OD_{260}/ml)	2*x*
0.5 M Tris pH 8.0	5
0.1 M $MgCl_2$	5
0.1 M 2-mercaptoethanol	5
^{32}P-ATP (10 μCi/μl)	1*x*
Polynucleotide kinase (10^4 u/ml)	1

Make up to total volume of 50 μl with water where *x* = number of vRNA samples

The labelling reaction mix is incubated at 37°C for 60 min. The phosphorylation mix is stopped by the addition of 75 μl 4 M ammonium acetate followed by the addition of 5 μl tRNA carrier (10 mg/ml) and 370 μl ethanol. Precipitation of labelled primer is carried out on dry ice for 15 min and is then pelleted at 13 000 rpm for 10 min (MSE Microcentaur). The pellet is resuspended and washed in 25 μl water, 1.3 μl 2 M NaCl and 75 μl ethanol and again precipitated on dry ice for 10 min followed by pelleting at 13 000 rpm for a further 10 min. The supernatant is discarded and the pellet is freeze-dried in readiness for resuspension in RTM (see over).

Preparation of vRNA samples

Purified whole genome vRNA is used as the source of template in the sequencing reactions. Samples of 1 μg vRNA in water are placed in 0.55 ml MCC tubes and chilled on dry ice prior to freeze-drying for approximately 30 min.

Preparation of reverse transcriptase mix

TABLE 2. *Preparation of RTM in a 1.8 ml MCC tube*

Reagent	Volume (μl)
0.5 M Tris pH 8.3	30.0*
0.1 M $MgCl_2$	30.0
0.2 M DTT	30.0
2.0 M KCl	7.5
10 mM dGTP	12.0
10 mM dATP	12.0
10 mM dCTP	12.0
10 mM dTTP	12.0
RNAsein (Amersham: 25 000 u/ml)	6.0

* This is made up to a total volume of 210 μl with water.

Sequencing reaction mix

The freeze-dried pellet of ^{32}P-labelled primer is resuspended in 210 μl of RTM and thoroughly vortex mixed (RTM/^{32}P-primer mix). Each vRNA mix is then resuspended in 16 μl of the RTM/^{32}P-primer mix.

For each vRNA sample, four reaction tubes (0.55 ml MCC tubes) each containing one class of dideoxynucleotide are set up as shown in Table 3.

TABLE 3. *Preparation of four classes of dideoxynucleotides in 0.55 ml MCC tubes*

	Volume of ddNTP (μl)			
Reagent	ddG	ddA	ddC	ddT
RTM/^{32}P-primer/vRNA	3.5	3.5	3.5	3.5
0.5 mM ddNTP	1.0	1.0	1.0	1.0
Reverse transcriptase AMV, Life Sciences (5 u/μl)	0.5	0.5	0.5	0.5

The MCC tubes are briefly microcentrifuged in order to collect the contents into the bottom, followed by brief vortex mixing. The reactions are then incubated at 42°C for 60 min, after which the reactions are stopped by the addition of 5 μl formamide–dye stop solution (see Appendix) and boiling for 3 min. At this stage the samples may be directly loaded onto the electrophoresis gel or may be stored at −20°C until required.

Primer Extension Sequencing of Flavivirus Genes

The flaviviruses comprise a group of approximately 68 viruses of both veterinary and human importance, and are responsible for a range of diseases. The flaviviruses possess a host-derived lipid envelope with a single-stranded positive-sense RNA genome which encodes genes for three structural and seven non-structural proteins.

Molecular cDNA cloning of many representative flavivirus types has been reported previously, including yellow fever (Rice *et al.* 1985), tickborne encephalitis/Russian spring summer encephalitis (Mandl *et al.* 1988; Pletnev *et al.* 1986), and dengue (Mason *et al.* 1987; Deubel *et al.* 1986; Zhao *et al.* 1986).

The primer extension dideoxy termination technique has been used infrequently in sequencing flaviviruses, and has generally only found use as a quality control method (Trent *et al.* 1987) and in conjunction with conventional molecular cloning techniques, e.g. for determining the complete nucleotide sequence of two Japanese encephalitis (JE) strains (Sumiyoshi *et al.* 1987; Hashimoto *et al.* 1988).

In respect of the sequencing of flaviviruses, the primer extension technique has been largely avoided, owing to the difficulty of growing many of the flaviviruses to sufficiently high titres to enable extraction of high quantities of sufficiently pure RNA.

Sequencing of the flaviviruses requires a different regime to that used for the influenza viruses. This is necessary because of the different growth characteristics of the viruses and also because of the extensive RNA secondary structure observed within the flavivirus genome. The technique of choice for the sequencing of flaviviruses has been modified to reduce the secondary structure of the template vRNA and to increase the efficiency of the primer-RNA hybridization and chain extension reactions.

Growth of flaviviruses

Virus seed stocks with infectious titres of 10^5–10^7 pfu/ml are inoculated (100 μl volumes) onto subconfluent monolayers (approximately 80%) on SW13 (human adenocarcinoma) cells in 175 cm^2 plastic tissue culture flasks, and incubated at room temperature for 30 min to allow for virus adsorption. Then

20 ml of media is added to the flasks, which are incubated at 37°C for up to 3–5 days, depending upon the virus strain. The cultures are best maintained at a neutral pH by the addition of sodium bicarbonate on a daily basis. The incubation time given to infected cultures was previously determined by establishing growth curves, although maximum titres of virus, determined by plaque assays, were usually achieved following the observation of greater than 80% cytopathic effect (cpe).

Clarification and concentration of virus

The first step in the purification of the virus is the clarification of the cell culture medium. The infected medium contains large quantities of dead and dying cells along with other debris, which are most readily removed by centrifugation. The tissue culture medium from each of the infected flasks is decanted and pooled into 250 ml centrifuge pots with sealable lids and centrifuged at 10 000 rpm for 30 min at 4°C.

Pelleting of virus from large volumes of medium is impractical by conventional ultracentrifugation. The virus is, however, readily aggregated by the addition of polyethylene glycol (PEG) 8 000 and pelleted by high speed centrifugation. The supernatant is therefore transferred to a suitably large vessel situated in an ice bath placed on a magnetic stirrer, to which is added PEG 8 000 at 7 g/100 ml fluid, and gently stirred for a further 3 h. Pelleting of virus is achieved by centrifugation of the PEG medium at 10 000 rpm for 30 min at 4°C, following which the supernatant is discarded. If an angled rotor is used, the virus pellet is usually found distributed on the outer apex of the centrifuge pot. The virus pellet is best removed and resuspended by washing with 1–2 ml TNE buffer and mechanical abrasion with a disposable soft plastic Pasteur pipette.

Further purification of the virus is achieved by ultracentrifugation on tartrate–glycerol gradients. Isopycnic gradients are prepared from 30% glycerol (v/v) in TNE buffer and 45% (w/v) potassium tartrate in TNE buffer. Linear gradients are prepared by using a dual channel gradient maker with potassium tartrate forming the heavy portion of the gradient, and an equal volume of the lighter glycerol to a total volume of 11 ml. The virus sample is carefully loaded onto the top of the gradient with a Pasteur pipette and is overlaid with liquid paraffin to prevent generation of aerosols. The loaded gradients are centrifuged at 39 000 rpm (200 000 *g*) for 3 h in a SW40 rotor with minimum acceleration.

The position of the virus band is identified by projecting a narrow field of light into the centrifuge tube against a dark background. Removal of gradient bands can be undertaken by means of either narrow gauge needles or Pasteur pipettes, however, whichever method is decided upon, great care must be taken not to disturb the integrity of the gradient.

On removal of the virus band, it is further purified to equilibrium by overnight centrifugation on a second 30–45% glycerol–potassium tartrate gradient at 22 000 rpm (62 000 *g*) in the SW40 rotor. Detection and removal of virus is carried out as with the first gradient. At the end of the virus purification, any remaining contaminating gradient materials must be removed from the sample before RNA extraction is undertaken. The virus band (or gradient fraction) is diluted in TNE buffer up to a maximum volume of 36 ml and is pelleted overnight at 25 000 rpm (85 000 *g*) in a SW28 rotor. The supernatant is decanted and the centrifuge tube allowed to drain, the virus pellet is covered with 200 μl TNE buffer, and the pellet allowed to swell for a minimum of 3 h at 4°C.

Extraction of RNA

Following the above purification procedure, the virus suspension is dispensed into 200 μl amounts in presiliconized 1.8 ml plastic MCC tubes. The first step in extraction of the vRNA involves the disruption of viral lipid membranes by detergent, followed by enzymic digestion of any extraneous protein material. The membranes are initially disrupted by the addition of 20 μl Triton X-100 (BDH Ltd) followed by rapid vortexing at which time 40 μl (1 mg) autodigested proteinase-K (Pharmacia or Sigma) is added with incubation at 37°C for 30 min. At approximately 10 min intervals, the tubes should be vortexed to ensure thorough mixing.

The nucleic acids are separated from the contaminating protein and lipid material by organic solvent phase extraction at room temperature. To each sample tube 260 μl phenol:chloroform:isoamylalcohol solution (see Appendix) is added, followed by swift vortexing and phase separation by centrifugation at 13 000 rpm for 3 min (MSE Microcentaur). The upper aqueous phase containing the nucleic acid is most easily removed by means of a proprietary automatic pipettor (e.g. Gilson Pipettman) adjusted to approximately 25 μl. It is advisable to use the smallest adjusted volume practicable since there is less chance of disturbing the interface. Likewise, removal of the aqueous phase down as far as the interface should be avoided for the same reason. Once removed, the aqueous phase should be temporarily stored on wet ice whilst the organic phase is 'back-extracted'. This involves the addition of a further 50 μl of TE buffer to the organic phase followed by brief vortex mixing and separation by centrifugation. The aqueous phase, once separated, is then pooled with the main aqueous phase stored on ice. During the extraction process a white precipitate may form at the solvent interface: it is important that any take-up of this material is avoided and subsequent extractions of the aqueous phase with equal volumes of phenol:chloroform should be continued until the precipitate is no longer present and a clear distinct interface is obtained. The final aqueous phase is given two washes in buffer-saturated diethyl ether (1:1 stored at −20°C), an equal volume of the ether is added to the aqueous phase followed by a brief vortex mix, the phases being allowed to separate under normal conditions, and the upper ether phase being removed with a small volume automatic pipettor following each wash. It is particularly important at this stage to ensure that the MCC tubes offer a good tight fit as samples can easily leak following mixing with ether. An additional precaution at this stage is to ensure a tighter seal by binding the tube tops with parafilm or nescofilm.

The nucleic acid is precipitated from the final aqueous phase by following the protocol as described for influenza viral RNA.

The RNA content and purity are estimated by determining optical density (OD) measurements at 280 nm, 260 nm and 230 nm (Spectrophotometer, Pye Unicam) for the 5 μl sample in 1 ml total volume with ultrapure water, where OD_{260} 1.0 is equivalent to 40 mg RNA and an OD_{260}:OD_{280} ratio of 2.0 represents optimum purity.

Primer extension sequencing

Kinase reaction

Sufficient synthetic oligonucleotide primer for five reactions is 5′ end-labelled with γ ^{32}P with the reaction carried out as shown in Table 4.

TABLE 4. *Preparation of synthetic oligonucleotide primer in a 0.55 ml MCC tube*

Reagent	Volume (μl)
10 × kinase buffer	0.65
Primer at OD_{260} = 1.0	0.5
^{32}P-ATP (10 μCi/μl)	2.5
Polynucleotide kinase	0.5
HPLC-ultrapure water	2.35

The reactions are carried out for 30 min at 37°C in either a waterbath or dri-block and are stopped by heat inactivation of the enzyme by incubation for 5 min at 65°C, at which time a further 3.5 μl of HPLC-pure water is added and then temporarily stored on wet ice until required.

Secondary structure denaturation

Flavivirus template RNA contains high degrees of secondary structures which have to be denatured before primer extension can begin. We have found that this is best achieved by heating the required quantity of RNA at 90°C in a waterbath for 3 min and subsequently cooling on wet ice for 30 s. To each of the RNA samples a total volume of 1.1 μl 200 mM methyl mercury hydroxide is added and incubated at room temperature for a further 15 min. The methyl mercury hydroxide is inactivated by the addition of an excess of 2-mercaptoethanol, 0.8 μl of a 7 M solution, with cooling on wet ice for 2 min, following which 6 μl of 10 mM Tris, 250 mM KCl solution is added to each of the samples.

RNA–primer hybridization

To each of the denatured RNA samples, 2 μl of the radiolabelled primer is added, and is then incubated for 3 min at 80°C. Hybridization of primer to template vRNA is undertaken by incubating for 45 min at 5°C lower than the dissociation temperature of the primer, calculated from the respective numbers of nucleotides present (G, C, A and T), and as defined by the following equation:

$$4(G+C) + 2(A+T) = x°C$$

where x = dissociation temperature of the oligonucleotide.

Reverse transcription

The generation of single DNA copies of varying chain lengths is obtained under the following conditions.

TABLE 5. *Reverse transcriptase mix sufficient for 5 vRNA reactions*

Reagent	Volume (μl)
0.5 M Tris	3.6
0.1 M $MgCl_2$	12.0
0.5 M DTT	1.2
10 mM dATP	3.0
10 mM dCTP	3.0
10 mM dTTP	3.0
10 mM dGTP	6.0
Actinomycin D (1 mg/ml in water)	3.75
RT (RAV-2, Amersham International 20 u/μl)	6.9
HPLC-pure water	33.05

Reverse transcription reaction

In all of the following reactions, 0.55 ml MCC tubes are used. For each of the primer–vRNA template samples a series of four tubes are prepared, each of which will contain only one of the following dideoxynucleotide bases:

5 mM ddG	2.0 μl,		or
5 mM ddA	2.0 μl,		or
5 mM ddC	2.0 μl,		or
5 mM ddT	1.4 μl	+	0.6 μl HPLC-ultrapure water.

To each of the 4 tubes in the series, 2.0 μl of the hybridized primer-mix is added along with 3.3 μl of the RTM. The contents should be collected to the bottom of each tube by briefly pulsing in a microcentrifuge for a few seconds followed by thorough vortex mixing. The RT reactions are incubated at 50°C for 45 min and stopped by the addition of 4 μl formamide–dye stop solution and boiling for 3 min.

On reaching this stage, the samples are relatively stable and can either be loaded directly onto the gel system or can be stored short term at −20°C. Storage of samples beyond the half-life of the radioisotope used is not recommended, and any stored samples should be boiled for 3 min before loading onto the gel.

Polyacrylamide gel electrophoresis

Electrophoresis is carried out in 20 × 40 cm PAGE tanks, which allows 5 × 4 reaction samples to be easily run although this can be increased to a maximum of 6 × 4 reaction samples if required.

Gel moulds are prepared from hardened float glass plain backplates with a notched front plate. The front plates should previously be coated with a suitable aqueous repellent, such as 2% dimethyl dichlorosilane (BDH Ltd), whilst to ensure successful binding of the gel to the backplate, this should be coated with a suitable attractant, such as Wackers solution (see Appendix). The mould is formed from a suitably thin material, such as 0.2 mm plasticard side spacers between the plates, although ultrathin strips may be prepared from strips of X-ray film to provide the spacers. The gel mould is initially held together with bulldog clips whilst the gel is tightly sealed with water-proof plastic tape (electrical tape, Sellotape Ltd).

The choice of gel system is dependent upon both the virus to be sequenced and individual preference. The sequencing samples should be divided into two samples, one to be loaded onto a short-run gel in order to visualize the sequence close to the primer and the other to be loaded onto the long-run gel in order to visualize the sequence of regions more distal to the primer.

Gel solutions are deionized for 2 h during mixing in the presence of amberlite bed resin, 4% (w/v), which is subsequently removed by filtration (see Appendix).

The regime in Table 6 has been successfully used to sequence influenza and flavivirus genes.

TABLE 6. *Sequencing of influenza and flavivirus genes*

		Short-run		Long-run	
Virus	Current	Percentage	Length of run	Percentage	Length of run
Influenza	32 mA	8	1 h 45 min	6	4 h
Flavivirus	24 mA	6.3	2 h 30 min	6.3	5 h

The above times are for guidance only and the necessary run times are variable under a variety of conditions including the use of different gel tanks and local temperature variations. A useful guide for determining the required electrical current and run time for a given gel and apparatus is by observation of the two dye fronts. The 'short' gel run should be terminated when the leading bromophenol blue band has approached the base of the gel. The time given for the longer-run gel is determined by a more trial and error approach, since a suitable overlap with the short-run gel is required, and the length of this will be dependent upon the quality of the sequence at the top of this shorter gel. A useful estimate can, however, be based on allowing approximately twice the time given for the short-run gel.

On completion of the gel run, the plates are removed from the tank and carefully separated with a scalpel blade. The efficiency of radiolabelling and extension can be estimated at this stage by the use of a hand held B-monitor. The gel is fixed *in situ* on the backplate by immersing it in a 2 litre bath of 10% glacial acetic acid for a minimum of 15 min, followed by washing in continuously running tap water for a further 15 min. The gels are dried, attached to the backplates, in an oven at 80°C for 45 min or, alternatively, if this is not available, overnight at 37°C. On cooling, the gels should be exposed with a suitable X-ray film (such as Fuji Rx X-ray film) and placed in light-tight containers (e.g. Harmer cassettes) for the required period of time. The exposure time depends upon the relative radioactive levels of the gels, but is generally in the order of 1–14 days. If weak signals are obtained, the intensity of the sequence ladders can be enhanced by using intensifying screens and storing at −70°C.

Comments

The primer extension technique provides a useful method for examining nucleotide sequence data within a relatively short period of time, and is particularly pertinent when specific short regions of the genome of a number of closely related viruses need to be examined. Although the techniques described relate to the sequencing of vRNA, there is no reason why they should not be used for sequencing other RNA species with a minimum of modification.

There is one drawback to the technique and that is the large quantity of RNA required for the procedure. Once it has been used in reactions it can not be re-used and, unlike molecular cloning, the product of the reactions cannot then be used in further experiments.

The primer extension technique generally requires the presence of the purest vRNA available, and there is no doubt that the best results are obtained with the best available vRNA. The protein present in any sample should be removed by the proteinase-K digestion, followed by phenol extraction. However, the removal of any contaminating DNA presents a more difficult problem. The removal of the DNA by enzymic digestion, for example with DNase 1, presents one approach, although there is some concern over the potential presence of RNase, which is reputed to be present in commercial preparations of the enzyme and the associated difficulties inherent in ensuring its absence (Maniatis *et al.* 1982). The presence of the DNA can, however, be largely ignored since the technique has been developed to inhibit the non-specific polymerization by the use of actinomycin D in the reverse transcription reactions. Similarly, the presence of any other species of cellular specified RNA is negligible due to both the high specificity of the synthetic oligonucleotide primers and the relatively large excess of vRNA in the reaction.

Although the AMV enzyme produced good quality sequence with influenza virus and was the preferred enzyme, the authors have, with the sequencing of the flaviviruses, demonstrated that the origin of the RT enzyme is a crucial factor. RT isolated from several sources, e.g. avian myeloblastosis virus (AMV) and Moloney cloned RT, produced poor sequences with unpredictable extensions and an unacceptable degree of transcription errors. However, in our hands, the RT enzyme isolated from Rous associated virus II (Amersham International) produced sequence data of optimum quality with flavivirus vRNA. It is apparent therefore that the choice of a particular RT enzyme must be based upon individual selection criteria. Whatever the source of the RT enzyme, the relatively high error rates recorded for this enzyme (1/600–1/5 000) compared to the values reported for DNA polymerases (1/3 000–1/300 000) should be borne in mind (reviewed by Perbal 1988).

The genomes of flaviviruses are composed of regions of extensive RNA secondary structure, resulting in a greatly reduced binding of the primer to the complementary regions on the template vRNA. Where regions of the genome are believed to possess a high degree of such structures as hairpin turns, the primer extension resulting in premature termination of the cDNA chain due to fall-off of the primer from the RNA template which is visualized by electrophoresis as continuous banding in all four channels with the abrupt end of any banding a short distance from the primer (Fig. 1). The specificity of the conditions for the removal of secondary structure is such that primer attachment and denaturation by both heat treatment and methyl mercury hydroxide is required, in conjunction with hybridization at the correct incubation temperatures, in order to produce readable sequence data (Fig. 2). Such treatment was not required for the sequencing of the segmented

FIG. 1. Autoradiograph of polyacrylamide gel electrophoresis showing dideoxy chain termination sequencing of non-denatured flavivirus template RNA. (G = guanine; A = adenine; C = cytosine; T = thymine.)

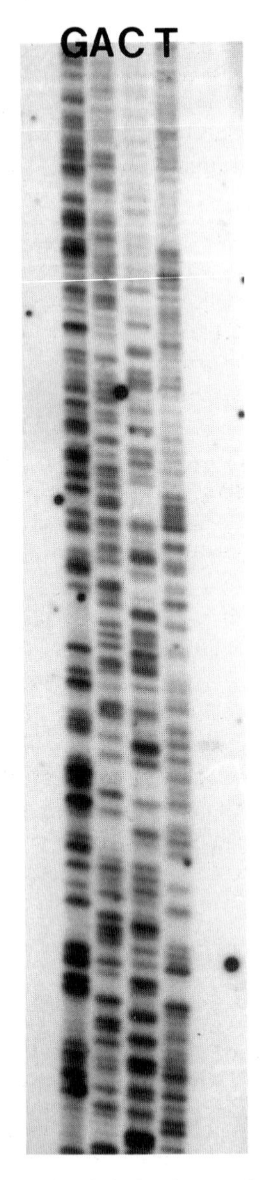

FIG. 2. Autoradiograph of polyacrylamide gel electrophoresis showing dideoxy chain termination sequencing of heat and methyl mercury denatured flavivirus template RNA. (G = guanine; A = adenine; C = cytosine; T = thymine.)

genomes of influenza virus, indicating the absence of complex secondary structure. Other premature terminations may occur throughout the length of the sequence due to the presence of protein contamination.

The maximum distance that it is able to be read from the primer and the intensity of these bands is dependent upon the ratio of concentrations of deoxynucleotides to dideoxynucleotides. The greater the ratio of dideoxynucleotides in the reactions the shorter the chain length will be, due to the more rapid incorporation of the dideoxynucleotides accompanied by an increase in band intensity. Conversely, increasing the relative proportion of deoxynucleotides in the reaction results in an increase in the average chain lengths of the cDNA transcripts.

In the examination of a sequencing autoradiograph, many apparent artefacts may be observed. These may occur when the primer-template annealing temperature is too low, or, as can be observed when partially purified material is sequenced, many crossbands are observed when bands are present in all four lanes and at multiple positions within the sequence.

When conditions are such that a quality sequencing 'ladder' is observed with minimal crossbanding, but of low intensity, then a number of factors may apply. The efficiency of binding of primer to template may be low, requiring a reassessment of annealing conditions or the ratios of template and primer. Alternatively, the estimation of the quantity of vRNA present may be inaccurate due to the presence of unknown amounts of other contaminating RNA species. Of equal consideration is the chosen method of radioactive labelling. The authors have routinely used 5′-end labelling of primers with γ^{32}P-labelled ATP, which results in autoradiographs of acceptable quality. However, in our hands, alternative methods utilizing internal labelling of the cDNA chains with α-^{32}P or ^{35}S produced less satisfactory results.

References

Deubel, V., Kinney, R.M. & Trent, D.W. 1986. Nucleotide sequence and deduced amino acid sequence of the structural proteins of dengue type 2 virus. *Virology* **155**, 365–377.

Hashimoto, H., Nomoto, A., Watanabe, K., Mori, T., Takazawa, T., Aizawa, C., Takegami, T., & Hiramatsu, K. 1988. Molecular cloning and complete nucleotide sequence of the genome of Japanese encephalitis virus Beijing-1 strain. *Virus Genes* **1**, 305–317.

Mandl, C.W., Heinz, F.X. & Kunz, C. 1988. Sequence of the structural proteins of tick-borne encephalitis virus (Western subtype) and comparative analysis with other flaviviruses. *Virology* **166**, 197–205.

Maniatis, T., Fritsch, E.F. & Sambrook, J. 1982. *Molecular Cloning: A Laboratory Manual.* Cold Spring Harbor, New York: Cold Spring Harbor Laboratory.

Mason, P.W., McAda, P.C., Mason, T.L. & Fournier, M.J. 1987. Sequence of the dengue 1 virus genome in the region encoding the three structural proteins and the major non structural protein NS1. *Virology* **161**, 262–267.

Perbal, P. 1988. *A Practical Guide to Cloning.* London: John Wiley & Sons.

PLETNEV, A.G., YAMASHCHIKOV, V.F. & BLINOV, V.M. 1986. Nucleotide sequence of the genome region encoding the structural proteins and the NS1 protein of the tick-borne encephalitis virus. *FEBS Letters* **200**, 317–321.

RICE, C.M., LENCHES, E.M., EDDY, S.R., SHIN, S.J., SHEETS, R.L. & STRAUSS, J.H. 1985. Nucleotide sequence of yellow fever virus: implications for flavivirus gene expression and evolution. *Science, New York* **229**, 726–733.

SAIKI, R.K., GELFAND, G.H., STOFFEL, S., SHARF, S., HIGUCHI, S.J., MULLIS, K., HORN, G. & ERLICH, H.A. 1988. Primer-directed enzymic amplification of DNA with a thermostable DNA polymerase. *Science, New York* **239**, 487–491.

SANGER, F., NICKLEN, S. & COULSON, A.R. 1977. DNA sequencing with chain determinating inhibitors. *Proceedings of the National Academy of Sciences of the United States of America* **74**, 5463–5467.

SUMIYOSHI, H., MORI, C., FUKE, I., MORITA, K., KUHARA, S., KIKUCHI, Y., NAGAMATU, H. & IGARISHI, A. 1987. Complete nucleotide sequence of Japanese encephalitis virus genome. *Virology* **161**, 497–510.

TRENT, D.W., KINNEY, R.M., JOHNSON, B.J.B., VORNDAM, A.V., GRANT, J.A., DEUBEL, V., RICE, C.M. & HAHN, C. 1987. Partial nucleotide sequence of St Louis encephalitis virus RNA: Structural proteins, NSI, ns2a and ns2b. *Virology* **156**, 293–304.

ZHAO, B., MACKOW, E., BUCKLER-WHITE, A., MARKOFF, L., CHANOCK, R.M., LAI, C-J. & MAKINO, Y. 1986. Cloning full-length dengue 4 viral DNA sequences: Analysis of genes coding for structural proteins: *Virology* **155**, 77–88.

Appendix: Buffers and Reagents

PBS

Phosphate buffered saline tablets (Oxoid), 1 tablet in 100 ml MilliQ water.

dNTPS

Deoxynucleotides, supplied as 100 mM solutions (Pharmacia), diluted to 10 mM working solutions.

ddNTPS

Dideoxynucleotides, supplied as 50 mM solutions (Pharmacia), diluted to 5 mM working solutions.

Virus infection medium

10 × Eagles MEM 10 ml, penicillin (10^5 u/ml) and streptomycin (1% w/v) 2 ml, 200 mM glutamine 2 ml, fetal calf serum 4 ml, made up to 1 100 ml with MilliQ water.

10 × kinase medium

0.5 M Tris–HCl pH 7.6, 0.1 M $MgCl_2$, 50 mM dTT, 1 mM spermidine, 1 mM EDTA.

DTT (dithiothreitol)

1 M stock made up as 3.09 g of dithiothreitol in 20 ml 0.01 M sodium acetate pH 5.2. Sterilize by filtration and store at −20°. Dilute to 0.2 M as required.

Sodium acetate stock solution

3 M sodium acetate. Adjust pH to 5.2 with glacial acetic acid.

TE buffer 10 ×, pH 7.6

100 mM Tris–HCl, 10 mM EDTA pH 8.0, pH 7.6.

TNE buffer 10 ×

100 mM Tris–HCl pH 8.0, 1 M NaCl, 10 mM EDTA pH 8.0.

Gradient solutions

1 30% glycerol (w/w) in 50 mM Tris–HCl, 1 mM EDTA, pH 7.5.
2 45% potassium tartrate (w/v) in 50 mM Tris–HC1, 1 mM EDTA, pH 8.8.

Formamide–dye stop solution

0.3% xylene cyanol, 0.3% bromophenol blue, 0.37% EDTA–Na_2 pH 7.0 in deionized formamide.

10 × Tris–borate buffer

Tris 218.06 g, boric acid 111.3 g, EDTA–Na_2 18.62 g. Make up initially to 1.5 litres, adjust pH to 8.3 and make up to final volume of 2 litres.

Tris–saline buffer

10 mM Tris–HCl, 150 mM NaCl, pH 7.6.

Wacker's solution

Ethanol 4 ml, freshly prepared 10% glacial acetic acid 120 μl, γ(methacryloxy)–propyl trimethoxysilane (Sigma) 12 μl.

Phenol:chloroform

Melt crystalline ultrapure phenol (BRL) at 60°C in a waterbath. Thoroughly mix equal volumes of molten phenol and HPLC-pure water and allow to separate into two phases. Remove the upper phase by aspiration, add another equal volume of HPLC-pure water and repeat the above steps twice. Remove the majority of the aqueous (upper) phase, leaving a small aqueous layer covering the phenol. Store the phenol at +4°C in a container covered with silver foil. When required for RNA extraction, thoroughly mix the required amount of phenol with an equal quantity of a 24:1 mixture of chloroform: isoamylalcohol.

Polyacrylamide gel solutions

Acrylamide and bisacrylamide (BDH, Electran grade) are at the required percentage in the ratio of 20:1 and mixed with a magnetic stirrer along with 16.8 g urea (BDH, Analar) and 4 ml 10 × TBE buffer, made up to a total volume of 40 ml with deionized distilled water. The gel solution is deionized by mixing it with 2 g amberlite bed resin which is removed by filtration before the addition of the accelerator catalysts. Before pouring and polymerization, add 330 μl of 10% ammonium persulphate and 35 μl of TEMED (*N,N,N,N'*-tetramethylethylenediamine, Kodak Ltd).

Forensic Applications of DNA Profiling

B.H. Parkin
Metropolitan Police Forensic Science Laboratory,
109 Lambeth Road, London SE1 7LP, UK

In crimes of violence, the forensic biologist often has to determine the identity of an assailant by grouping bloodstains on his or her clothing and comparing the results with the blood of the victim. Similarly, in sexual assault cases, the rapist may be identified by comparing semen on the victim's vaginal swabs or clothing with the blood of a suspect. Until 1988, such examinations were carried out in the Metropolitan Police Forensic Science Laboratory by means of grouping systems involving the identification of antigens by serological techniques or polymorphic proteins by electrophoresis. This evidence was evaluated by estimating the frequency in the population of the combination of groups obtained. The problems with this approach were:

1 The frequencies obtained were relatively high. Hence, in large populations, there could be a number of individuals with the same groups.

2 The body fluid groups degraded rapidly and most could not be identified after 2–3 months of the stain being made.

3 Few groups could be determined from semen stains. In addition, masking of the rapist's semen type could occur by groups found in the victim's vaginal secretions, e.g. semen of group O would be masked by body fluids of group A from the female.

4 Many semen enzymes lost their activity in the vagina 3–4 h post intercourse.

A possible solution to these problems was evident when Wyman & White (1980) reported the first demonstration that hypervariable regions (HVR) occurred in human DNA. This chance discovery was followed by reports of a number of other HVRs including three in the region of the α-globin gene cluster (Higgs *et al.* 1981; Proudfoot *et al.* 1982; Goodbourn *et al.* 1983; Jarman *et al.* 1986). Bell *et al.* (1982) reported a HVR 5′ to the human insulin gene and a further such region was found 3′ to the c-Ha-*ras* 1 oncogene (Capon *et al.* 1983).

In structure, each of these HVRs consist of tandem repeats of a short core sequence of the DNA, or 'minisatellite' and the polymorphism of these

0-632-02926-9

regions is due to the variability in the number of tandem repeats of the core DNA sequence. This results in a large number (between 50 and 100 in some cases) of 'alleles' consisting of different lengths of DNA, with the heterozygous frequency of many of these being between 95 and 100%.

Based on the work of Knowlton *et al.* (1986), the number of such HVRs in the human genome has been estimated to be as high as 1 500.

Multilocus Probes

Weller *et al.* (1984) reported the discovery of a small minisatellite composed of four tandem repeats of a 33 bp sequence within one of the introns of the human myoglobin gene. This minisatellite cross-hybridized weakly with multiple loci, and Jeffreys *et al.* (1985) cloned a random selection of eight of these loci, which they found to be minisatellites. Of these, four showed variable number tandem repeat (VNTR) polymorphism.

Hybridization probes made from different core sequences of two of these HVRs were made single-stranded, labelled with ^{32}P and hybridized to Southern blots of *Hin*fI digests of DNA from a random sample of British Caucasians and from selected members of a large British Asian family. The probes cross-hybridized at low stringency conditions to a large number of DNA fragments. The resulting autoradiograph demonstrated a complex set of large and highly variable DNA fragments and was termed a DNA 'fingerprint' (Fig. 1).

These two probes, λ33.6 and λ33.15, detect almost completely different sets of hypervariable loci. Each probe produces individually specific patterns, except in the case of monozygotic twins where the patterns will be identical.

The DNA fragments detected act as single heterozygous Mendelian characteristics and are therefore transmitted from a parent to approximately half the offspring.

Jeffreys claims that about 36 bands greater than 3 kb in size may be seen in each individual by using both λ33.6 and λ33.15 probes, with a calculated band sharing between individuals about 25%, giving the conservative probability that all 36 fragments in one individual will be present in an unrelated randomly selected second individual at 2×10^{-22}.

The potential for the technique to be used in forensic science was obvious. Gill *et al.* (1985, 1987) developed techniques of extracting and purifying human DNA from the small quantities of blood and semen in liquid or dried stains that are found in forensic casework. They also demonstrated that DNA could be extracted from hair roots. Further, they developed a method for preferentially lysing the vaginal epithelial cells present in most semen stains and swabs examined after rape. This allowed DNA profiling of uncontaminated spermatozoa DNA.

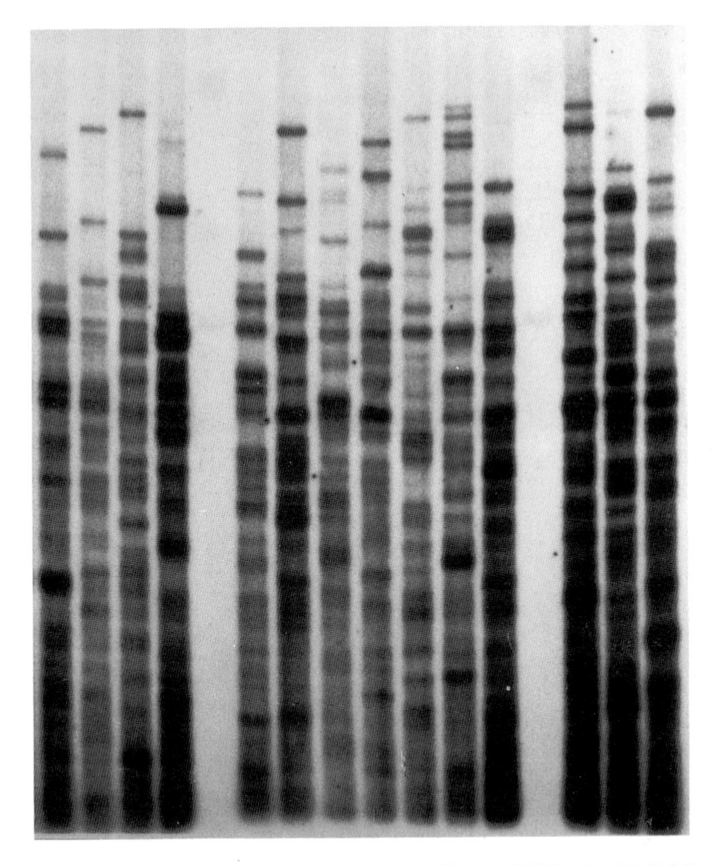

FIG. 1. Human DNA 'fingerprint' detected by digestion of human DNA with *Hin*fI followed by Southern blot hybridization with DNA minisatellite probe λ33.15. This figure shows the diversity of the larger DNA fragments from person to person.

In laboratory trials the same authors reported that about 60 μl of blood and 5 μl of semen was enough to give sufficient high molecular weight DNA to obtain DNA profiles after hybridization with ^{32}P labelled probes λ33.15 and λ33.6. Results were obtained from body fluid stains up to 4 years old, demonstrating that the test is well suited for use in the forensic laboratory.

Single Locus Probes

Although the use of multilocus probes has been successful in a large number of forensic cases, single locus probes offer several advantages:

1 They are more sensitive.

2 They enable mixtures of body fluids from more than one person to be disclosed.
3 The problem of comparing partial profiles in weak samples is largely overcome.
4 The results are more easily computerized.

Wong *et al.* (1986, 1987) isolated 6 large hypervariable fragments from DNA fingerprints. These were purified by preparative gel electrophoresis, cloned and, under high stringency conditions, these fragments acted as single locus probes, detecting, after restriction and Southern blotting, a maximum of two DNA fragments (alleles) per individual.

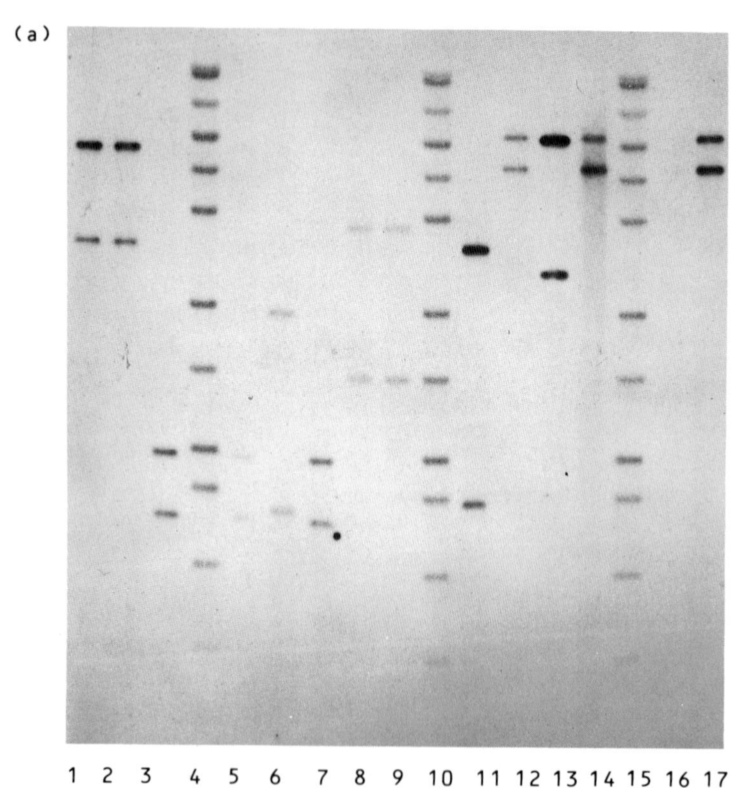

FIG. 2. DNA profiles of human blood samples and forensic body fluid stains detected by digestion with *Hin*fI followed by Southern blot hybridization with single locus probes λMS1 (a); λMS31 (b) and λMS43A (c). The probes were used sequentially, stripping the membrane after each hybridization with 0.4 M sodium hydroxide solution at 45°C. Matching 'alleles' from body fluid stains and control blood samples are in lanes 1 and 2; 3, 5 and 7; 8 and 9; 12, 14 and 17. DNA size marker 'ladders' are in lanes 4, 10 and 15.

(b)

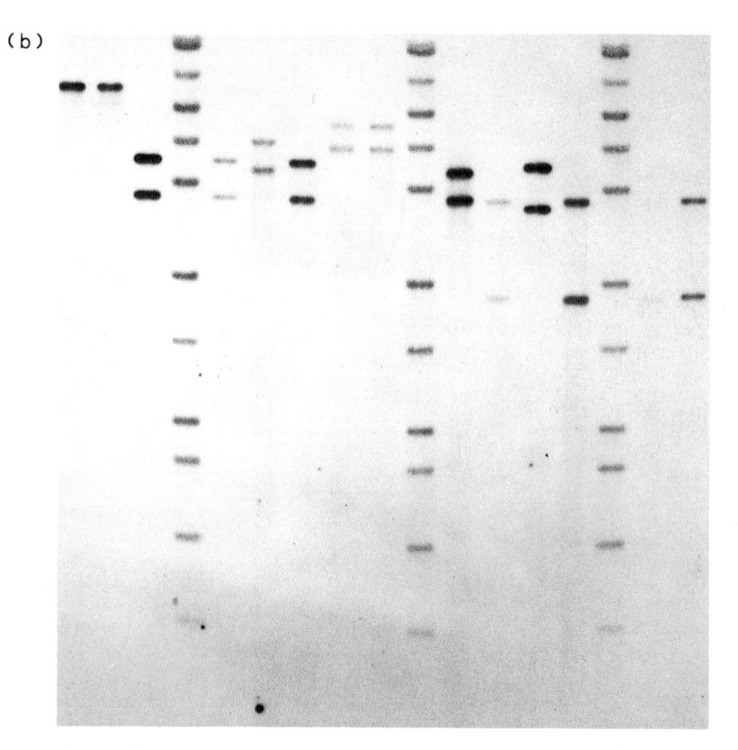
1 2 3 4 5 6 7 8 9 10 11 12 13 14 15 16 17

(c)

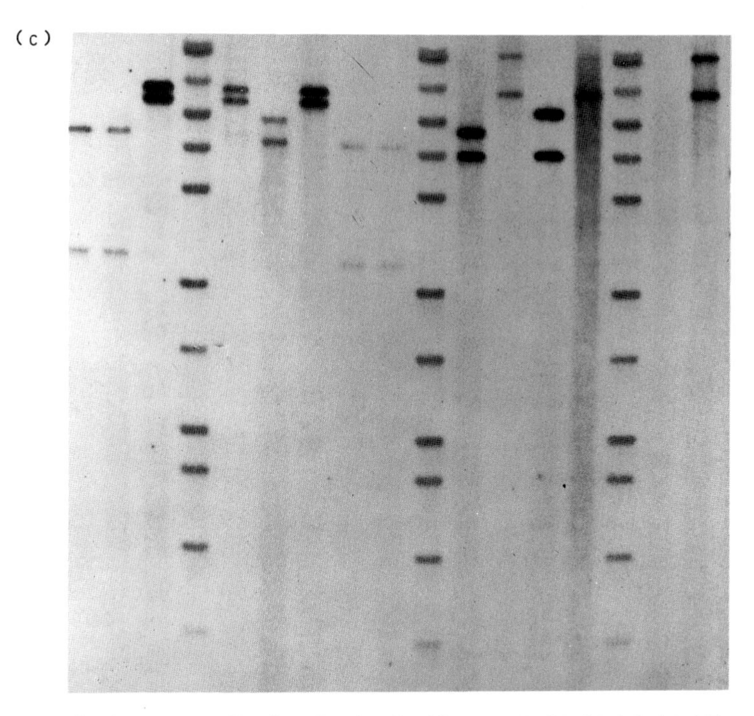
1 2 3 4 5 6 7 8 9 10 11 12 13 14 15 16 17

The probes were termed λg3, λMS1, λMS8, λMS31, λMS32 and λMS43 respectively. Each of these loci consisted of minisatellites with tandem repeat units varying from 9 to 45 bp in length and, under high stringency conditions, acted as powerful single locus probes with heterozygocities varying from 90 to 99%. The alleles revealed by these probes are spread throughout the population varying in size from about 1.5 to a maximum of 28 kb, although most are within a much narrower range of about 3–10 kb.

In addition to these single locus probes, others are rapidly becoming available, particularly those developed by White and Nakamura (Nakamura *et al.* 1987). The forensic scientist thus has an increasingly large number of single locus probes to select for use in crime cases.

At the Metropolitan Police Forensic Science Laboratory, London, this technique has been in use since December 1988, during which time over 400 cases have been examined by the sequential use of three probes λMS1, λMS31 and λMS43A (Fig. 2).

Methods

Extraction

DNA is extracted from 150 μl blood, 20 μl semen or the equivalent amount of dry stain of these fluids by a mixture containing proteinase-K (10 mg/ml), 1 M dithiothreitol (DTT) and 2% sodium dodecyl sulphate (SDS).

When it is necessary to examine semen DNA from a vaginal swab, the vaginal epithelial cells and any white blood cells present may be lysed by incubation with proteinase-K and SDS followed by washing and pelleting the intact sperm cells by centrifugation (Gill *et al.* 1985). The sperm DNA may then be liberated by addition of DTT which disrupts the sperm membranes.

Purification

The DNA is purified by phenol–chloroform extractions and precipitated by the addition of 0.1 volumes of 2 M sodium acetate and 2 volumes of ethanol. The pelleted DNA is redissolved and assayed by fluorimetry using Hoechst dye (H33258) as a fluorophore.

Southern blot hybridization

Aliquots of 500 ng of DNA are removed from each sample and restricted by *Hin*fI. The restricted DNA fragments are separated by submarine electrophoresis in a 0.7% agarose gel and transferred to a nylon membrane by vacuum blotting, which speeds up this process from about 16 h to 45 min.

The membranes are prehybridized at 65°C for 5 min in 0.5 M sodium phosphate, 7% SDS, 1% bovine serum albumin (BSA), 1 mM EDTA (Church & Gilbert 1984) and hybridized overnight with 0.5 ng/ml ^{32}P-labelled probe DNA in the same solution.

After hybridization, the membranes are washed at 65°C in 2 × SSC containing 1% SDS followed by a high stringency wash at 65°C in 0.1 × SSC (15 mM sodium chloride, 1.5 mM trisodium citrate, pH 7.0) containing 0.1% SDS.

The membranes are autoradiographed from 1 to 10 days at −70°C in the presence of an intensifier screen. At the completion of autoradiography, the membranes are stripped and rehybridized with further single locus probes.

DNA Profiling in Forensic Casework

DNA profiling is used in forensic casework in the same way as conventional grouping. A body fluid stain relevant to a crime is compared with the bloods of the suspect, victim and any possible third person to ascertain its origin. Most samples required to be examined are those which originated from crimes of violence, particularly sexual assault.

In these cases, vaginal, anal or oral swabs are taken from the victim, by a forensic medical examiner, as the reported circumstances require. These, together with items of clothing from the victim, are searched for semen stains. Blood samples, taken in EDTA, are used for control purposes to compare with the relevant semen stains. The victim's husband or boyfriend should also give a blood sample for elimination purposes.

In other crimes of violence, bloodstained weapons or clothing are examined and the results compared with blood samples from the suspect or victim. On occasions, other human tissues are necessarily examined, e.g. muscle from a cadaver, small fragments of flesh on a knife or hairs removed from a weapon.

Compared with conventional grouping, DNA profiling yields results of much greater discriminating power, allowing for more positive statements of guilt or innocence. In additon, the test is more versatile in that it can be carried out on a greater range of body fluids and tissues. Further, because of its greater stability, DNA can be examined in body fluid stains of up to about 5 years old when stored under ideal conditions. Not all stains, however, can be profiled after this length of time. Since the test relies on the isolation and purification of high molecular weight DNA, conditions causing DNA to fragment obviously render the stain less likely to give a result. Of these, high humidity has been shown to degrade DNA rapidly (Gill *et al.* 1987). Also, because of nuclease activity, bacterial attack will also destroy DNA, thus, items submitted for DNA analysis are stored below 0°C.

The size of stain required depends on the body fluid to be examined. Hence a bloodstain on cotton material of about 20 mm diameter, or a semen stain of about 7 mm is necessary.

Much of the semen DNA profiling is carried out on vaginal swabs taken from rape victims after the offence. Laboratory trials have shown that the time after intercourse in which DNA profiling is successful varies considerably with the female concerned, but swabs taken between 20 and 30 hours post intercourse have regularly been shown to give the semen profile (Gill *et al.* 1987).

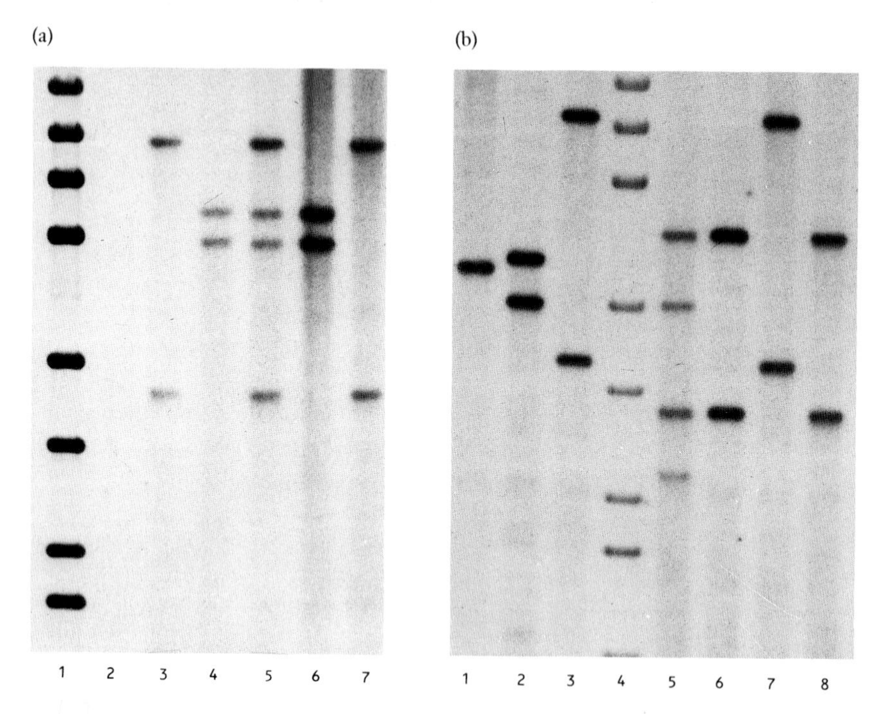

FIG Profiles of DNA extracted from control samples of blood and vaginal swabs taken from rape victims. Vaginal epithelial cells were removed preferentially so only semen DNA is present. The DNA was digested with *Hin*fI followed by Southern blot hybridization, with single locus probe λMS1:

(a) The vaginal swab (lane 5) shows four DNA 'alleles' demonstrating that semen from two men are present. Two of the 'alleles' match the husband's DNA profile (lanes 4 and 6) showing that his semen is probably present. The remaining 'alleles' match the suspect's control sample (lanes 3 and 7);

(b) the vaginal swab (lane 5) shows four DNA 'alleles' demonstrating that semen from two men is present. Two 'alleles' match those of the husband (lanes 6 and 8) showing that his semen is probably present. The remaining 'alleles' do not match those of the suspect (lanes 3 and 7) showing that he did not commit the offence.

If a blood sample cannot be taken for control purposes, a buccal swab or DNA combined from about 10 plucked head hairs will suffice.

On occasions, sufficient cellular material has been found on cigarette ends for DNA profiling.

By means of single locus probes used sequentially, mixtures of body fluids from more than one individual may be detected. This is particularly useful in cases where the rapist's semen may be mixed with that of the victim's husband or boyfriend, or when an act of multiple rape has occurred. In such circumstances, it is often possible to say that such a pattern could have been produced by a mixture of the semen of the two males in question by com-

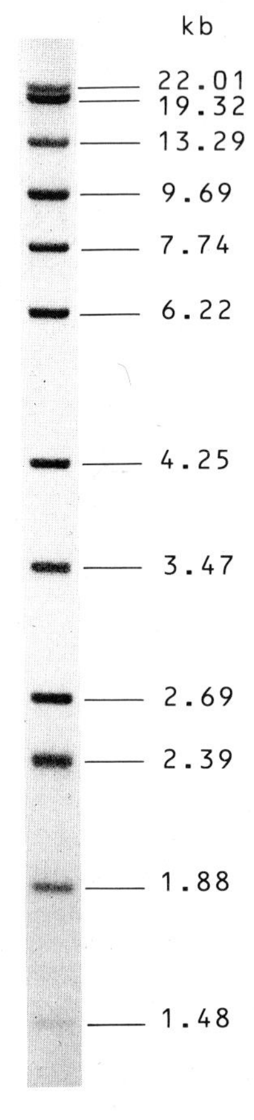

FIG. 4. DNA size marker 'ladder' labelled with ^{35}S (Amersham). Sizes of the DNA fragments are given in kb.

parison with the DNA profiles obtained from their control blood samples (Fig. 3a). By this method it is also possible to eliminate a suspect from the enquiry (Fig. 3b).

In presenting the results of DNA profiling to a Court of Law, the probability of occurrence of the combination of alleles produced by the three single locus probes used is calculated with reference to calibration lanes on each autoradiograph (Fig. 4). Even conservative estimates show that the degree of characterization provided by a DNA profile is such that the probability of obtaining a similar matching profile from an individual other than the person whose DNA profile matches that of the suspect sample is between 1 in 1 million and 1 in 50 million of the population. This probability of occurrence is made by reference to relevant data bases prepared from Caucasian (Fig. 5), Afro-Caribbean and Asian populations.

DNA profiling has revolutionized forensic body fluid grouping, giving evidence of very high probability of guilt and readily showing innocence. In order to improve this powerful technique, future developments will enable smaller body fluid stains to be profiled and improve the time taken to obtain results.

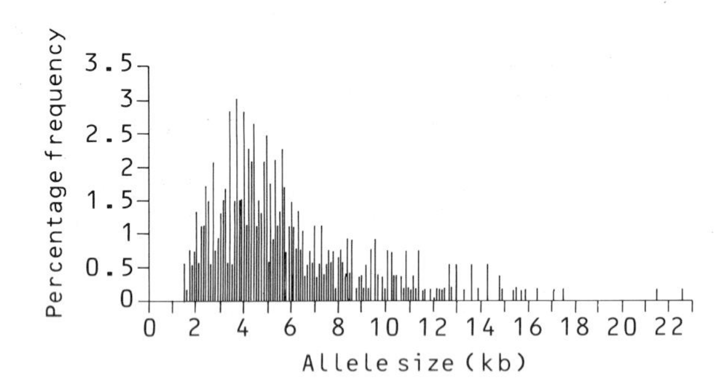

FIG. 5. Histogram showing the distribution of allele sizes, in 100 base-pair steps, in a population of 311 unrelated Caucasians. The DNA was digested with *Hinf*I and, after Southern blot hybridization, was probed with λMS1. At least 120 'alleles' were resolved, all of which are rare.

References

BELL, G.I., SELBY, M.J. & RUTTER, W.J. 1982. The high polymorphic region near the human insulin gene is composed of simple tandemly repeating sequences. *Nature* **295**, 31–35.

CAPON, D.G., CHEM, E.Y., LEVINSON, A.D., SEEBURG, P.H. & GOEDDEL, D.V. 1983. Complete nucleotide sequence of the T24 human bladder carcinoma oncogene and its normal homologue. *Nature* **302**, 33–37.

CHURCH, G.M. & GILBERT, W. 1984. Genomic sequencing. *Proceedings of the National Academy of Sciences of the United States of America* **81**, 1991–1995.

Gill, P., Jeffreys, A.J. & Werrett, D.J. 1985. Forensic applications of DNA 'fingerprints'. *Nature* **318**, 577–579.

Gill, P., Lygo, J.E., Fowler, S.J. & Werrett, D.J. 1987. An evaluation of DNA fingerprinting for forensic purposes. *Electrophoresis* **8**, 38–44.

Goodbourn, S.E.Y., Higgs, D.R., Clegg, J.B. & Weatherall, D.J. 1983. Molecular basis of length polymorphism in the human ζ-globin gene complex. *Proceedings of the National Academy of Sciences of the United States of America* **80**, 5022–5026.

Higgs, D.R., Goodbourn, S.E.Y., Wainscoat, J.S., Clegg, J.B. & Weatherall, D.J. 1981. Highly variable regions flank the human α-globin genes. *Nucleic Acids Research* **9**, 4213–4214.

Jarman, A.P., Nicholls, R.D., Weatherall, D.J., Clegg, J.B. & Higgs, D.R. 1986. Molecular characterisation of a hypervariable region downstream of the human α-globin gene cluster. *EMBO Journal* **5**, 1857–1863.

Jeffreys, A.J., Wilson, V. & Thein, S.L. 1985. Individual-specific 'fingerprints' of human DNA. *Nature* **316**, 76–79.

Knowlton R.G., Brown, V., Barman, J.C., Barker, D., Schumm, J.W., Murray, C., Takvorian, T., Ritz, J. & Donnis-Keller, H. 1986. Use of highly polymorphic DNA probes for genotypic analysis following bone marrow transplantation. *Blood* **68**, 378–385.

Nakamura, Y., Leppert, M., O'Connell, P., Wolff, R., Holm, T., Culver, M., Martin, C., Fujimoto, E., Hoff, M., Kumlin, E. & White, R. 1987. Variable number of tandem repeat (VNTR) markers for human gene mapping. *Science* **235**, 1616–1622.

Proudfoot, N.J., Gil, A. & Maniatis, T. 1982. The structure of the human zeta-globin gene and a closely linked, nearly identical, pseudogene. *Cell* **31**, 553–563.

Weller, P., Jeffreys, A., Wilson, V. & Blanchetot, A. 1984. Organisation of the human myoglobin gene. *EMBO Journal* **3**, 439–446.

Wong, Z., Wilson, V., Jeffreys, A.J. & Thein, S.L. 1986. Cloning a selected fragment from a human DNA 'fingerprint': isolation of an extremely polymorphic minisatellite. *Nucleic Acids Research* **14**, 4605–4616.

Wong, Z., Wilson, V., Patel, I., Povey, S. & Jeffreys, A.J. 1987. Characterisation of a panel of highly variable minisatellites cloned from human DNA. *Annals of Human Genetics* **51**, 269–288.

Wyman, A. & White, R. 1980. A highly polymorphic locus in human DNA. *Proceedings of the National Academy of Sciences of the United States of America* **77**, 6754–6758.

Application of Nucleic Acid Probes to the Identification of Bacterial Enteric Pathogens

B. WREN, H. KLEANTHOUS, C.L. CLAYTON, P. MULLANY AND SOAD TABAQCHALI
Department of Medical Microbiology, St Bartholomew's Hospital, London EC1A 7BE, UK

All known organisms contain nucleic acids which constitute their genotype. Specific nucleic acid probes (either DNA or RNA) are therefore of potential use in the identification of micro-organisms from pathological specimens. The use of nucleic acid probes in the identification of pathogens is based on a combination of gene cloning, nucleic acid hydridization technology and, more recently, the polymerase chain reaction (PCR) amplification of target DNA.

Diarrhoea is one of the most infectious diseases world-wide with over 10 billion cases per year, resulting in several million deaths. Most enteric pathogens exert their harmful effects via the production of enterotoxins which cause massive losses of fluids from the body, resulting in dehydration. Despite much progress in recent years in the identification of enteric pathogens, the causative agents of diarrhoea in many cases cannot be identified. Without being able to identify the causative agent, it is difficult to study the epidemiology of the infectious agent and to develop effective vaccines to prevent diarrhoeal disease.

Pathogenic *Escherichia coli* is a major cause of diarrhoeal disease in man and farm animals throughout the world. There are four major categories of diarrhoeagenic *E. coli*: enterotoxigenic (ETEC; a major cause of traveller's diarrhoea and infant diarrhoea in developing countries), enteropathogenic (EPEC; an important cause of infant diarrhoea), enteroinvasive (EIEC; a cause of dysentery) and enterohaemorrhagic (EHEC; associated with haemolytic uraemic syndrome (HUS) and haemorrhagic colitis).

The major impediment to the study of diarrhoea caused by *E. coli* has been the difficulty in differentiating enterotoxigenic strains from non-toxigenic strains and the normal gut flora. The use of gene probes in the identification

Genetic Manipulation
0–632–02926–9

of pathogenic *E. coli* strains from patient specimens represents an advantage over current methods available in the clinical laboratory which are time consuming and expensive; for example, the use of animals to demonstrate the presence of enterotoxins.

In this chapter, experimental details for the identification of ETEC strains from diarrhoeal specimens by means of heat-labile enterotoxin gene probes (LTI and LTII) will be presented as an example of the application of nucleic acid probes in the clinical laboratory. The role of gene probes for the identification of other enteric pathogens will also be described, as well as the potential use of PCR as a method for the *in vitro* amplification of small amounts of target DNA from clinical specimens.

Identification of Enterotoxigenic *Escherichia coli* from Faecal Specimens

The following protocol represents a generalized procedure which has been used successfully to identify ETEC strains from patients with travellers' diarrhoea and infants with diarrhoea in developing countries, using LTI gene probe (Echeverria *et al.* 1985). The protocol has also been used in our department to identify ETEC strains which contain the LTII enterotoxin (Seriwatana *et al.* 1988).

Treatment of faecal samples

Faecal samples from diarrhoeal specimens are resuspended in an equal volume of sterile phosphate buffered saline (PBS) in the original specimen container. The sample can then be tested for the presence of heat-labile toxin (LTI or LTII) either directly by spot hybridization, or by colony hybridization if isolation of the ETEC strain is required.

1 *Direct spot hybridization*: spot in triplicate 10 μl of faecal suspension directly onto gridded nylon membranes (Hybond N 82 mm; Amersham International). Up to 90 spots can be loaded on the gridded membrane, which should include LTI/LTII positive and negative controls.

2 *Colony hybridization*: prepare serial dilutions of faecal suspension and spread 0.1 ml of each dilution on MacConkey agar plates and incubate for at least 6 h at 37°C. The remaining faecal suspension can be spun down and the supernatant filtered through a Millipore filter (0.45 μm) and tested for cytotoxicity on a monolayer of Chinese hamster ovary (CHO) tissue culture cells in microtitre plates, or for enterotoxicity in animals.

Preparation of samples for filter hybridization

1 For colony hybridization samples, choose suitable MacConkey plates with about 500 colonies and place at 4°C for 15 min.
2 Remove plates from 4°C and overlay with a nylon membrane for 15 min. Set up four dishes with Whatman 3MM paper saturated with denaturing solution (1.5 M NaCl, 0.5 M NaOH) neutralizing solution (1.5 mM NaCl, 0.5 M Tris–HCl pH 7.2, 0.001 M EDTA) × 2 and SSC solution (0.3 M NaCl, 0.03 M sodium citrate).
3 Peel the nylon membranes from the plates and transfer all membranes, including spot hybridization samples, colony side up, to the 3MM paper saturated with denaturing solution for 5 min. Take care not to cover the upper surface of the membranes with solution.
4 Transfer the membranes to the 3MM paper saturated with neutralizing solution for 2 min.
5 Repeat neutralization step.
6 Transfer the membranes to SSC solution for 2 min.
7 Place membranes on 3MM paper for 15 min to dry in air.
8 Wrap the membranes in cling film and place on u.v. transilluminator (colony side down) for 3 min.
Batches of membranes can be stored for several months at this stage before hybridization with gene probe. This can be particularly useful in developing countries where the DNA-fixed membranes can be sent to reference laboratories for hybridization with the relevant probe.

Hybridization of ETEC samples with LTI/LTII toxin gene probe(s)

1 Place membranes (up to 50) with 20 ml hybridization solution in a sealed plastic freezer bag. Incubate for at least 4 h at 42°C.

The hybridization solution consists of 10 ml deionized formamide; 4 ml 5 × P buffer (1% BSA, 1% polyvinylpyrolidone, 1% Ficoll, 250 mM Tris pH 7.5, 0.5% sodium pyrophosphate, 5.0% SDS); 4 ml dextran sulphate (50% solution); 0.1 ml denatured salmon sperm DNA (Sigma Chemical Co., Poole, Dorset); 1.9 ml sterile water; 1.16 g sodium chloride. The formamide, 5 × P buffer, and salmon sperm DNA can be dispensed and stored at −20°C.
2 Denature the ^{35}S or ^{32}P-labelled LT toxin probe(s) by heating for 2 min at 100°C, cool on ice and add directly to hybridization solution. Mix the probe in thoroughly, avoiding air bubbles, and incubate at 42°C on a shaker for at least 4 h or overnight.
3 After the hybridization is complete, discard the hybridization solution and wash with solution A (0.3 M NaCl; 60 mM Tris–HCl, pH 8.0; 2 mM EDTA)

for 30 min at room temperature. Make up solution B (solution A including 1% SDS) and pre-incubate at 68°C.
4 Wash membranes twice with solution B at 68°C for 30 min.
5 Wash membranes with solution C (1 in 10 dilution of solution A) for 30 min at room temperature.
6 Dry membranes in air on Whatman 3MM paper at room temperature. Tape the membranes (colony side up) onto the 3MM paper and expose to X-ray film at −70°C for at least 90 min (^{32}P) or 16 h (^{35}S) at room temperature.
7 The autoradiographs of the colony hybridizations are examined together with the original plates so that colonies that hybridize with the LT toxin probe(s) can be identified. The colony hybridization results are compared with the spot hybridization results for confirmation as well as the tissue culture results, if available.
8 If positive colonies can be identified, these probe-positive strains can be serotyped and identified biochemically.

Notes

1 The authors have found the random primer method of Feinberg & Vogelstein (1983) for radiolabelling DNA to be highly sensitive and easy to use. The Multiprime kit from Amersham is convenient and an incorporation time of 1 h is sufficient.
2 The authors have found that the hybridization conditions in the above protocol yield reproducible results. The conditions are for well matched hybrids (T_m −20°C) and the use of 10% dextran sulphate, which increases the reassociation rate, allows for more rapid hybridization. A detailed account of hybridization conditions and procedures has been given by Anderson & Young (1985).

Gene Probes for other Bacterial Enteric Pathogens

The protocol detailed above should be suitable for the detection of most Gram-negative enteric pathogens where specific gene probes have been isolated for the organism. *E. coli* is also the host strain in many cloning experiments. Therefore, the protocol should prove useful in screening potential clones by colony hybridization.

Toxin gene probes against the two heat stabile toxins, STI (Maas *et al.* 1985) and STII (Echeverria *et al.* 1984), have also been used successfully in colony hybridization experiments to identify ETEC strains from diarrhoeal specimens. A Vero cytotoxin (also called shiga-like toxin) has been cloned from EHEC (Willshaw *et al.* 1985) and has proved useful in the identification of EHEC strains in diarrhoeal samples from HUS patients. DNA probes against entero-adhesive factors and entero-invasive factors have been used to specifically identify EPEC (Nataro *et al.* 1985) and EIEC (Wood *et al.* 1986) strains respectively. Toxin gene probes have also been isolated from *Vibrio cholerae* (Pearson & Mekalanos 1982), *Aeromonas hydrophilia* (Chakraborty *et al.* 1984), *Clostridium perfringens* (Iwanejko *et al.* 1989), *Staphylococcus aureus* (Bentley & Mekalanos 1988) and *Shigella flexneri* (Brazil *et al.* 1988) which have potential use in the identification of the organisms from diarrhoeal specimens. In the case of enteric pathogens where no readily identifiable virulence determinants are known, e.g. *Salmonella* and *Campylobacter* species, species-specific probes can be used to identify the organism. A species-specific oligonucleotide probe has been used to identify enteric *Campylobacter* species and is commercially available as part of a kit from DuPont.

Harsher treatment of the bacterial cells is often required to release DNA from Gram-positive enteric pathogens before hybridization with gene probes. For example, the authors have found that, in hybridization studies with *Clostridium difficile* (the causative agent of pseudomembranous colitis), it is necessary to include a preliminary step of soaking the nylon membrane filters in a 10% SDS solution prior to denaturation. Better results have been obtained by repeating the denaturation and neutralization steps. In these experiments, a gene fragment has been constructed from the cloned enterotoxin, toxin A (Wren *et al.* 1987) to identify toxigenic *C. difficile* strains in patient specimens which often contain non-toxigenic *C. difficile* strains. Hybridization studies have shown that the toxin A gene is absent in non-toxigenic strains, and therefore the lack of toxin activity is not due to non-expression of the gene. This observation is an important consideration before embarking on hybridization studies with virulence determinant gene probes.

Prospects for Nucleic Acid Probes in Clinical Microbiology

Synthetic oligonucleotides as nucleic acid probes offer several advantages over cloned double-stranded DNA probes and they have been successfully used to detect ETEC (Murray *et al.* 1987), *Yersinia entercolitica* (Murray *et al.* 1987) and *Campylobacter* species (Mikotis *et al.* 1989; Freir *et al.* 1988). Firstly, synthetic oligonucleotides are very pure probes and have no contaminating DNA sequences, in contrast to cloned DNA probes which often contain vector DNA. Secondly, oligonucleotide probes can be conveniently chemically coupled with colorimetric enzymes such as alkaline phosphatase. Thirdly, being single-stranded, it is unnecessary to denature the probe prior to hybridization.

The role of nucleic acid probes in the diagnostic microbiology laboratory remains unclear. Ideally, a battery of DNA probes with different colorimetric substrates would be used directly on clinical samples. For example, a faecal sample could initially be screened for the three major enteric pathogens, *Campylobacter jejuni, Salmonella enteritidis* and *E. coli.*

The two main problems with present day nucleic acid probe technology are sensitivity (hence difficulty in using probes directly on clinical specimens) and the necessity for detecting homologous DNA using potentially hazardous radioisotopes. By increasing the sensitivity of nucleic acid probes the latter problem could be resolved by using a less sensitive non-radioactive assay. Sensitivity of the probes can be overcome by using specific probes targeted to a sequence that occurs as multiple copies in the cell. This has been achieved using ribosomal RNA probes (Wilson *et al.* 1988) or probes against DNA on multicopy plasmids (Totten *et al.* 1983). An alternative approach would be to amplify the target DNA which can now be conveniently achieved using the PCR technique.

Polymerase Chain Reaction Amplification of DNA

The polymerase chain reaction (PCR) takes advantage of a DNA polymerase that uses a defined segment in a strand of DNA as a template for assembling a complementary strand. The PCR reaction vessel contains a mixture of buffers, nucleotides, primers, DNA polymerase and a small amount of the clinical specimen of interest. The PCR requires a three step cycling process.

1 *Denaturation of double-stranded DNA (1 min).* Heating the sample to 95–100°C is usually sufficient, although lower temperatures of around 80°C may be preferable under certain circumstances.

2 *Annealing of primers to dissociated DNA strands (2 min).* It is necessary to know the sequence of the target DNA to enable the correct synthesis of oligonucleotides to act as primers. The choice of primers can be crucial, and it is desirable to choose oligonucleotides which will anneal upstream and downstream of unique restriction enzyme sites. The primers are present in such vast molar excess that they are more likely to anneal to the dissociated strands than the strands are to reanneal to each other.

3 *Primer extension and synthesis of new DNA strand* (*3 min*). DNA polymerase adds nucleotides complementary to those in the unpaired DNA strand onto the annealed primer, resulting in a copy of the DNA strand containing the target DNA. The DNA polymerase of choice is one isolated from *Thermus aquaticus* (*Taq*) which has revolutionized PCR. This thermostable enzyme permits automation of the procedure because all reaction components can be combined at the beginning and processed in a thermal cycler without the need

to replenish heat-inactivated enzyme. *Taq* polymerase has also improved specificity, yield, sensitivity and length of the target DNA that can be amplified.

A typical cycle takes 3–5 min and the number of DNA strands doubles on completion of each cycles. After 30 cycles, a single copy of DNA can be increased up to one million copies. The amplified sequences of target DNA can be cut with a suitable restriction enzyme(s) that will yield a DNA fragment of defined length. If enough amplified DNA is present, it can be seen after gel electrophoresis and ethidium bromide staining. Alternatively, final detection can be made with an oligonucleotide probe which will hybridize to a portion of the amplified DNA product.

The experimental conditions under which PCR are run can be extremely variable and need to be determined empirically. Most commmercial thermal cyclers permit at least 24 simultaneous reactions in one run, which enables a rapid determination of the optimal conditions to be made.

Variable factors to optimize PCR experiments

Mg^{2+} concentration (1–5 mM): this is the most important variable and should be tried first. If samples contain EDTA or other chelators, increase the Mg^{2+} concentration.

Temperature of denaturation (80–100°C): it is recommended that samples be heated for 5 min at 95°C before the addition of *Taq* in order to inactivate any bacterial proteases in the sample.

Extension time (2–5 min): longer extension times are required if large segments of DNA are to be amplified.

Taq polymerase enzyme (1–5 units): more units are required for larger reaction volumes.

Number of cycles (10–40): depends on concentration of target DNA.

Length of oligonucleotide (0.2–1 μM of 17–30 bp): the G + C percentage of primers should be near 50% to maximize specificity. Avoid AT and GC rich regions and complementary oligonucleotides which may form primer dimers. Lowering the concentration of oligonucleotides will reduce primer dimers.

Volume of reaction (25–100 μl): mineral oil can be added to sample tubes to prevent evaporation.

Annealing temperature (45–65°C): higher temperatures generally result in much more specific products.

The major problem with PCR is that the technique is a victim of its own success. It is often too sensitive and may lead to the amplification of a non-specific DNA. All PCR experiments should have appropriate controls including a water sample as a negative control.

It is also recommended that pipettes, solutions and work areas should be set aside specifically for PCR work. However, with due care and the development of convenient methods for the identification of amplified DNA, there can be little doubt that PCR will play a significant role in clinical microbiology research and routine diagnosis in the future.

References

ANDERSON, M.L. & YOUNG, B.D. 1985. Quantitative filter hybridisation. In *Nucleic Acid Hybridisation; A Practical Approach.* eds Hames, B.D. & Higgins, S.J. Oxford: IRL Press.

BENTLEY, M.L. & MEKALANOS, J.J. 1988. Nucleotide sequence of the Type A staphylococcal enterotoxin gene. *Journal of Bacteriology* **170**, 34–41.

BRAZIL, G.M., CLAYTON, C.L., SEKIZAKI, T. & TIMMIS, K.N. 1988. Development of DNA probes for cytotoxin and enterotoxin genes in enteric bacteria. *Experimentia* **44**, 848–853.

CHAKRABORTY, T., MONTENEGRO, M.A., SANYAL, S.C., HELMUTH, R., BULING, E. & TIMMIS, K.N. 1984. Cloning of enterotoxin gene from *Aeromona hydrophila* provides conclusive evidence of production of a cytotoxic enterotoxin. *Infection and Immunity* **46**, 435–441.

ECHEVERRIA, P., SERIWATANA, J., PATAMAROJ, U., MOSELEY, S.L., MCFARLAND, A., CHITYOTHIN, O. & CHIACUMPA, W. 1984. Prevalence of heat-stable II enterotoxigenic *Escherichia coli* in pigs, water and people at farms in Thailand as determined by DNA hybridization. *Journal of Clinical Microbiology* **19**, 489–491.

ECHEVERRIA, P., SERIWATANA, J., TAYLOR, D.N., TIRAPAT, C., CHAICUMPA, W. & ROWE, B. 1985. Identification of DNA hybridisation of enterotoxigenic *Escherichia coli* in a longitudinal study of villages in Thailand. *Journal of Infectious Diseases* **151**, 124–130.

FEINBERG, A.P. & VOGELSTEIN, B. 1983. A technique for radiolabelling DNA restriction endonuclease fragments to high specific activity. *Analytical Biochemistry* **132**, 6–13.

FREIR, S.M., ROSZAK, D.B., RISEN, L.A., MARICH, J.E. & WONG, D.F. 1988. A non-radioactive, highly sensitive oligonucleotide probe to detect *Campylobacter. DuPont Biotechnology Update* **4**, 14–16.

IWANEJKO, L., ROUTLEDGE, M.N. & STEWART, G.S. 1989. Cloning in *Escherichia coli* of enterotoxin gene from *Clostridium perfringens* Type A. *Journal of General Microbiology* **135**, 903–909.

MAAS, R., SILVA, R.M., GOMES, T.A.T., TRABULSI, L.R., & MAAS, W.K. 1985. Detection of genes for heat-stable enterotoxin I in *Escherichia coli* strains isolated in Brazil. *Infection and Immunity* **49**, 16–51.

MIKOTIS, M.D., GALEN, J.E., KAPER, J.B. & GLEN-MORRIS, J. 1989. Development and testing of a synthetic oligonucleotide probe for the detection of pathogenic *Yersinia* strains. *Journal of Clinical Microbiology* **27**, 1667–1670.

MURRAY, B.W., MATHEWSON, J.J., DUPONT, H.L. & HILL, W.E. 1987. Utility of oligodeoxyribonucleotide probes for detecting enterotoxigenic *Escherichia coli. Journal of Infectious Diseases* **155**, 809–811.

NATARO, J.P., BALDINA, M.M., KAPER, J.B., BLACK, R.E., BRAVO, N. & LEVINE, M.M. 1985. Detection of an adherance factor of entero-pathogenic *Escherichia coli* with a DNA probe. *Journal of Infectious Disease* **152**, 560–565.

PEARSON, G.D.N. & MEKALANOS, J.J. 1982. Molecular cloning of *Vibrio cholerae* enterotoxin genes in *Escherichia coli* K12. *Proceedings of the National Academy of Sciences of the United States of America* **79**, 2976–2980.

SERIWATANA, J., ECHEVARRIA, P., TAYLOR, D.N., RASRINAUL, L., EDWARD-BROWN, J., PEIRIS, J.S.M. & CLAYTON, C.L. 1988. Type II heat-labile enterotoxin-producing *Escherichia coli* isolated from animals and humans. *Infection and Immunity* **56**, 1158–1161.

TOTTEN, P.A., HOLMES, K.K., HANDSFIELD, H.H., KNAPP, J.S., PERINE, P.L. & FALKOW, S.

1983. DNA hybridization technique for the detection of *Neisseria gonorrhoeae* in men with urethritis. *Journal of Infectious Diseases* **148**, 462–471.

WILLSHAW, G.A., SMITH, H.R., SCOTLAND, S.R. & ROWE, B. 1985. Cloning of genes determining the production of Vero cytotoxin by *Escherichia coli*. *Journal of General Microbiology* **131**, 3047–3053.

WILSON, K., BLITCHINGTON, R., HINDENACH, B. & GREENE, R.C. 1988. Species-specific oligonucleotide probes for rRNA of *Clostridium difficile* and related species. *Journal of Clinical Microbiology* **26**, 2484–2488.

WOOD, P.K., MORRIS, J.G. JR., SMALL, P.L.C., SETHABUTR, O., TOLEDO, M.R.E., TRABULSI, L. & KAPER, J.B. 1986. Comparison of DNA probes with the Sereny test for identification of invasive *Shigella* and *Escherichia coli* strains. *Journal of Clinical Microbiology* **24**, 298–500.

WREN, B.W., CLAYTON, C.L., MULLANY, P. & TABAQCHALI, S. 1987. Molecular cloning and expression of *Clostridium difficile* toxin A in *Escherichia coli* K12. *FEBS Letters* **225**, 82–86.

Quantification of Collagen mRNA Levels in Mammalian Cells

J.S. Campa, J.E. Bishop and G.J. Laurent
Biochemistry Unit, National Heart and Lung Institute, Dovehouse Street, London SW3 6LY, UK

Most studies of cell gene expression begin with observations of phenotypic changes due to enhanced production of a specific protein. The mechanisms involved in changing the production of the protein can then be assessed at several levels: nuclear run-off assays to measure transcription rates; analysis of steady-state mRNA levels to examine the regulation of events occurring between transcription and translation; studies of protein appearance to assess translation and post-translational control. A convenient starting place for analysis of regulation of protein production is at the mRNA level. The study of steady-state mRNA levels, which reflect the balance of mRNA production and degradation after experimental manipulation, can be used as a guide to differentiate important pretranslation regulatory mechanisms. For example, increased protein production accompanied by increases in the mRNA of that protein suggest pretranslation regulation and nuclear run-off assays or assessment of mRNA half-life may be the next experimental approach. Alternatively, if no changes in mRNA accumulation are noted, regulation must occur at a translation or post-translation level. Complete analysis will necessitate measurements at all levels, each of which can be divided even further in order to look at the many events of which they are comprised.

Collagens, the most abundant proteins in mammals, account for about 30% of all proteins and belong to a gene family whose products share a basic structure and certain chemical properties. There are more than ten distinct types of collagen recognized to date, encoded by at least 18 genes. A protein can be classified as a collagen if it contains triple-helical domains, repeats of the triplet glycine–X–Y (X and Y indicating any amino acid), the modified amino acids hydroxyproline and hydroxylysine and has the ability to form supramolecular structures such as fibrils, filaments or networks. Each collagen molecule is formed by three polypeptide chains that can be identical (homo-

0–632–02926–9

trimers) or different (heterotrimers). Collagen types are named by Roman numerals and within each type, chains are identified by a number accompanying a letter α.

Collagens, however, exhibit a striking diversity of function, as revealed by the differential regulation of their expression, both in time and space. They are predominant constituents of extracellular matrices of nearly all kinds of tissues, where they contribute to the preservation of tissue shape and cell integrity. In addition, collagens are involved in embryonic development, cell adhesion and cell mobility.

Our interest has been mainly with the interstitial or fibrillar collagens type I and III and their metabolism in normal development and pulmonary and cardiovascular disorders. We have developed methods to determine hydroxyproline, and thus collagen levels and production rates both *in vivo* (McAnulty & Laurent 1987) and *in vitro* (Campa *et al.* 1990). This allows the assessment of translational and post-translational control. We have more recently turned to RNA analysis to examine the control between transcription and translation.

RNA analysis classically involves the isolation of intact mRNA from cells and/or tissue, size fractionating these transcripts through an agarose gel and identifying the specific mRNA of interest by DNA–RNA hybridization with a labelled probe. We can then determine relative levels of RNA accumulation in our samples and begin to understand the transcriptional mechanism of gene expression control, and regulation of translation.

Preparation of Total RNA from Cells in Culture

There are many different procedures described in the literature for the extraction of eukaryotic RNA. All have common features:

1 Prevention of ribonuclease activity in order to obtain intact RNA.
2 The separation of RNA from protein and DNA.
3 The separation of certain mRNA species from other mRNAs and other classes of RNA.

Certain precautions must be taken in the isolation of RNA. Ribonuclease contamination, which will degrade RNA present in samples, must be avoided by using pure reagents: most companies now sell molecular biology grade reagents. Additionally, all glassware must be baked at 250°C for 4.5 h and all solutions and water should be treated with 0.2% diethyl pyrocarbamate (DEPC) to inactivate these enzymes. To treat solutions with DEPC, they should be adjusted to 0.2% in DEPC and shaken thoroughly or stirred vigorously for 10 min to disperse DEPC. The solution should then be autoclaved (liquid cycle for 15 min). Specific ribonuclease inhibitors may be needed for samples in which high levels of endogenous ribonucleases are expected. This will be dependent on the cell type or tissue used. The spleen and pancreas are tissues with high levels of endogenous ribonucleases. All personnel should wear plastic or latex gloves when handling samples, glassware or reagents to avoid contamination with the large amount of ribonuclease present on skin.

In our laboratories, we have adopted the single-step guanidinium thiocyanate–phenol–chloroform extraction procedure which can be completed in a minimum of 4 h (Chomczynski & Sacchi 1987). The most widely used technique for isolating total RNA from whole cells is the guanidinium thiocyanate/caesium chloride procedure (Chirgwin *et al.* 1979). This technique takes advantage of different buoyancies of nucleic acids in a CsCl cushion and requires overnight ultracentrifugation.

Reagents

Guanidinium thiocyanate.
0.75 M sodium citrate pH 7.0.
10% sodium sarcosinate.
β-mercaptoethanol.
2 M sodium acetate pH 4.0.
Phenol.
Chloroform: isoamylalcohol (49:1 v/v).
Isopropanol.
70% ethanol (v/v).
0.5% sodium dodecyl sulphate (SDS).

Guanidinium thiocyanate is an efficient protein denaturing agent and is thus an effective nuclease inactivator. The addition of denaturation enhancers, such as SDS and β-mercaptoethanol, produce a solution which, following RNA dispersion, meets the first two criteria for RNA isolation.

The guanidinium solution, Solution A: 4 M guanidinium thiocyanate, 25 mM sodium citrate pH 7, 0.5% sarcosinate, 0.1 M β-mercaptoethanol, is prepared by adding 0.36 ml β-mercaptoethanol to 50 ml of stock solution, this can be stored for 1 month at room temperature. The denaturing stock solution is prepared by dissolving 100 g guanidinium thiocyanate in 194.3 ml of water, 7.1 ml of sodium citrate pH 7, and 10.6 ml 10% sodium sarcosinate at 65°C. This can be stored for 3 months at room temperature.

Fibroblasts were grown to confluence in 100 mm diameter petri dishes and incubated for 24 h under optimal conditions for collagen production: Dulbecco's modified Eagle's medium (DMEM) containing 2% newborn calf serum (NCS), 100 u/ml penicillin, 100 μg/ml streptomycin, 4 mM glutamine, 50 μg/ml ascorbic acid and 0.2 mM proline. After this period the medium was replaced with media as above but supplemented with the factor(s) under investigation and allowed to act on the cells for a further 24 h before the assessment of collagen protein levels by high pressure liquid chromatography (Campa *et al.* 1990) and the isolation of total RNA as follows:

1 Remove medium and wash cells three times with ice-cold phosphate buffered saline (PBS).

2 Add 2 ml of Mg/Ca-free ice-cold PBS, and stand on ice.

3 Scrape off cell sheet into PBS, transfer to 15 ml polypropylene tube and wash dish with a further 2 ml of PBS.

4 Centrifuge the 4 ml PBS at 2000 ***g*** for 5 min at 4°C.

5 Aspirate PBS and add 100 μl of solution A per 10^6 cells, pass three times through a 25-gauge needle for complete cell disruption and shearing of DNA.

6 The solution can be frozen (−70°C) or used immediately.

7 Add sequentially, with inversion mixing after each reagent:
0.1 volume of 2 M sodium acetate pH 4;
1 volume of phenol (water saturated);
0.2 volume of chloroform: isoamylalcohol (49:1).

8 Shake final suspension vigorously for 15 s and cool on ice for 15 min.

9 Centrifuge samples at 10000 ***g*** for 20 min at 4°C.

10 The aqueous phase containing the RNA is transferred to another tube, leaving the DNA and protein in the interphase and phenolic phase respectively.

11 Precipitate the RNA with 1 volume of isopropanol for at least 1 h at −20°C. For maximal recovery allow precipitation to continue overnight.

12 Centrifuge samples at 10000 ***g*** for 20 min at 4°C.

13 Redissolve the RNA pellet in 0.3 volume of solution A and transfer to an Eppendorf tube.

14 Reprecipitate with 1 volume of isopropanol at −20°C for 1 h.

15 Centrifuge in a microcentrifuge for 10 min at 4°C.

16 Wash RNA pellet in 70% ethanol, microcentrifuge and remove supernatant, dry under vacuum for 15 min.

17 Finally the pellet is dissolved in 50 μl 0.5% SDS at 65°C for 10 min.

18 The RNA preparation can be enriched for poly $(A)^+$ mRNA by oligo(dT) chromatography, but this is unnecessary for the high copy number collagen genes.

The amount and concentration of RNA obtained can be estimated by measuring the absorbance at 260 nm. An OD of 1 corresponds to approximately 40 μg/ml of RNA. In addition, a purity check is performed by determining the 260/280 absorbance ratio. The ratio for each sample should be between 1.5 and 2.0. Ratios less than 1.5 signify a significant protein contamination and the preparation should be reprecipitated. A ratio of 2.0 is considered a virtually pure nucleic acid sample but does not imply an intact sample.

Quantification of Specific mRNA Levels

The total RNA isolated contains all species of RNA with ribosomal RNA being the most abundant. To quantitate single transcripts from our gene of interest, very sensitive detection assays are required. This is accomplished by hybridization techniques involving labelled complementary DNA (cDNA) strands.

Fractionation of RNA

A gel electrophoresis step, under denaturing conditions, allows separation of RNA according to molecular weight. This technique has the advantage of giving a transcript size, separate multiple or cross-reactive RNAs and a check on the quality of isolated RNA. A number of methods of denaturation are in common use; for example, the use of formaldehyde (Lehrach *et al.* 1977), methyl mercuric hydroxide (Bailey & Davidson 1976) and glyoxal (McMaster & Carmichael 1977). Only the glyoxal procedure will be presented here since the authors have found this to be as effective as the other techniques and it precludes the use of toxic compounds.

The RNA is denatured by incubation at 50°C with glyoxal and dimethylsulphoxide (DMSO). The high temperature and DMSO combine to disrupt hydrogen bonding which allows glyoxal to interact with the RNA. It modifies guanine residues to form a covalent adduct which is stable at neutral or acidic pH. After electrophoresis and transfer to a suitable membrane, the glyoxalation reaction is reversed by a high temperature incubation at pH 8.0 and the specific RNA is detected by hybridization with a radioactively-labelled probe.

Reagents

Glyoxal, a 40% aqueous solution.
Dimethylsulphoxide, molecular biology grade.
10 mM NaH_2PO_4/Na_2HPO_4, pH 7.0.
Agarose, ultrapure electrophoresis grade.
Sample buffer, 0.05% bromophenol blue, 50% glycerol, 10 mM sodium phosphate.
Mixed-bed ion exchange resin, AG-501-X8(D), with blue/gold indicator dye.

Glyoxal deionization: commercial aqueous solutions of glyoxal contain hydrated forms and oxidation products, such as glyoxylic acid, glycolic acid and formic acid. Unless the solution is deionized, these products will cause RNA degradation during denaturation. Deionization is achieved firstly by stirring the glyoxal solution with mixed-bed resin AG-501-X8(D) and subsequently by passing the suspension through a column of fresh resin. AG-501-X8(D) contains a blue dye that turns yellow when the exchange capacity is exhausted, thus treatment of the glyoxal solution must continue with fresh resin until the indicator remains blue. Deionized glyoxal should be stored in completely filled and capped 250 μl–1.5 ml Eppendorf tubes at −20°C.

1 To denature add:
4 μl of deionized glyoxal;
3 μl of 80 mM sodium phosphate buffer pH 6.5;
12 μl of DMSO;
to 20 μg of RNA sample contained in 5 μl of 0.5% SDS.

2 RNA sample in 1 M glyoxal, 50% DMSO, 10 mM sodium phosphate buffer pH 6.5 is incubated in a tightly sealed plastic tube at 50°C for 1 h.

3 Cool on ice and add 5 μl of sample buffer.

4 Prepare a horizontal 1.1% agarose gel in 10 mM sodium phosphate. In this procedure we used a IBI model MPH gel apparatus with gel dimensions of 10 × 14 cm and gel depth 3 mm.

5 A sample range of 2–20 μg of RNA and glyoxalated RNA markers are loaded and the gel is run in 10 mM sodium phosphate buffer pH 7.0 at 55 V for 6 h (4.5 V/cm gel current should not exceed 45 mA). Once samples have moved a few mm in the gel, recirculation of buffer is started. Buffer should remain at pH 7.0 or below pH 8.0 to prevent dissociation of glyoxal. This can be checked periodically with narrow range pH indicator paper.

6 Quality of RNA preparation can be checked by staining a gel with ethidium bromide. This gel may not be suitable for subsequent transfer as staining has been shown to reduce the efficiency of transfer.

Transfer of RNA

Once size-fractionated, the RNA must be immobilized on a solid phase to facilitate its detection. Currently, RNA can be transferred to a membrane in several different ways: vacuum, electrically or passively. In the absence of data to suggest more efficient transfer by other techniques we have used the passive techniques as it requires no specialized equipment. For passive transfer, the gel is placed in direct contact with the membrane and a transfer buffer is allowed to pass through it. By capillary action, buffer is drawn through the gel and the membrane: RNA species migrate out of the gel and are immobilized on the membrane.

Reagents

20 × SSC, 3 M NaCl, 0.3 M trisodium citrate.
Whatman 3MM paper.

Paper towels.
20 mM Tris–HCl, pH 8.0.
1 Place two sheets of 3MM paper soaked in 20 × SSC (the same width as the gel but longer, to act as wicks) on a clean glass plate and allow the ends to drape into a shallow reservoir of 20 × SSC.
2 After electrophoresis, without prior treatment of any kind, place the gel well-side down onto the 3MM paper, carefully removing any air bubbles.
3 A piece of membrane wetted in water, slightly smaller that the gel, is smoothly placed on top of the gel, removing trapped air. It should not be moved once the gel and membrane are in contact. Care must be taken to ensure that the membrane and the 3MM paper do not come into direct contact as this will short-circuit the flow of buffer through the gel. Parafilm or clingfilm can be placed along each edge of the gel, flush with the edge, to ensure no contact.
4 Two sheets of 20 × SSC soaked 3MM paper, the same size as the membrane, are placed on top of the membrane with removal of any trapped air.
5 A third sheet of dry 3MM is added along with paper towels, 2–3 mm smaller than the size of the gel, to a height of 5–7 cm.
6 A glass plate is placed over the stack of towels and covered to minimize evaporation. The sandwich is compressed by a 0.5–1 kg weight and transfer allowed to occur overnight.
7 Blots are carefully removed, after orientation marking has been added and air-dried. The membrane is finally baked, in a 'tent' of 3MM paper, *in vacuo* at 80°C for a maximum of 2 h.
8 Baking does not effectively dissociate the glyoxal adduct from the RNA, so baked blots must be treated with 20 mM Tris–HCl pH 8.0, at 100°C for 5 min. Place blots in 200 ml of 20 mM Tris–HCl pH 8.0, at 100°C and allow to cool to room temperature. The blots are air-dried and stored desiccated, sealed in bags at 4°C.

Detection of RNA

A complementary DNA strand is used to form the DNA–RNA hybrids and allows a specific RNA of interest to be detected. The probe or the DNA strand used to hybridize with the RNA, is the complementary DNA (cDNA) of the gene. cDNA represents a reverse transcription product of the RNA under examination. The cDNA is not the gene, as it lacks introns and may also lack important 5′ and 3′ flanking regions, but has the exon sequences which were transcribed into the RNA. This cDNA will hybridize through hydrogen bonding only to the RNA from which it was transcribed. Owing to the large number of interactions which must correctly occur, this binding is very specific. cDNAs are usually inserted into plasmids which can be propagated by transforming them into bacteria, growing the bacterial culture and isolating the plasmids with cDNA inserts. The cDNA is labelled to a high specific activity with ^{32}P, and provides a specific and sensitive assay for probing the fractionated, immobilized RNA of interest. Random primer or nick translation are the most popular methods for the generation of radiolabelled cDNA. (For details see chapters by Sanders *et al.*, this volume.)

After the RNA has been fractionated and immobilized, the non-specific binding sites on the membrane are first blocked by prehybridizing the membrane in a solution containing denatured salmon sperm DNA. This is then removed and replaced by the same solution containing the radiolabelled probe. These steps are usually carried out at 42–65°C, depending on DNA structure and choice of hybridization buffers. Formulas are available to help in choosing the correct temperature (Anderson & Young 1985). Hybridization is allowed to occur for 16–24 h: at the end of this period the membrane is removed and washed in low salt solutions at room temperature and at 50–65°C. These washes remove any weak interactions between the probe and RNA on the membrane which is not of interest. After the washes are complete, autoradiography is performed and the signal on the X-ray film identifies the DNA–RNA hybrid of interest and the relative amount.

A large number of cDNA probes for the ten different collagen types are currently available. The authors have used pα1R1 and pα2R2 (Genovese *et al.* 1984) which are rat collagen probes containing type I α1 and α2 collagen-specific sequences respectively, as well as pMCS1 mouse probe (Liau *et al.* 1985) which is specific for collagen type III. The following conditions are for a random primer ^{32}P labelled cDNA probe.

Reagents

Formamide, deionize by stirring in the presence of 5 g/100 ml of AG-501-X8(D) for 1 h at room temperature, filter through Whatman 1MM paper, snap-freeze single-use amounts and store at −70°C.

20 × SSC: 3 M NaCl, 300 mM trisodium citrate.

1 M potassium phosphate pH 7.4.

10% sodium dodecyl sulphate (SDS).

100 mg/ml denatured salmon sperm DNA.

50 × Denhardt's solution: 2.5 g Ficoll, 2.5 g polyvinylpyrrolidone, 2.5 g bovine serum albumin (BSA) to 250 ml water. Filter sterilize and store in aliquots at −20°C.

1 Incubate the membrane in a heat-sealable plastic bag with 0.1 ml/cm^2 warmed prehybridization buffer at 42°C overnight or at least 2 h. Prehybridization buffer consists of:

25% formamide;

0.75 M NaCl, 75 mM sodium citrate (5 × SSC);
25 mM potassium phosphate pH 7.4;
100 μg/ml denatured salmon sperm DNA;
5 × Denhardt's solution.

2 The probe is denatured by heating for 10 min in boiling water, placed immediately on ice and added directly to the prehybridization buffer. The prehybridized filters are hybridized for 24 h at 42°C in 0.1 ml/cm^2 of the above solution containing 10–15 ng/ml of the cDNA random primer labelled to a specific activity of 6 to 9×10^8 cpm/μg.

3 Wash hybridized filters twice in 400 ml of 2 × SSC, 1% SDS at room temperature for 15 min, once in 0.4 × SSC, 0.2% SDS at 56°C for 30 min, and once 0.2 × SSC, 0.1% SDS at 56°C for 30 min. Background signal can be reduced by increasing the strigency of the washes. This is achieved by decreasing the salt concentration and/or by increasing the temperature and time of washes.

4 Wrap a slightly damp membrane blot in a plastic food seal bag, and place in a cassette with an X-ray film, Kodak X-Omat AR, and an intensifying screen. Place at −70°C for 24 h or up to a week, depending on radioactivity of the blot.

5 Relative quantitation is achieved by scanning densitometry or, alternatively, filters can be cut into pieces and segments counted in a liquid scintillation counter.

6 Probes bound to nylon membranes such as Genescreen or Hybond N can be stripped off by placing the membranes in boiling water and allowing them to cool to room temperature or by washing them in 5 mM Tris–HCl pH 8.0, 0.2 mM EDTA, 0.05% pyrophosphate, 0.1 × Denhardt's solution for 2 h at 65°C. Membranes can then be stored and subsequently rehybridized. This stripping procedure should be limited to a maximum of 3–4 times.

The hybridization pattern shown in Fig. 1 for the different collagen types comprises several signals of various lengths. These multiple forms arise from the appearance of several polyadenylation consensus sequences of the 3′ untranslated region of the collagen genes.

Slot blots

Variations on the above Northern analysis are those of dot/slot blots and cytoplasmic dot blots. In the latter, the whole cell lysate, including the RNA, is placed directly onto the membrane. The technique is rapid but has the disadvantage of high backgrounds.

Slot blot techniques allow the quantitation of RNA as serial dilutions, by the direct spotting of the total RNA on the membrane and hybridizing with required probe. This avoids the need to fractionate the RNA, but in so doing

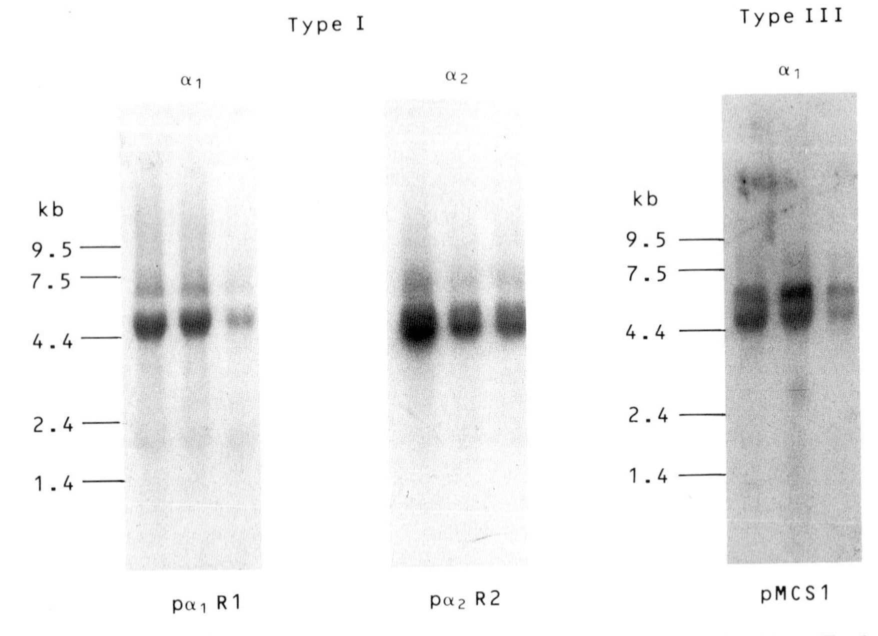

FIG. 1. Northern blot analysis of total mRNA obtained from cultured IMR 90 fibroblasts. Total RNA was isolated by guanidinium thiocyanate–phenol–chloroform extraction, then separated by gel electrophoresis under denaturing conditions. Hybridization was carried out after blotting to Genescreen filters, using probes for collagen type I α_1 and α_2 and collagen type III α_1.

the ability to ascertain if RNA is intact is lost nor can it distinguish between multiple or cross-reactive RNA species. Thus slot blot analysis is only useful when a complete Northern analysis has been completed for the particular message and probe under investigation.

Method

1 Isolate total RNA by guanidinium thiocyanate–phenol–chloroform extraction as detailed above.
2 Denature RNA, 5–10 μg, in 1 M deionized glyoxal, 10 mM sodium phosphate for 1 h at 50°C. Omit the DMSO as this will dissolve the membrane.
3 Prepare the required serial dilutions of the sample in 0.1% SDS. Final volume should not exceed 5 μl.
4 The membrane is floated on water, placed in 20 × SSC for 5 min, dried on 'concertina' foil under a heat lamp, then clamped in the slot blot apparatus.

5 Samples are applied to the wells and the membrane air-dried and baked for a maximum of 2 h at 80°C.
6 Remove residual glyoxal by placing filter in 20 mM Tris–HCl, pH 8.0 at 100°C and allow to cool to room temperature.
7 Hybridize and wash filters as described for Northern blots and quantitate by scanning densitometry.

The methods outlined above permit the elucidation of post-transcriptional events occurring within the cell and, coupled with other techniques, lead to a better understanding of how genes are controlled.

References

ANDERSON, M.L.M. & YOUNG, B.D. 1985. Quantitative filter hybridisation. In *Nucleic Acid Hybridisation, A Practical Approach*. eds Hames, B.D. & Higgins, S.J. ch. 4 pp. 73–111. Oxford: IRL Press.

BAILEY, J.M. & DAVIDSON, N. 1976. Methylmercury as a reversible denaturing agent for agarose electrophoresis. *Analytical Biochemistry*, **70**, 75–85.

CAMPA, J.S., MCANULTY, R.J. & LAURENT, G.J. 1990. Application of high pressure liquid chromatography to studies of collagen production by isolated cells in culture. *Analytical Biochemistry* **186**, 257–263.

CHIRGWIN, J.M., PRZYBYLA, A.E., MACDONALD, R.J. & RUTTER, W.J. 1979. Isolation of biologically active ribonucleic acid from sources enriched in ribonuclease. *Biochemistry* **18**, 5294–5299.

CHOMCZYNSKI, P. & SACCHI, N. 1987. Single-step method of RNA isolation by acid guanidinium thiocyanate–phenol–chloroform extraction. *Analytical Biochemistry* **162**, 156–159.

GENOVESE, C., ROWE, D. & KREAM, B. 1984. Construction of DNA sequences to Rat α1 and α2 collagen mRNA and their use in studying the regulation of type I collagen synthesis by 1,25-dihydroxyvitamin D. *Biochemistry* **23**, 6210–6216.

LEHRACH, H., DIAMOND, D., WOZNEY, J.M. & BOEDTKER, H. 1977. RNA molecular weight determinations by gel electrophoresis under denaturating conditions, a critical re-examination. *Biochemistry* **16**, 4743–4751.

LIAU, G., YAMADA, Y. & DE CROMBRUGGHE, B. 1985. Coordinate regulation of the levels of type III and I collagen mRNA in most but not all mouse fibroblasts. *Journal of Biological Chemistry* **260**, 531–536.

MCANULTY, R.J. & LAURENT, G.J. 1987. Collagen synthesis and degradation *in vivo*. Evidence for rapid rates of collagen turnover with extensive degradation of newly synthesised collagen in tissues of the adult rat. *Collagen and Related Research* **7**, 93–104.

MCMASTER, G.K. & CARMICHAEL, G.G. 1977. Analysis of single and double-stranded nucleic acids on polyacrilamide and agarose gels by using glyoxal and acridine orange. *Proceedings of the National Academy of Sciences of the United States of America* **74**, 4835–4838.

Detection of Foodborne Bacterial Pathogens by a Colorimetric DNA Hybridization Method

M. Mozola, D. Halbert, S. Chan, H-Y. Hsu, A. Johnson, W. King, S. Wilson[1], R.P. Betts, Pamela Bankes and J.G. Banks[2]

[1] *GENE-TRAK Systems, 31 New York Avenue, Framingham, Massachusetts 01701, USA*;

[2] *Campden Food and Drink Research Association, Chipping Campden, Gloucestershire GL55 6LD, UK*

Historically, isolation and identification of foodborne bacterial pathogens and indicator organisms has been achieved through 'conventional microbiology', that is, enrichment and isolation of bacteria from the test sample using selective/differential culture media and subsequent identification of isolates on the basis of their biochemical and serological characteristics. Whilst alternative, non-traditional methods of analysis (e.g. radioimmunoassay, enzyme immunoassay, automated blood culture and microbial identification systems) have been in routine use in the clinical microbiology laboratory for many years, it is only recently that these technologies have been introduced to the food microbiologist. These emerging rapid alternative methods offer many advantages, including labour savings, increased accuracy, and decreased time of analysis. The development of methods for the microbiological analysis of food, based on these advanced technologies, is currently the focus of considerable effort, and the reader is referred to recent reviews which summarize developments in the field (Pierson & Stern 1986; Hartman *et al.* 1990).

Among the rapid diagnostic technologies recently applied to the microbiological analysis of foods is DNA hybridization. Detailed discussions of the theory of DNA hybridization technology, its diagnostic possibilities, and strategies for the isolation and design of specific DNA probes are presented in other chapters of this volume and in reviews available in the literature (Tenover 1988; Hartman *et al.* 1990).

In the area of detection of pathogenic bacteria in foods, DNA hydridization procedures have been described for the identification of *Salmonella* species (Fitts 1985; Curiale *et al.* 1986), *Listeria* species (Klinger *et al.* 1988; Klinger

0–632–02926–9

& Johnson 1988), *Yersinia enterocolitica* (Hill *et al.* 1983 b; Jagow & Hill 1986), *Listeria monocytogenes* (Datta *et al.* 1988), enterotoxigenic *Escherichia coli* (Hill *et al.* 1983 a, 1986), and *Vibrio vulnificus* (Morris *et al.* 1987). These developments have recently been reviewed (Hartman *et al.* 1990).

In 1985, the first commercially available DNA probe-based diagnostic system for food analysis was introduced (GENE-TRAK® Salmonella Assay, GENE-TRAK Systems, Framingham, Massachusetts). This test system employs *Salmonella*-specific DNA probes directed against chromosomal DNA to detect *Salmonella* in enriched food samples. The test format involves hybridization between target DNA bound to a membrane filter and ^{32}P-labelled probes (Fitts 1985; Curiale *et al.* 1986). Following approximately 40–44 h of sample enrichment in non-selective and selective media, the hybridization procedure requires 4–5 h, resulting in a total analysis time of approximately 48 h, or about half that required to obtain a presumptive result by conventional culture procedures (US Food and Drug Administration (FDA), 1984). The *Salmonella* hybridization method has been extensively evaluated in comparative and interlaboratory collaborative studies, and has been proved to be equivalent and, in some cases superior, in accuracy to standard culture methods (Flowers *et al.* 1987 a,b). A hybridization assay for *Listeria* species in a similar format has also been developed and introduced commercially (Klinger *et al.* 1988; Klinger & Johnson 1988).

Notwithstanding the advantages in analysis time and accuracy offered by the hybridization methods, some potential users have objected to the use of radioisotopes required by the procedures, and thus it was clear that a hybridization method employing a non-isotopic detection system would be highly desirable. This chapter outlines the development and evaluation of a second-generation hybridization assay using an enzyme-mediated colorimetric detection system and its application to the determination of *Salmonella, Listeria*, and *E. coli* in foods. A preliminary description of the test system has recently been published (Chan *et al.* 1989).

Colorimetric DNA Hybridization Assay System

Test chemistry and format

The first-generation hybridization assays rely on the use of radioactively labelled probes directed against organism-specific nucleic acid sequences. The target in the case of the *Salmonella* assay, for example, is chromosomal DNA. This target is present in low copy number (1–4 copies) in each cell. In order to achieve a colorimetric format for the *Salmonella* assay, it was necessary to develop probes against nucleic acid targets that are present in much higher copy number, while retaining a high degree of specificity. Thus, probes that are targeted against regions of ribosomal RNA unique to the genus *Salmonella* have been developed. Ribosomes are ubiquitous components of the cellular protein synthesis machinery and are present in an estimated 5 000–20 000 copies in every bacterial cell. Ribosomal RNAs are the nucleic acid components of ribosomes, and are therefore represented in the same high copy number. In order to develop specific probes directed against these high copy number targets, ribosomal RNAs from a wide variety of salmonellae and non-salmonellae were isolated and the rRNA sequences were determined (Lane *et al.* 1985). Regions unique to the *Salmonella* genus were identified, and synthetic oligonucleotide probes were developed and tested for specificity. Once probes with the appropriate specificity were identified, they were modified and incorporated into the second-generation hybridization format as described below. Similar procedures were followed for the development of rRNA-specific probes for the genus *Listeria* and for *E. coli*.

The colorimetric hybridization assay is based on a liquid hybridization reaction between the target nucleic acid (rRNA) and two oligonucleotide probes (capture probe and reporter probe) specific for the organism of interest. The capture probe is enzymatically tailed at the 3′ end with a homopolymer comprised of approximately 100 deoxyadenosine monophosphate (dA) residues, in order to allow capture of the probe onto a poly(deoxythymidylic acid) (dT)-coated solid support (Morrissey & Collins 1989). The reporter probe is chemically labelled with a fluorescein hapten to allow detection with the enzyme horseradish peroxidase coupled to an anti-fluorescein antibody.

A subsample of the enriched food sample is added to a test tube, and the organisms are lysed either by the addition of NaOH followed by neutralization (*Salmonella* and *E. coli* assays) or by enzyme pretreatment and chemical lysis with guanidinium thiocyanate (*Listeria* assay). A solution containing capture and reporter probes is then added, and hybridization is allowed to proceed. If the specific target nucleic acids (rRNA) are present in the sample, the probes will hybridize with them. This solution containing the target–probe complex is then brought into contact with a solid surface (plastic dipstick) containing bound dT homopolymer, allowing hybridization of the poly(dA) tail on the capture probe to the poly(dT) on the dipstick, and capture of the target–probe complex onto the surface of the dipstick. Unhybridized nucleic acids and cellular debris are washed away, and an anti-fluorescein antibody coupled to horseradish peroxidase is added and allowed to bind to the fluorescein on the reporter probe. Excess conjugate is removed by washing, and the bound complex is detected by the addition of the enzyme substrate H_2O_2 and the chromogen tetramethylbenzidine (TMB). Following a specified colour development period, a stop solution is added, and the colour is measured photometrically. This reading is compared to a negative control and a cut-off value to determine a positive or negative test result. The assay chemistry is shown schematically in Fig. 1. Procedural details are discussed below (see *Assay procedure*).

Applications

In theory, the colorimetric hybridization technology can be applied to the detection of any variety of target pathogenic or indicator bacteria in foods provided that three conditions are satisfied:

1 An enrichment scheme must be defined that is capable of recovering low numbers of potentially injured cells and amplifying this population, in the possible presence of high numbers of competing bacteria, to the detectable titre threshold of 10^5-10^6 cells/ml of final enrichment broth culture.

2 An efficient means of lysing the target organism and releasing its rRNA component must be developed.

3 The rRNA complement of the target organism must contain regions of unique nucleotide sequence such that probes can be produced which exhibit the desired specificity, i.e. the ability to react with all members (species, biotypes, etc.) of the target group while not reacting with genetically related non-target bacteria.

To date, three test methods employing the colorimetric hybridization format have been developed; namely, those for (a) *Salmonella* species in all food product types (Wilson *et al.* 1989, 1990), (b) *Listeria* species in dairy products, meats, poultry, seafood, and environmental samples (King *et al.* 1989 a,b) and (c) *E. coli* in all food product types (Hsu *et al.* 1989). In addition, a method for the determination of *Yersinia enterocolitica* in foods has been developed to the prototype stage (Chan *et al.* 1988). While it is not possible here to discuss the procedural details, potential applications, specificity and performance characteristics of each method in detail, examples will be given in the aforementioned areas to illustrate the basic approach, advantages, and limitations of this technology.

Enrichment methods

In the microbiological analysis of foods, low numbers of organisms must be recovered from the test samples. In the case of known pathogens such as *Salmonella* or *Listeria*, contamination levels as low as 1 cell/25 or 50 g of sample are of significance. All currently available analytical techniques for food analysis thus require a period of cultural enrichment of the test sample in

FIG. 1. Colorimetric DNA hybridization assay chemistry. Cross-hatched area denotes polystyrene dipstick solid support, . . . AAA . . . = polydeoxyadenylic acid, . . . TTT . . . = polydeoxythymidylic acid, FL = fluorescein, anti-FL = anti-fluorescein antibody, HRP = horseradish peroxidase.

a) Sample lysis

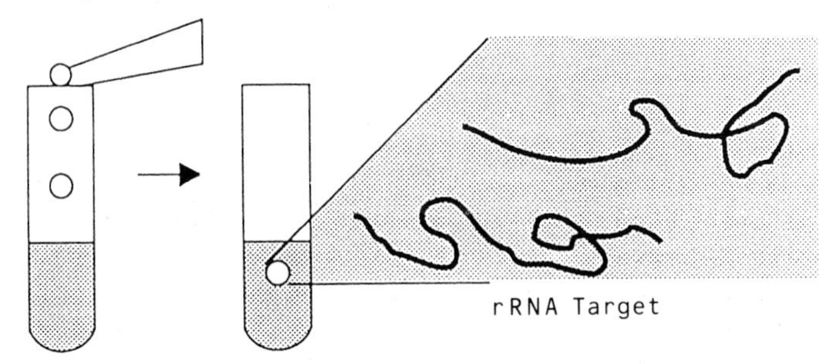

b) Hybridization

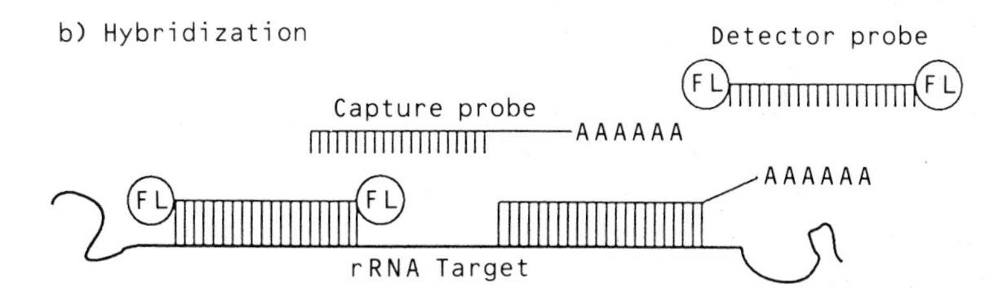

c) Capture

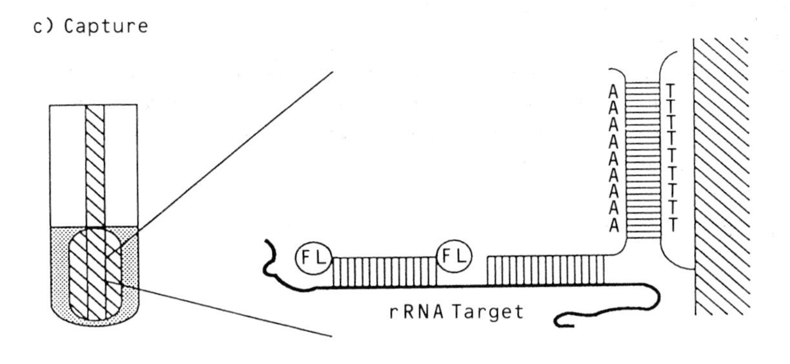

d) Detection

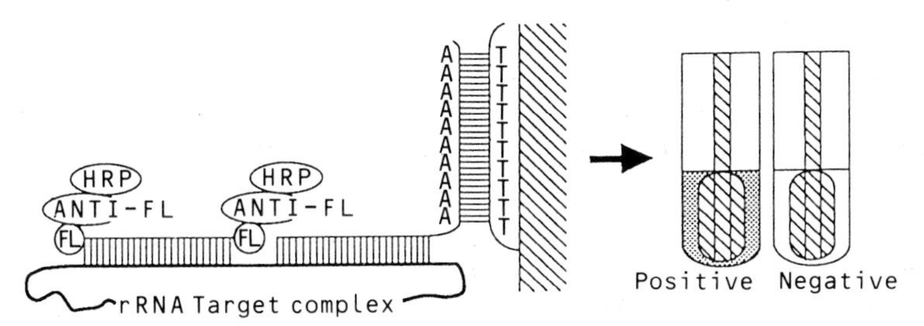

order to allow multiplication of these low numbers of target bacteria to detectable levels. Enrichment procedures used in conjunction with rapid detection methods such as immunoassay or DNA hybridization are usually derived from those of conventional microbiological isolation procedures.

The enrichment schemes used in conjunction with the colorimetric hybridization methods are adapted, in large part, from those recommended by regulatory agencies in the USA such as the Food and Drug Administration (US FDA 1984). Enrichment is usually accomplished in a two or three stage procedure with selective broth media. In cases such as *Salmonella* analysis, where low moisture foods may contain injured organisms requiring a recovery period, a non-selective primary enrichment is used before exposure of the sample to selective media. Recommended enrichment protocols for use with the colorimetric hybridization assays are summarized in Table 1.

The most novel enrichment procedure is that developed for the *Listeria* assay, in which the secondary enrichment is performed on a solid agar selective medium (LPM agar) rather than in broth culture. In comparative studies on artificially and naturally contaminated foods, approximately 16% more positives were detected by hybridization assays conducted on cell resuspensions from secondary LPM plate enrichments than from hybridization assays performed on traditional secondary broth culture enrichments (unpub-

TABLE 1. *Recommended enrichment procedures for use with the colorimetric hybridization assays*

	Enrichment stage		
Determination	Primary	Secondary	Tertiary
E. coli	Lauryl sulphate typtose broth, 35°C, 24 h	Lauryl sulphate typtose broth, 35°C, 24 h	—
Salmonella	Lactose broth*, 35°C, 24 h	Tetrathionate broth and selenite cystine broth, 35°C, 6 h†	Gram-negative broth, 35°C, 12–18 h‡
Listeria	Modified *Listeria* enrichment broth or UVM broth§ 35°C, 24 h	LPM plate, 35°C 24 h	—

* Or other pre-enrichment medium as appropriate for sample type (US FDA 1984).
† 16–18 h for raw meat and raw milk products.
‡ 6 h for raw meat and raw milk products.
§ Modified *Listeria* enrichment broth (MEB) = LEB + morpholinopropane sulphonic acid (Klinger *et al.* 1988). MEB is used for primary enrichment of dairy product samples, UVM is used for primary enrichment of all other sample types.

lished results). It should be noted that while the recommended enrichment procedures outlined in Table 1 have been validated and shown to be effective, many alternatives are possible; for example, the use of buffered peptone water rather than lactose broth as a pre-enrichment medium for *Salmonella*, or the use of alternative *Salmonella* selective enrichment media or different incubation temperatures. As mentioned previously, the primary requirement that must be met by any enrichment scheme proposed for use in conjunction with the colorimetric hybridization method is that it must be capable of recovering low numbers of target organisms from the test sample and allowing multiplication of the target organisms to a final titre of 10^5-10^6 cells/ml. The titre of competitor organisms in the final culture is not important and can be in vast excess to that of the target organism as long as the 10^5-10^6/ml titre requirement for the target organism is satisfied.

Assay procedure

The colorimetric hybridization assays use a common test format, with only relatively minor procedural variations incorporated to satisfy the unique requirements of the various determinations (e.g., different requirements for cell lysis, hybridization conditions, etc.). The reader is referred to the individual assay package inserts for detailed information regarding reagent formulations and test procedures. As an example, the test procedure for the *Salmonella* assay is given below.

1 Add 0.5 ml test sample (final enrichment broth(s)) to a 12 × 75 mm glass test tube. Run one positive control and one negative control (supplied as formaldehyde-inactivated preparations of *S. typhimurium* and *Citrobacter freundii*) with each set of samples.

2 Add 0.1 ml lysis solution. Mix, and incubate for 5 min at room temperature.

3 Add 0.1 ml neutralization solution. Mix.

4 Cover the tubes with aluminum foil and incubate statically for 15 min in a waterbath at 65°C.

5 Add 0.1 ml *Salmonella* probe solution. Mix, then cover the tubes with foil again and incubate for 15 min at 65°C.

6 Place dipsticks into the sample tubes and incubate for 1 h at 65°C.

7 Remove the dipsticks from the sample tubes and wash the dipsticks sequentially for 1 min each, first in wash solution at 65°C, then in wash solution at room temperature.

8 Blot the dipsticks on absorbent paper, then place the dipsticks into a second set of 12 × 75 mm glass test tubes containing 0.75 ml enzyme conjugate. Incubate for 20 min at room temperature.

9 Remove the dipsticks from the enzyme conjugate and wash them sequen-

tially for 1 min each in two basins of wash solution at room temperature.

10 Blot the dipsticks on absorbent paper, then place the dipsticks into a third set of 12 × 75 mm glass test tubes containing 0.75 ml substrate-chromogen. Incubate for 20 min at room temperature.

11 Remove the dipsticks from the tubes and discard.

12 Add 0.25 ml stop solution to each tube containing substrate-chromogen (include a blank tube containing substrate-chromogen and stop solution). Mix.

13 Determine the absorbance at 450 nm of the positive and negative controls and each test sample using the GENE-TRAK photometer or other appropriate instrument. Read the positive and negative controls against the substrate-chromogen–stop solution blank. Read the test samples against the negative control.

14 *Interpretation*: the A_{450} of the positive control should be ≥ 1.00. The A_{450} of the negative control should be ≤ 0.15. A test sample $A_{450} > 0.10$ indicates a positive assay result.

15 *Confirmation*: positive hybridization assays may be confirmed, if desired, by streaking the final enrichment broth culture(s) onto *Salmonella* selective/differential agar media and continuing with biochemical and serological identification of presumptive *Salmonella* isolates using standard procedures (US FDA 1984).

The test procedure for the *E. coli* assay is virtually identical to that of the *Salmonella* test. Most reagents are common to both systems, and procedural variations are limited to some minor differences in incubation times and wash temperatures. More extensive variations to the basic procedure are required for the *Listeria* assay. Complete lysis of Gram-positive bacteria cannot be accomplished simply with the type of alkaline denaturation used in the *Salmonella* and *E. coli* assays, and thus an alternative lysis procedure incorporating enzymatic pretreatment is used. The lysis reagents used in the *Listeria* assay contain the enzymes lysozyme, mutanolysin and proteinase-K, which degrade the bacterial cell wall and membrane structures. Following this pretreatment, complete lysis is accomplished by the addition of the chaotropic salt, guanidinium thiocyanate, contained in the *Listeria* probe solution. Guanidinium thiocyanate has the effect of destabilizing the hydrogen bonding between the complementary strands of double-stranded nucleic acid molecules, and thus inclusion of this denaturing agent in the hybridization mixture requires that the hybridization reaction be conducted at 37°C rather than 65°C in order to maintain the necessary degree of stringency. Steps in the *Listeria* assay following hybridization are essentially the same as those for the other two tests, again with the exception of some minor variations in incubation times and wash temperatures.

Test performance

Probe specificity characteristics

Any effective diagnostic method must, of course, exhibit the desired specificity characteristics, that is, the ability to detect all members within the target group of organisms while not producing positive reactions with organisms that do not belong to the target group. During the development of the colorimetric hybridization assays, individual probes were screened by colony blot hybridization (Grunstein & Hogness 1975) against a large collection of target and non-target bacterial strains to provide a preliminary assessment of their specificity. This was followed by testing the final probe set against a similar collection of organisms in the actual assay. Results of specificity testing of the probe set used in the *Listeria* hybridization assay have been reported. The assay produced positive results with all 291 strains of *Listeria* tested and negative results with all 65 non-*Listeria* strains examined, including representatives of genetically related bacterial genera such as *Brochothrix*, *Jonesia*, *Enterococcus* and *Streptococcus* (King *et al.* 1989 a,b). An interesting example of the potential of the hybridization method is provided by examining the specificity characteristics of the colorimetric *E. coli* hybridization assay. Results of specificity testing are shown in Table 2. In this experiment, comparative data

TABLE 2. *Specificity testing of the* E. coli *hybridization assay and two alternative methods*

Biotype category	Number of isolates	Number of positives		
		Hybridization*	Gas from lactose†	MUG reaction‡
Typical	158	158	148	145
Enterotoxigenic	16	16	14	11
Enteroinvasive	6	6	6	5
Haemorrhagic (0157:H7)	11	11	11	0
Other enteropathogenic	16	16	16	16
Total	207	207	195	177

* Test strains were inoculated into LB-broth and incubated at 35°C for 16–24 h. Hybridization assays were performed on 1:100 dilutions of the overnight cultures (approximately 10^7 cells). Assays producing an A_{450} greater than 0.2 are considered positive.

† Test strains were inoculated into LST-MUG-broth, incubated at 35°C for 48 h and read for gas production.

‡ Test strains were inoculated into LST-MUG-broth, incubated at 35°C for 48 h and read for fluorescence.

were also generated for two alternative *E. coli* methods, a standard culture procedure which uses the production of gas from lactose fermentation as the primary indicator reaction (US FDA 1984), and a rapid method based on the conversion of a substrate to a visually discernible fluorescent end-point by the action of the enzyme β-glucuronidase (MUG assay; Feng & Hartman 1982; Moberg *et al.* 1988). The hybridization method produced positive reactions with all 207 *E. coli* strains tested, whereas the standard culture method and MUG assay produced negative results with several strains (Table 2). *E. coli* exhibit considerable strain-to-strain variability in their biochemical profiles; some do not ferment lactose with the subsequent production of gas (Ewing 1986) and others apparently do not possess active β-glucuronidase (Kilian & Bulow 1976; Doyle & Schoeni 1984). Such isolates will, *a priori*, go undetected by methods based on these indicator reactions. It is particularly noteworthy that isolates of haemorrhagic *E. coli* serotype 0157:H7, a serious human foodborne pathogen (Doyle & Padhye 1989), are consistently negative in the MUG assay (Doyle & Schoeni 1984). In addition, other pathogenic strains of *E. coli* exhibit atypical biochemical behaviour (reviewed in US FDA 1984; Doyle & Padhye 1989), and thus methods based on biochemical indicator reactions may not be appropriate for screening of foods in which the possible presence of pathogenic biotypes of *E. coli* is of concern.

Analysis of foods

Considerable data have been generated on the performance of the colorimetric hybridization methods relative to those of standard culture procedures. The reader is referred to several reports describing these findings (King *et al.* 1989 a,b; Chan *et al.* 1989, 1990; Wilson *et al.* 1990; Curiale *et al.* 1990). The hybridization methods for *Salmonella* and *Listeria* have undergone preliminary evaluation at the Campden Food and Drink Research Association and the results of these studies are summarized here.

Table 3 presents results of a comparison of the *Salmonella* hybridization method and a conventional European culture procedure. A variety of sample types were analysed and several *Salmonella* serovars were used as inocula. The enrichment procedures used in conjunction with the hybridization method were essentially those recommended (see Table 1) except that the selective enrichment period was 18 h and the post-enrichment period in Gram-negative broth was 6 h for all sample types. In addition, buffered peptone water was used as the pre-enrichment medium instead of lactose broth for some samples. Results of the analysis showed equivalence of the hybridization method and the standard culture procedure; each method detected 56 positive samples although there were some discrepancies in the number of positives obtained by the two methods within individual sample categories. The use of buffered

TABLE 3. *Performance of the* Salmonella *hybridization method and a standard culture procedure*

Sample type	Number of samples	Number inoculated*	Number of positives	
			Hybridization	Culture†
Skim milk powder	25	21	21	18
Raw chicken	15	13	12	10
Formed chicken	21	18	11	16
Raw egg	12	9	9	9
Chocolate	4	3	3	3
Other	4	0	0	0
Total	81	64	56	56

* Inoculum strains were *S. ealing*, *S. typhimurium*, *S. enteritidis*, *S. napoli*, *S. pullorum*, *S. senftenberg*, *S. infantis*, and *S. virchow*. Inoculation levels ranged from 1×10^0 to 3.7×10^5 cells per 25 g sample.

† Pre-enriched samples were inoculated into Muller-Kauffmann tetrathionate broth, selenite cystine broth and Rappaport-Vassiliadis broth and incubated at 37°C (MK, Sel) or 42°C (RV) for 24 and 48 h. Selective enrichment cultures were streaked to xylose lysine desoxycholate, bismuth sulphite and Hektoen enteric agars and incubated at 37°C for 24 h.

peptone water as an alternative pre-enrichment medium proved to be effective (data not shown).

Table 4 shows the results of a preliminary study comparing the performance of the *Listeria* hybridization assay with that of a standard culture method. Four product types were examined and a variety of *Listeria* species were employed as inocula. Common primary enrichments were used for the hybridization and culture analysis; these were performed essentially as described in Table 1. Two selective agars were evaluated as secondary enrichment media for use in conjunction with the hybridization method (LPM and Oxford medium) and two selective media were compared in the conventional culture analysis (modified McBride agar and Oxford medium). Results showed that the hybridization and standard culture methods were equivalent in performance. The Oxford medium provided for a significant enhancement of performance of both the hybridization and culture methods compared with the other selective media evaluated. It must be noted that these results are preliminary, but it would appear that a rigorous assessment of the use of the Oxford medium in conjunction with the hybridization assay is warranted.

Conclusions

A rapid DNA hybridization method incorporating a colorimetric detection system has been developed and applied to the determination of bacterial

TABLE 4. *Performance of the* Listeria *hybridization method and a standard culture procedure*

Sample type	Number of samples	Number inoculated*	Number of positives			
			Hybridization		Culture†	
			LPM	Oxford	McBride	Oxford
Brie cheese	13	12	4	8	4	8
Raw milk	4	3	0	4	1	4
Raw chicken	7	6	1	7	0	7
Cooked chicken	4	3	3	3	3	3
Total	28	24	8	22	8	22

* Inoculum strains were *L. ivanovii*, *L. monocytogenes* serotype 4b, *L. monocytogenes* serotype 1/2a, *L. monocytogenes* serotype 1/2b and *L. innocua*. Inoculation levels ranged from 1×10^0 to 3.6×10^4 cells per 25 g sample.

† Modified *Listeria* enrichment broth was used as the primary enrichment medium for brie cheese and raw milk, UVM broth was used for raw and cooked chicken. Incubation was at 35°C for 24 h.

pathogens and indicator organisms in foods. The assay system utilizes solution-phase hybridization of synthetic oligonucleotide probes directed against ribosomal RNA of the target organism, solid-phase capture of probe target hybrids, and an enzyme-mediated detection system resulting in a colorimetric end-point that is determined photometrically. The assay chemistry and test format are generic in that, given the appropriate enrichment conditions for the target groups of interest and the appropriate specific DNA probes, the test system can in theory be applied to any bacterial foodborne pathogen or indicator organism of interest. Hybridization methods for *Salmonella*, *Listeria*, and *E. coli* have been developed from the basic test procedure. The specificity characteristics of the probes used in the assays have been established, and the performance of the methods with food samples has been assessed relative to that of conventional culture procedures. The hybridization assays have been shown to be at least equivalent in accuracy to standard culture methods, and offer the significant advantage of decreased time of analysis compared to the requirements of microbiological culture procedures. Future work with the colorimetric hybridization test system will focus on increasing the sensitivity of the assay chemistry and defining more effective enrichment schemes for use with the various methods, with the ultimate goal of further improvements in performance and reduction of analysis time.

References

CHAN, S., PITMAN, T., SHAH, J., KING, W., LANE, D. & LAWRIE, J. 1988. Non-radioactive DNA assay for detection and identification of foodborne *Yersinia*. *Abstracts of the 88th Annual*

Meeting of the American Society for Microbiology. p. 280.

Chan, S.W., Wilson, S., Hsu, H.-Y., King, W., Halbert, D.N. & Klinger, J.D. 1989. Model non-isotopic hybridization systems for detection of foodborne bacteria: preliminary results and future prospects. In *Proceedings of the International Symposium on Biotechnology and Food Quality*, ed. Kung, S.-D. Boston: Butterworth.

Chan, S.W., Wilson, S., Vera-Garcia, M., Whippie, K., Ottaviani, M., Whilby, A., Shah, A., Johnson, A., Mozola, M.A. & Halbert, D.N. 1990. Comparative study of a colorimetric DNA hybridization method and the conventional culture procedure for detection of *Salmonella* in foods. *Journal of the Association of Official Analytical Chemists* **73**, 419–424.

Curiale, M.S., Flowers, R.S., Mozola, M.A. & Smith, A.E. 1986. A commercial DNA probe-based diagnostic for the detection of *Salmonella* in food samples. In *DNA Probes: Applications in Genetic and Infectious Disease and Cancer*, ed. Lerman, L.S. pp. 143–148. Cold Spring Harbor, New York: Cold Spring Harbor Laboratory.

Curiale, M.S., Klatt, M.J. & Mozola, M.A. 1990. Colorimetric deoxyribonucleic acid hybridization assay for rapid screening of *Salmonella* in foods: collaborative study. *Journal of the Association of Official Analytical Chemists* **73**, 248–256.

Datta, A.R., Wentz, B.A., Shook, D. & Trucksess, M.W. 1988. Synthetic oligodeoxyribonucletide probes for detection of *Listeria monocytogenes*. *Applied and Environmental Microbiology* **54**, 2933–2937.

Doyle, M.P. & Padhye, V.V. 1989. *Escherichia coli*. In *Foodborne Bacterial Pathogens*, ed. Doyle, M.P. pp. 235–281. New York: Marcel Dekker Inc.

Doyle, M.P. & Schoeni, J.L. 1984. Survival and growth characteristics of *Escherichia coli* associated with hemorrhagic colitis. *Applied and Environmental Microbiology* **48**, 855–856.

Ewing, W.H. 1986. *Edwards and Ewing's Identification of Enterobacteriaceae*. 4th edition, New York: Elsevier Science Publishing Co. Inc.

Feng, P.C.S. & Hartman, P.A. 1982. Fluorogenic assays for immediate confirmation of *Escherichia coli*. *Applied and Environmental Microbiology* **43**, 1320–1329.

Fitts, R. 1985. Development of a DNA-DNA hybridization test for the presence of *Salmonella* in foods. *Food Technology* **39**, 95–102.

Flowers, R.S., Mozola, M.A., Curiale, M.S., Gabis, D.A. & Silliker, J.H. 1987 a. Comparative study of a DNA hybridization method and the conventional culture procedure for detection of *Salmonella* in foods. *Journal of Food Science* **52**, 842–845.

Flowers, R.S., Klatt, M.J., Mozola, M.A., Curiale, M.S., Gabis, D.A. & Silliker, J.H. 1987b. DNA hybridization assay for detection of *Salmonella* in foods: collaborative study. *Journal of the Association of Official Analytical Chemists* **70**, 521–529.

Grunstein, M. & Hogness, D.S. 1975. Colony hybridization: method for the isolation of cloned DNAs that contain a specific gene. *Proceedings of the National Academy of Sciences of the USA* **72**, 3961–3965.

Hartman, P.A., Swaminathan, B., Curiale, M.S., Eden, R., Sharpe, A.N., Cox, N.A., Fung, D.Y.C. & Goldschmidt, M.C. 1990. Rapid methods and automation. In *Compendium of Methods for the Microbiological Examination of Foods*, 3rd edn, ed. Vanderzant, C. Washington D.C.: American Public Health Association. (In press).

Hill, W.E., Madden, J.M., McCardell, B.A., Shah, D.B., Jagow, J.A., Payne, W.L. & Boutin, B.K. 1983a. Foodborne enterotoxigenic *Escherichia coli*: detection and enumeration by DNA colony hybridization. *Applied and Environmental Microbiology* **45**, 1324–1330.

Hill, W.E., Payne, W.L. & Aulisio, C.C.G. 1983b. Detection and enumeration of virulent *Yersinia enterocolitica* in food by colony hybridization. *Applied and Environmental Microbiology* **46**, 636–641.

Hill, W.E., Wentz, B.A., Jagow, J.A., Payne, W.L. & Zon, G. 1986. DNA colony hybridization method using synthetic oligonucleotides to detect enterotoxigenic *Escherichia coli*: collaborative study. *Journal of the Association of Official Analytical Chemists* **69**, 531–536.

HSU, H-Y., SOBELL, D.I., CHAN, S.W., MCCARTY, J., PARODOS, K., LANE, D.J. & HALBERT, D.N. 1989. A colorimetric DNA hybridization method for the detection of *Escherichia coli* in foods. *Abstracts of the 89th Annual Meeting of the American Society for Microbiology*. p. 322.

JAGOW, J. & HILL, W.E. 1986. Enumeration by DNA colony hybridization of virulent *Yersinia enterocolitica* colonies in artifically contaminated food. *Applied and Environmental Microbiology* **51**, 441–443.

KILIAN, M. & BULOW, P. 1976. Rapid diagnosis of Enterobacteriaceae. I. Detection of bacterial glycosidases. *Acta Pathologica et Microbiologica Scandinavica* **84b**, 245–251.

KING, W., RAPOSA, S., WARSHAW, J., JOHNSON, A., HALBERT, D. & KLINGER, J.D. 1989a. A new colorimetric nucleic acid hybridization assay for *Listeria* in foods. *International Journal of Food Microbiology* **8**, 225–232.

KING, W., RAPOSA, S.M., WARSHAW, J.E., JOHNSON, A.R., LANE, D., KLINGER, J.D. & HALBERT, D.N. 1989b. A colorimetric assay for the detection of *Listeria* using nucleic acid probes. In *Proceedings of the Comprehensive Conference on* Listeria monocytogenes, ed. Miller, A.J. pp. 117–124. Society for Industrial Microbiology.

KLINGER, J.D. & JOHNSON, A.R. 1988. A rapid nucleic acid hybridization assay for *Listeria* in foods. *Food Technology* **42**, 66–70.

KLINGER, J.D., JOHNSON, A., CROAN, D., FLYNN, P., WHIPPIE, K., KIMBALL, M., LAWRIE, J. & CURIALE, M. 1988. Comparative studies of a nucleic acid hybridization assay for *Listeria* in foods. *Journal of the Association of Official Analytical Chemists* **71**, 669–673.

LANE, D.J., PACE, B., OLSEN, G.J., STAHL, D.A., SOGIN, M.L. & PACE, N.R. 1985. Rapid determination of 16S ribosomal RNA sequences for phylogenetic analyses. *Proceedings of the National Academy of Sciences of the United States of America* **82**, 6955–6959.

MOBERG, L.J., WAGNER, M.K. & KELLEN, L.A. 1988. Fluorogenic assay for rapid detection of *Escherichia coli* in chilled and frozen foods: collaborative study. *Journal of the Association of Official Analytical Chemists* **71**, 589–602.

MORRIS, J.G., WRIGHT, A.C., ROBERTS, D.M., WOOD, P.K., SIMPSON, L.M. & OLIVER, J.D. 1987. Identification of environmental *Vibrio vulnificus* isolates with a DNA probe for the cytotoxin-hemolysin gene. *Applied and Environmental Microbiology* **53**, 193–195.

MORRISSEY, D.V. & COLLINS, M.L. 1989. Nucleic acid hybridization assays employing dA-tailed capture probes. Single capture methods. *Molecular and Cellular Probes* **3**, 189–207.

PIERSON, M.D. & STERN, N.J., eds. 1986. *Foodborne Microorganisms and their Toxins: Developing Methodology*, New York: Marcel Dekker Inc.

TENOVER, F.C. 1988. Diagnostic deoxyribonucleic acid probes for infectious diseases. *Clinical Microbiology Reviews* **1**, 82–101.

U.S. FOOD AND DRUG ADMINISTRATION 1984. *Bacteriological Analytical Manual*, 6th edn, Association of Official Analytical Chemists, Arlington, Virginia.

WILSON, S., CHAN, S., DE ROO, M., VERA-GARCIA, M., LANE, D., NIETUPSKI, R., PITMAN, T., SHAH, J. & HALBERT, D.N. 1989. A rapid, colorimetric DNA hybridization method for the detection of *Salmonella* in foods. *Abstracts of the 89th Annual Meeting of the American Society for Microbiology*. p. 323.

WILSON, S.G., CHAN, S., DE ROO, M., VERA-GARCIA, M., JOHNSON, A., LANE, D. & HALBERT, D.N. 1990. Development of a second generation colorimetric nucleic acid hybridization method for detection of *Salmonella* in foods and a comparison with conventional culture. *Journal of Food Science* (In press).

Use of a DNA Probe for a Viroid in Plant Pathology

P.S. Harris and C.M. James
Department of Agriculture and Fisheries for Scotland, Agricultural Scientific Services, East Craig, Edinburgh EH12 8NJ, UK

Viroids are small, circular, extensively base-paired single-stranded RNA pathogens with no gene product. They infect only higher plants and can seriously damage propagation of coconut, avocado, citrus and potato. Chrysanthemum, hops and vine propagation can be affected (Singh 1984). Potato spindle tuber viroid (PSTV) is efficiently transmitted in potato true seed and is known to contaminate world potato genetic resources. There is an obligation, under an EC decision, and a strong technical case for protecting UK Potato breeding and clean stock programmes. Tests for PSTV have been conducted by the Department of Agriculture and Fisheries for Scotland (DAFS) for two decades (Harris & James 1987).

Viroid concentration in tissue can vary and maximum sensitivity is required for avocado sunblotch viroid (ASBV) (Barker *et al.* 1985) and for PSTV under some conditions (Harris & James 1987). A concentration range of $\times$ 10 000 for PSTV has been estimated (Schumacher *et al.* 1986). Reliable detection methods have been sought. Bioassay is still useful (Grasmick & Slack 1985) and biological amplification in tomato after inoculation of sample nucleic acids (NA) can be combined with subsequent polyacrylamide gel-electrophoresis (PAGE) NA visualization to provide a sensitive test of composite samples (Morris & Smith 1977; Harris & Miller-Jones 1981). Return gel-electrophoresis offers sensitive direct detection of viroid in individual samples (Schumacher *et al.* 1986) if high sensitivity silver staining is used.

Viroids have been an early target for NA probes (Owens & Diener 1981). High sensitivity has been reported for oligonucleotide primed high activity radiolabelled bacteriophage M13 probes for ASBV with full length viroid DNA inserts (Barker *et al.* 1985). Apparently much lower sensitivity has been reported for probes for PSTV (Macquaire *et al.* 1984; Singh & Boucher 1988). At DAFS we have used a PSTV DNA probe obtained from the US Department of Agriculture (USDA) in 1983 (Harris *et al.* 1988).

In 1987 a modified method for use of a DNA probe superseded previous

Genetic Manipulation
0–632–02926–9

methods and further modifications are described here. Hybridizing conditions suitable for high activity radiolabelled probing of samples on nylon filters have been devised and sensitivities reported here are comparable with those for M13 probes for ASBV (above) and RNA probes for PSTV (Lakshman *et al.* 1986). Nylon has had a significant advantage over nitrocellulose for probing viroid in nucleic acid extracts. Methods are described for reducing serious signal blocking caused by sap and nucleic acid extracts.

Ideally, for large-scale testing, sample manipulation should be kept to a minimum. Electrophoresis requires sample extraction and, even for NA probing, samples must be extracted to reduce non-specific bonding and to increase signal strength. Full NA extraction of samples followed by resuspension in small volumes allows a concentration step which increases potential sensitivity. We report the use of a NA fractionation method (Skrzeczkowski *et al.* 1985) which increases the signal both by concentration and by reduction of signal blocking. Composite samples can be tested without compromising sensitivity and larger throughputs of tests, up to 100 × 25 potato samples, may be performed in a week by one person.

Materials and Methods

Phytosanitary arrangements, the sources of viroid and inoculation and growing of tomato and potato sources are essentially as previously described (Harris & Miller-Jones 1981).

Sample sap extraction

Clarified sap extracts are prepared and denatured as follows:

1 Leaf discs or microplant pieces (0.4 g) are thoroughly ground with 0.6 ml AMESS buffer (Palukaitis *et al.* 1985) in a mortar and pestle with a trace of acid-washed sand (AMESS buffer, stored at −18°C, requires warming before use: 0.5 M sodium acetate, pH 6.0; 10 mM $MgCl_2$; 20% (v/v) ethanol; 3% sodium dodecyl sulphate (SDS); 1.0 M NaCl. The authors use 90% ethanol (methylated spirit)).

2 Chloroform (0.8 ml) at room temperature is stirred in and the whole sample is transferred to a 1.5 ml microfuge tube. Incubation is omitted.

3 Centrifugation is in a microfuge for 5 min at room temperature.

4 Formaldehyde (37%) is mixed 2:3 with 20 × standard saline citrate (3 M NaCl; 0.3 M sodium citrate, pH 7.0) and stored at room temperature.

5 An aliquot of supernatant (50 μl) is transferred to a fresh tube containing 50 μl of formaldehyde/SSC and mixed by vortexing.

6 Denaturing is at a final concentration of 7.5% formaldehyde and 6 ×

SSC at 60°C for 15 min followed by snap-cooling in ice water (Skrzeczkowski *et al.* 1985: following White & Bancroft 1982).

Smaller tissue quantities are prepared in a 1.5 ml microfuge tube by grinding in proportional quantities of buffer with a spatula-ended glass rod which has been ground to fit the bottom of the tube.

Sample nucleic acid (NA) extraction

Nucleic acid extraction and a 2-step precipitation yield a low molecular weight RNA fraction (Skrzeczkowski *et al.* 1985). Extraction of 1 or 2 g of fresh tissue is as follows (volumes are adjusted accordingly):

1 Tissue (1 g) is squashed quickly with 0.1 ml 10% SDS with a trace of acid-washed sand in a mortar and pestle.

2 Rapidly, 0.9 ml buffer (3.5 M LiCl, 0.3 M glycine, pH 9.5) is added and the tissue is finely-ground.

3 Phenolics (2 ml) (500 g phenol, 500 ml chloroform, 20 ml octanol) is stirred in.

4 The sample is transferred to a 16 ml polypropylene tube for centrifugation at 4°C for 15 min at 10 000 rpm in a Sorval SM24 rotor.

5 A 1.0 ml sample of supernatant is transferred to a 1.5 ml microfuge tube containing enough finely-ground PEG 6 000 to achieve a final concentration after mixing of 10% (w/v) (for speed the authors measure PEG by volume with a calibrated scoop made from the bottom of a microfuge tube, delivering about 0.12 g).

6 Tubes are stored at 4°C for 30–60 min before centrifugation for 15 min at room temperature.

7 The supernatant is retained and made to more than 20% (w/v) PEG at final concentration and the pellet, collected by centrifugation as above, is carefully washed with ethanol/diethyl ether at −20°C (ethanol is mixed 50:50 with ether and stored at −20°C).

A fine-nosed plastic 'pastette' is used to deliver and retrieve the wash. The pellet of low molecular weight RNA may detach without problem from the tube during the wash. All viscous material must be removed by the washes.

8 The pellet is dried *in vacuo* and stored at −20°C or resuspended in 25 μl water or buffer for use.

A single-stage PEG precipitation is used to provide a 'total NA' extract by raising PEG to 20% in the first instance and collecting and washing the pellet as above and resuspending it in 50 μl water or buffer.

Small quantities of tissue (0.3 g) are finely-ground with buffer in microfuge tubes using the glass rods described above. Phenolics are added, the tubes capped and then vortexed.

Sample transfer to nylon filter

Samples are transferred to charge-modified nylon 66 (Gelman Biotrace RP) and 2 μl spots are placed at a density of 96 spots per 5 × 5 cm filter; no vacuum or pretreatment of the filter is made. The filter is air-dried and subjected to u.v. irradiation on the surface of a Trans-illuminator for 4–5 min. A large batch of identical small filters has been prepared with dilution series in ten-fold steps of sap extracts from infected and non-infected tomato tissue and stored dry at 4°C. One of these is included with all filter hybridizations.

Probe preparation

M13 DNA with a full length 359 nucleotide insert of PSTV DNA was provided by R.A. Owens, USDA, in 1983, and used to transform *E. coli* JM101. A large-scale preparation was made in 1986 (Harris *et al.* 1988; following Barker *et al.* 1985) and is still in use.

The DNA is labelled across the insert and denatured before use by boiling for 5 min followed by quick chilling in ice water:

1 Single-stranded M13 preparation (12 μl) of DNA (250 ng) is annealed with 4 μl of primer ($TC_3AGTCACGACGT$ (Boerhringer) at 2.5 mg/ml); 4 μl anneal buffer (100 mM Tris–Cl pH 8.0–8.5, 100 mM $MgCl_2$) and 20 μl water at 45°C for 1 h.

2 The mixture is slow-cooled in the dri-block to 35°C and then further mixed with 16 μl of 1 mM dATP, 1 mM dGTP, 1 mM dTTP, 12 μl [Alpha ^{32}P]dCTP, 3 000 Ci/mmol (up to 4.5 MBq), 1.2 μl Klenow fragment of DNA polymerase-I (4.5–6 units/μl; Pharmacia), 11 μl water and heated at 45°C for 1 h.

3 The labelled probe is separated on a Sephadex G-50 (medium grade) spun column in a swing out rotor (4 min 1 600 ***g***). It is important not to use fine grade (DNA grade) G-50 on a spun column.

Prehybridization and hybridization

Protocols for prehybridization and hybridization are derived from methods of K.J. Hardy, California Medical School (published by Hoeffer Instruments) with major differences in the molarity and concentration of dextran sulphate in the hybridization buffer. Prehybridization is in 1 M sodium chloride and

1% sodium dodecyl sulphate for 3 h or overnight at 65°C. Hybridization is done in 5% dextran sulphate, 0.2 M sodium chloride, 1% SDS; 0.1 mg/ml boiled and sonicated herring sperm DNA for 20 h at 70°C at 20 ml for every four 5 × 5 cm filters. Washings are done in saline phosphate EDTA (SSPE): 1 × SSPE is 150 mM sodium chloride, 10 mM sodium phosphate, 1 mM EDTA, pH 7.4. The low stringency wash is 2 × SSPE, 1% SDS and the high stringency wash is 0.1 × SSPE, 1% SDS. Washing consists of one rapid swirl of high stringency wash at room temperature in the hybridizing dish after probe removal, followed by one low stringency wash for 15 min at 70°C and two high stringency washes for 15 min at 70°C. Typically, 10 ml of each wash is used for each 5 × 5 cm filter.

Prehybridization, hybridization and washing are performed in glass dishes sealed with a doubled plastic (Saran) wrap. Smaller 5 × 5 cm filters are placed in less buffer in disposable square plastic weigh boats fitted in the glass dishes. Access for insertion and withdrawal of hybridizing buffer and washes is made through a slit in the plastic wrap which is sealed with sticky tape. Insertions and withdrawals are made with 5 ml transfer pipette disposable tips. Up to four glass dishes are laid in a plastic dish in a New Brunswick G24 bench incubator shaker (modified to withstand 70°C) and shaking is achieved at low speed.

Autoradiography is typically for 20 h or 68 h at −30°C with Kodak X-O mat film and two intensifying screens.

Results

Probe activity

After incorporation of 60% of 4.5 MBq on 250 ng of M13PSTV DNA, we estimate a maximum specific activity of 6.5×10^8 cpm/μg DNA. At this probe activity, weekend exposure began to raise filter background on the autoradiograph even at −30°C; the filter background is otherwise satisfactory.

Detection of M13PSTV DNA

DNA concentrations estimated by A_{260} measurement and calculated thereafter from the dilution series are detected on a nylon filter to a limit of 0.5 pg per 2 μl spot. Absorbance ratio A_{260} to A_{280} is close to 1.8 for M13PSTV DNA.

Detection in clarified sap extracts

A standard filter is included with all routine and experimental probings. Clarified sap extracts provide two dilution series in ten-fold steps; one series diluted in water, the other in healthy clarified sap (viroid infection diluted, sap concentration held constant). A denatured replicate of the two series is also provided. Signal strength varied with probe specific activity, within limits defined by less than one ten-fold step in the dilution series but varied more with the blocking effect provided by plant sap extracts. Calculations suggest that if denatured, a 2 μl spot at the start of a dilution series contains about 1 part per 1 500 of the viroid in 1 g of fresh tissue, if it is assumed that viroid is solubilized with high efficiency by sap extraction buffer.

Detection in clarified sap extracts diluted in water

At maximum probe activity, detection is to the dilution of 10^{-4} after autoradiographic exposure for 20 h. Following estimates (Schumacher *et al.* 1986) of about 8 μg viroid per g fresh tissue as a peak concentration, we are detecting from hundreds of femtogram to a few picogram per spot of sap in water. Denaturing does not increase the signal from low concentrations of sap in water.

Detection in clarified sap extract diluted in healthy clarified sap extract

This dilution simulates low concentrations of viroid in tissue: the addition of healthy sap blocks over 90% of the signal at lower viroid concentrations. Denaturing reduces the blocking and increases signal about three times but the dilution end-point rarely approaches the 10^{-4} end-point which is achieved by the same sap diluted in water. Estimates of the concentrations detected are approximate but probably close to those quoted for tissue with one or two copies of viroid per cell (Schumacher *et al.* 1986).

Detection in nucleic acid (NA) extracts

Nucleic acid extraction followed by resuspension in small volumes provides a concentration step; we estimate the 2-step PEG fractionated extracts allow for a 30-fold concentration compared with non-denatured sap extract. A 2 μl spot represents 1 part in 25 of the viroid in 1 g of tissue, assuming efficient extraction. Nucleic acid extracts for test are only partially purified.

Detection in low molecular weight NA extracts in water

Detection with high activity probes and long exposures approaches a dilution end-point of 10^{-6} for starting solutions equivalent to 1 part in 25 of viroid in 1 g fresh tissue. Denaturing does not increase signal from lower concentrations.

Detection in low molecular weight NA extracts diluted in healthy NA

Low concentrations of viroid generate less signal when embedded in healthy NA extracts, even in healthy low molecular weight fractions when, perhaps, typically 50–70% of signal is blocked. Detection is below the best sensitivity of 10^{-6} for the same peak concentration diluted in water and denaturation of low molecular weight fractions does not increase the signal. The addition of total NA extracts to a dilution series of infected low molecular weight NA (total NA constant, viroid fraction diluted) has a greater blocking effect, probably proportional to the extra NA on each spot.

Denaturing does improve detection of low concentrations of viroid in total NA extract but detection of small amounts of viroid in a total NA extract was ten times less sensitive than detection of the same amount in a low molecular weight NA extract. Additional experiments (× 3), where a low quantity of a viroid-rich fraction was kept as a constant, to which increasing quantities of either healthy sap or NA extracts were added, gave similar results. The viroid-rich fraction was obtained by two-step PEG fractionation (see p. 219) except that a narrower band 14–18% PEG reduced NA other than viroid.

Non-specific signals

Experience has been gained in volume testing of leaves, tubers, sprouts and microplants in culture: non-specific signal on spots has been low for NA on nylon, particularly when the low molecular weight fraction is probed. Non-specific signals are seen frequently on non-denatured sap extracts of microplants, sprouts and tubers sometimes to a significant level of 100 times greater than the background. Most of this non-specific signal is abolished by the sample denaturing step but sprouts and microplants occasionally gave low but discernible background, even after samples had been denatured.

Discussion

The use of nylon filters is important for our NA extracts because in the previous nitrocellulose-based system the small RNA target which had been baked to the filter was shown to migrate (Harris & James 1987). Spurious signals on healthy NA spots arose from the relocation of the migrating target.

The authors have noted previously that signals on nylon were weaker than those on nitrocellulose (Harris *et al.* 1988) confirming an earlier report (Barker *et al.* 1985). Hybridization conditions have been changed accordingly to suit the nylon. The current hybridization on Gelman Biotrace RP provides as strong, if not stronger, signals than those on nitrocellulose with the benefit of lower backgrounds.

It has helped routine work to avoid the use of plastic bags for hybridization. Our low-cost glass dishes are rapid and clean to use, can be adjusted to smaller sizes of filter and there is less exposure of an operator to radiation and less contaminated solid and fluid waste.

Although probing of viroids is a specialized procedure, there are some useful pointers to probing for other pathogen NA at low concentration. M13 probes are capable of detecting very low quantities of NA. In our case there is a possibility that concatenamerization is occurring and that particular hybridizing conditions favour that amplification (unpublished results).

The efficiency of the probe is much reduced if the target is embedded in high concentrations of cell constituents, particularly other nucleic acids. Methods for fractionating viroid NA extracts and thereby increasing the signal are not applicable to higher molecular weight NA of other pathogens. Generation of lower molecular weight fragments as a product by polymerase chain reaction is, however, a possible detection strategy for some pathogens at low concentration, for instance in soil. Fractionation of such an NA extract by molecular size could allow a much larger subsequent amplification of signal by an NA probe to the product.

Acknowledgements

We thank Dr R.A. Owens, USDA, Beltsville, USA for the gift of M13mp9 containing inserts and Dr L.J. Skrzeczkowski, Institute of Biochemistry and Biophysics, Polish Academy of Sciences, Warsaw, Poland for his method of fractionating NA extracts and for introducing us to viroid denaturing. We are grateful for the earlier support of E. Okely, Department of Molecular Biology, University of Edinburgh and members of Prof. C.J. Leaver's group in the Department of Botany, University of Edinburgh, and P. Kelly, Mercia Diagnostics Ltd.

References

BARKER, J.M., McINNIS, J.L., MURPHY, P.J. & SYMONS, R.H. 1985. Dot-blot procedure with [^{32}P] DNA probes for the sensitive detection of avocado sunblotch and other viroids in plants. *Journal of Virological Methods* **10**, 87–98.

GRASMICK, M.E. & SLACK, S.A. 1985. Symptom expression enhanced and low concentrations of Potato Spindle Tuber Viroid amplified in tomato with high light intensity and temperature. *Plant Disease* **69**, 49–51.

HARRIS, P.S. & JAMES, C.M. 1987. Exclusion of viroids from potato resources and the modified use of a cDNA probe. *OEPP/EPPO Bulletin* **17**, 51–60.

HARRIS, P.S., JAMES, C.M. & KELLY, P. 1988. Use of a cDNA probe for sensitive detection of potato spindle tuber viroid in potato quarantine. *Proceedings of International Seminar, Warsaw, Poland, 1986, in Viroids of Plants and their Detection* Warsaw: Warsaw Agricultural University Press.

HARRIS, P.S. & MILLER-JONES, D.N. 1981. An assessment of the tomato/polyacrylamide gel electrophoresis test for potato spindle tuber viroid in potato. *Potato Research* **27**, 399–408.

LAKSHMAN, D.K., HIRUKI, C., WU, X.N. & LEUNG, W.C. 1986. Use of [^{32}P] RNA probes for the dot hybridisation detection of potato spindle tuber viroid. *Journal of Virological Methods* **14**, 309–319.

MACQUAIRE, G., MONSION, C., MOUCHES, T., CANDRESSE, T. & DUNEZ, J. 1984. Spot hybridisation: application to viroid identification. *Annals of Virology (Institute Pasteur)* **135e**, 219–230.

MORRIS, T.J. & SMITH, E.M. 1977. Potato spindle tuber disease: procedures for the detection of viroid RNA and certification of disease-free potato tubers. *Phytopathology*, **67**, 145–150.

OWENS, R.A. & DIENER, T.O. 1981. Sensitive and rapid diagnosis of potato spindle tuber viroid disease by nucleic acid hybridisation. *Science* **213**, 670–672.

PALUKAITIS, P., COTTS, M. & ZAITLIN, M. 1985. Detection and identification of viral nucleic acids by 'dot-blot' hybridisation. *Acta Horticulturae* **164**, 109–118.

SCHUMACHER, J., MEYER, N., RIESNER, D. & WEIDEMANN, H.L. 1986. Diagnostic procedure for detection of viroids and viruses with circular RNAs by 'return-gel electrophoresis'. *Journal of Phytopathology* **115**, 332–343.

SINGH, R.P. 1984. Bibliography of viroid reviews through 1983. *Canadian Plant Disease Survey* **64**, 15–16.

SINGH, R.P. & BOUCHER, A. 1988. Comparative detection of mild strains of potato spindle tuber viroid from the dormant potato tubers by return-polyacrylamide gel electrophoresis and NA hybridisation. *Potato Research* **31**, 159–166.

SKRZECZKOWSKI, L.J., OKELY, E. & MACKIEWICZ, N. 1985. Improved PAGE and cDNA diagnosis of potato spindle tuber viroid in *Abstracts for Posters, AAB Virology Group*, Cambridge, UK, April 10–12.

WHITE, B.A. & BANCROFT, F.C. 1982. Cytoplasmic dot hybridisation. *Journal of Biological Chemistry* **257**, 8569–8572.

Analysis of Restriction Fragment Length Polymorphisms in the Study of Bacteria

N.A. Saunders
Division of Microbiological Reagents and Quality Control, Central Public Health Laboratory, 61 Colindale Avenue, London NW9 5HT, UK

The use of nucleic acid probes for the detection of inter or intraspecies DNA sequence differences by analysis of restriction fragment length polymorphisms (RFLPs) is of increasing importance in several distinct areas of biology. The method is based on the use of a probe able to hybridize to specific restriction fragments, the length of which is determined by the positions of the relevant endonuclease recognition sites in the DNA sequence. A well-known application of this technique is the use of probes for the hypervariable minisatellite sequences of human DNA in paternity testing and forensic medicine (Jeffreys *et al.* 1985 and the chapter by Parkin, this volume). In prokaryotes the detection of RFLPs is useful for epidemiological typing purposes (Eisenach *et al.* 1986; Tompkins *et al.* 1986; Irino *et al.* 1988; Musser *et al.* 1988; Saunders *et al.* 1989, 1990) and for strain identification to the species level (Grimont & Grimont 1986; Saunders *et al.* 1988a).

Epidemiological Typing

Epidemiological typing of bacterial strains is crucial in public health microbiology. The data obtained can be used to monitor trends in the occurrence of pathogenic strains or to identify possible sources of infection. The frequently used methods, for example, serogrouping with polyclonal anti-sera or monoclonal antibodies, phage typing and isoenzyme analysis all rely on the detection of interstrain differences in phenotypic characteristics and each has its own particular advantages and disadvantages. The analysis of RFLPs for typing is highly discriminatory since it is possible to distinguish between strains which differ by only one nucleotide residue in their DNA sequences, providing that

0-632-02926-9

the varying base is within an appropriate restriction site. In addition, DNA rearrangements, insertions and deletions can also alter the restriction fragment patterns observed.

The degree of pattern variation observed may be used as a crude measure of the overall sequence divergence between strains. The discrimination of RFLP analysis depends not only upon the degree of similarity between the base sequences of the strains to be compared but also on the specific probe used. This is because the tendency for base sequence drift in different parts of the genome during strain divergence varies considerably from the highly conserved rRNA genes through to those parts of the genome coding for gene products having highly variable amino acid sequences. RFLPs also arise from rearrangements of the genome and these events also vary in frequency at different genetic loci. Some repeated sequences, for example, are recombination 'hot spots' (Lin *et al.* 1984).

Bacterial identification

Measurements of DNA/DNA (Owen & Pitcher 1985) and rRNA/DNA (Mordarski 1985) homology are the 'gold-standard' in studies of the taxonomic relationships between closely and distantly-related bacteria respectively. Unfortunately the methods often require comparisons between the appropriate nucleic acids of many representative strains and are therefore rather time-consuming and specialized so that few laboratories are able to use them routinely. RFLP analysis with probes derived from rRNA provides a rapid alternative to these methods. The structural RNA genes of the eubacteria are clustered to form operons which vary in number according to the species. These consist of the rRNA genes, in the order 5′−16S, 23S, 5S−3′, followed by a variable number of tRNA genes with short intervening sequences which are generally less conserved than the genes for the structural RNAs themselves. Parts of the rRNA genes are very highly conserved and a single probe derived from these sequences will hybridize to the homologous polynucleotides of species from many different bacterial genera. In general, it appears that the variation of restriction fragment patterns revealed by rRNA probes is correlated to the degree of homology between the strains in DNA/DNA and rRNA/DNA hybridization studies. The pattern similarities observed when strains of the same species are examined can be employed as markers for the identification of other strains of that species. In most cases intraspecies pattern differences are also observed and these can be employed as strain-specific markers for typing.

Methods of RFLP Analysis

Bacterial cultivation

DNA can usually be extracted successfully from cells grown to stationary phase on any suitable liquid or solid culture media. Cultures should contain few dead cells and little cellular debris but it is advantageous to avoid using rapidly-dividing cells which have a high RNA/DNA ratio since RNA contamination of DNA preparations can be a disadvantage of the DNA isolation procedure described below.

DNA preparation

The preparation of high molecular weight DNA from bacterial strains by the classical procedures which require repeated extractions with phenol and chloroform (as, for example, Owen & Pitcher 1985) is time-consuming and inconvenient. The method described here is a modification of that described previously (Pitcher *et al.* 1989) and requires a minimum of technical effort. Most Gram-positive and Gram-negative bacteria can be lysed by treatment with a solution of 5 M guanidine thiocyanate; 0.1 M EDTA following lysozyme treatment. However, some species are resistant to lysis under these conditions and, for example, *Staphylococcus aureus* strains must be pretreated with lysostaphin rather than lysozyme (Pitcher *et al.* 1989). The method used by the author, which is detailed in Table 1, has been applied successfully to the extraction of DNA from the following bacteria: *Clostridium difficile*, *Listeria* (4 species), *E. coli*, *Legionella* (34 species), *Yersinia* (6 species), *Hafnia alvei*, *Proteus mirabilis*, *Citrobacter freundii*, *Salmonella typhimurium*, *Klebsiella pneumoniae*, *Pseudomonas* (2 species), *Neisseria meningitidis* and *Haemophilus influenzae*.

Deoxyribonuclease (DNase) is inactive under the conditions of extraction described here but studies of *Yersinia enterocolitica* strains showed that residual nuclease activity was present in preparations from these organisms as judged by fragmentation of the DNA during incubation at 37°C (results not shown). DNase activity was destroyed by heating the resuspended DNA at 80°C for 30 min.

If necessary, contaminating RNA can be removed from the DNA by treatment with RNase and proteinase-K as described in Table 3(i) and (j).

DNA solutions are stable for several years when frozen at −20°C in microcentrifuge tubes. However, repeated freezing and thawing of the DNA results in its fragmentation and should be avoided.

Theoretical considerations of restriction endonuclease digestion

Many restriction endonucleases are available from commercial sources and all have the potential to produce RFLPs which can be used as species or strain-

TABLE 1. *Protocol for DNA purification*

(a) Harvest bacterial cells into a 1.5 ml microcentrifuge tube and pellet by centrifugation for 2 min (10 000 g_{max}). (The pellet size should be between 25 and 50 μl. If only a smaller cell volume is available the volumes of reagents employed should be scaled down proportionately.)

(b) Resuspend the cells by vortexing in a solution of lysozyme (2 mg/ml) in sterile distilled water (100 μl) and incubate at room temperature for 10 min (Gram-negative) or 30 min (Gram-positive.

(c) Lyse the cells by the addition of a solution containing 5 M guanidine thiocyanate and 0.1 M EDTA pH 7.0 (200 μl) with gentle pipetting. (Clearing of the solution should occur rapidly on mixing but the lysate often remains slightly cloudy and this is not resolved by prolonging the guanidine thiocyanate treatment. However, the yield of DNA is usually satisfactory.)

(d) Add 7.5 M ammonium acetate (150 μl) and chloroform:isoamylalcohol, 24:1, (450 μl) and emulsify by vigorous shaking for 30 s. Separate the phases by centrifugation in a microfuge (2 min, 10 000 g_{max}) and remove the aqueous (top) phase.

(e) Precipitate the high molecular weight DNA in the aqueous phase by addition of 0.54 volumes of propan-2-ol and mix by inversion. Remove the precipitated DNA. (The DNA, which forms a dense stringy precipitate within a few seconds of mixing, should be separated from the liquid phase immediately with a pipettor tip. The DNA should not be collected by centrifugation for two reasons: first, centrifugation results in the formation of a compacted pellet which is difficult to resuspend and, secondly, a cloudy precipitate (probably consisting mainly of RNA and protein) forms in some preparations and this is collected with the DNA if centrifugation is used).

(f) Wash the DNA twice in 80 % ethanol (1 ml), dry under vacuum and dissolve in TE buffer (TE contains 10 mM Tris–HCl pH 7.0, 1 mM EDTA). Measure the nucleic acid concentration spectrophotometrically (a 1 mg/ml solution of double-stranded DNA gives an absorbance of 20 OD units at 260 nm).

specific markers. An important consideration is whether or not the endonuclease is able to cleave the DNA from all of the strains of interest. The failure of some enzymes to digest DNAs isolated from particular micro-organisms is usually due to methylation of nucleotides at the relevant restriction sites.

It is convenient to analyse restriction fragments in the size range 500–10 000 bp since such molecules are well-resolved in agarose gels of approximately 1%. Gels of this porosity have sufficient strength to be handled easily during the subsequent processing steps and to allow efficient blotting of the DNA onto a solid support. The number and size of the restriction fragments to be examined can be determined by the selection of a suitable endonuclease. Each restriction fragment results from cleavage of the original DNA sequence at two specific sites. Thus, for a given length of chromosomal DNA, the number and size of restriction fragments depends upon the frequency of occurrence of the enzyme sites. In general, DNA cleavage with restriction enzymes having hexanucleotide sites produces fewer restriction fragments than when the same sequence is digested with an enzyme with a tetranucleotide recognition sequence. The frequency of occurrence of enzyme recognition sequences is also affected by the overall base composition of the substrate DNA (as judged by the % G + C). For example, more sites for the enzyme *Hae*III, which has the recognition sequence 5′–GGCC–3′, should arise in DNAs with a high % G + C. Another factor, which may be of importance, is that certain sequences of nucleotides arise infrequently in particular DNAs. If necessary, the test DNA can be cleaved with more than one endonuclease to reduce the size of the restriction fragments produced. The disadvantage of this approach is that it may be difficult to detect partial DNA digests resulting in poor reproducibility.

Interstrain RFLPs can result from chromosomal rearrangements, insertions or deletions of one or more nucleotides at the relevant restriction sites resulting in the creation/deletion of a site of cleavage. Alternatively, sequence variations at other loci may result in changes in the length of a restriction fragment. However, in order to be recognized, the difference in restriction fragment length between two strains must be approximately 5% (or more) of the fragment length and consequently the loss, addition or substitution of single nucleotide bases remote from the restriction sites is unlikely to result in observable RFLPs. Within a group of strains, large restriction fragments will normally be more polymorphic than the small fragments. This is because alterations to the DNA sequence, which in the absence of special factors are random events, occur in proportion to the length of the polynucleotide. Thus, when DNA is treated with an enzyme which cuts infrequently to produce large restriction fragments, the ratio of RFLPs to the number of fragments is likely to be higher than when an enzyme active on many sites on the substrate DNA is employed.

The ratio of RFLPs to restriction fragments is also influenced by the number of bases comprising the endonuclease site. Thus, an individual endonuclease site consisting of six bases is more likely (because of its size) to be altered by a random base change than a four-base site. If all of these factors are taken into consideration it should be possible to predict the number and heterogeneity of fragments produced by cleavage of a particular DNA sample with a restriction endonuclease and therefore to select a 'short-list' of suitable enzymes from those available. Enzymes such as *Ava*I, *Ava*II, *Sin*I or *Nci*I seem to be particularly useful when the probe employed consists of approximately 20 kb of a non-repeated DNA sequence. These nucleases cleave DNA at sites comprised of five or six bases but only four positions are specific for a single nucleotide base and the remaining positions may be filled by one of two possible bases. The final choice of enzyme should be made after comparison of the results obtained with several different enzymes. This is necessary because specific enzyme recognition sites may be either especially conserved or particularly liable to interstrain variation.

Clearly, the number of interstrain RFLPs observed within a group of strains depends upon the quantity and size of fragments subjected to analysis in addition to the specific sequence of the target DNA and the restriction enzyme recognition site. An advantage of the development of a system which allows the simultaneous analysis of a large number of restriction fragments and hence interstrain RFLPs is that this results in a greater accuracy of strain identification or discrimination in typing than when small numbers of fragments are subjected to analysis. The number of restriction fragments to be examined is limited by the ability of the system to resolve between bands and to produce a pattern that can be readily evaluated.

Practical aspects of restriction endonuclease digestion

It is convenient to digest 5 μg amounts of substrate DNA in a 50 μl reaction volumes as follows:

1 DNA in 45 μl of TE buffer is made 1 × in reaction buffer by addition of 5 μl of the 10 × stock solution supplied by the manufacturer (alternatively, if no buffer is supplied, one of the restriction endonuclease buffers described by Maniatis *et al.* (1982) may be employed).

2 An excess (1.2–2 u/μg DNA) of enzyme is added, the mixture is incubated at the recommended temperature for 2 h and then kept on ice.

3 A sample (5 μl) is withdrawn and subjected to agarose gel electrophoresis for 1 h at 5 volumes/cm, as described below, in order to check that the digestion is complete. A sample of the undigested DNA should be run in an adjacent track as a control.

4 The gel is stained with ethidium bromide (0.5 μg/ml in water) and examined under u.v. transillumination.

The DNA fragments should migrate faster than the undigested DNA and it should be possible to distinguish a few stronger bands which stand out against a smear of weak bands. If the DNA is only partially digested these bands cannot be seen clearly. If digestion is not apparent the activity of the enzyme should be checked against another DNA sample (for example commercial λ phage or plasmid DNA). The possibility that the DNA is contaminated with a general inhibitor of endonuclease activity (for example chemicals used in the extraction process) can be investigated by attempting its digestion with another enzyme. Contaminants can often be removed by reprecipitation of the DNA from 66% ethanol containing 0.3 M sodium acetate, pH 6.0. If neither the enzyme nor the DNA is the cause of the problem it is likely that the relevant restriction sites are methylated. If incomplete digestion has occurred more enzyme should be added and incubation should be resumed for a further 2 h.

When the digestion is complete, 0.5 volumes of a solution containing

10% Ficoll and 0.05% bromophenol blue is added. The restriction fragments can then be stored at −20°C or below until required for analysis.

Agarose gel electrophoresis

Many different types of apparatus are available for running horizontal agarose slab gels submerged beneath the surface of the electrophoresis buffer and all should give satisfactory results. Good results are obtained with slots of 0.5 cm in width. The use of smaller slots has the advantage that more samples can be processed concurrently but the effect of trailing at the edges of the slots is greater and the results have less clarity.

The concentration of agarose determines the size range of fragments which can be resolved on a gel. Gels with an agarose concentration of between 0.8 and 1% are easy to handle and give reasonable resolution of fragments in the size range 0.5–10 kb even when the gel is run at a high voltage.

Many types of agarose are available and considerable variation between grades and batches occurs. Some agarose preparations contain impurities which interfere with the binding of DNA to Hybond-N filters in vacuum blotting. Fortunately, different batches of 'ultrapure' agarose (BRL) have all given satisfactory results.

Agarose gels 20 × 20 × 0.5 cm in TBE buffer containing 89 mM Tris; 89 mM borate and 2.5 mM EDTA Na_2, pH 8.3, are subjected to electrophoresis at either 1.5 volumes/cm (approximately 16 h) or at 6 volumes/cm (approximately 3 h) until the marker has migrated 75% of the length of the gel. The resolution of the high molecular weight bands is lower at high voltage but the bands obtained are sharper. One microgram of sample DNA in 15 μl is applied to each slot. Restriction fragments of known molecular size should be applied to slots adjacent to each sample. If the probe is a recombinant λ phage it is convenient to use a digest of commercial λ phage DNA as a standard. The fragments produced by separate digestion with the enzymes *Pst*I and *Eco*RI can be mixed (1 ng of each digest in a total volume of 10–15 μl) to produce markers of a useful size range. However, digestion of λ DNA with many other enzymes or combinations of enzymes will yield satisfactory marker fragments.

Southern blotting

The transfer of DNA fragments from agarose gels to a hybridization membrane can be accomplished satisfactorily by capillary (Southern 1975), vacuum or electro-blotting (see manufacturers' literature for descriptions of the equipment available). Vacuum-blotting is preferred because it is rapid and gives superior resolution of closely spaced bands.

Many of the different hybridization membranes available are suitable for use in blotting experiments. Hybond-N (Amersham) 'a nylon-based membrane' is convenient to use, since it is very strong and has high binding capacity for nucleic acids which can be fixed to it covalently by a short exposure to u.v. radiation.

A standard vacuum-blotting procedure is summarized in Table 2. The hybridization membrane should be handled as carefully as possible to avoid scratches and finger-marks on its surface. These marks often stain intensely when the blot is developed.

Cloning and selection of probes

A full description of the various methods available for cloning bacterial DNA and screening for sequences suitable for use as probes is beyond the scope of this chapter and the standard techniques required are the subject of some very helpful texts (for example see Maniatis *et al.* 1982; Karn *et al.* 1983; Kaiser & Murray 1985).

The use of bacteriophage λ vectors is particularly suitable for the preparation of cloned probes for RFLP analysis, having advantages over the M13, plasmid and cosmid alternatives. The advantages are:

1 Relatively large DNA fragments can be cloned with great efficiency. The small fragments readily cloned in M13 and plasmid vectors can be used as probes in RFLP analysis but such probes, if they represent unique sequences,

TABLE 2. *Transfer of DNA to hybridization membrane*

(a) Soak the gel in 2.5 gel volumes of 0.25 M HCl twice for 15 min with gentle agitation. (This treatment results in the partial depurination of the DNA which is then susceptible to alkaline hydrolysis (step b).)

(b) Denature the DNA by soaking the gel in 2.5 volumes of 0.5 M NaOH, 1.5 M NaCl twice for 15 min with gentle agitation.

(c) Neutralize in 2.5 volumes of Tris–HCl pH 8.0, M of NaCl twice for 15 min with gentle agitation. (If necessary the gel may be left in this buffer for several hours but gradual diffusion of the DNA will occur, resulting in reduced resolution of the bands.)

(d) Assemble the gel onto a vacuum blotting apparatus (suitable apparatus can be obtained from several manufacturers, for example the Vacublot (Anderman), and blot the restriction fragments onto Hybond-N using 10 × SSPE as transfer buffer (1 × SSPE is 0.15 M NaCl, 2 mM EDTA, 10 mM NaH_2PO_4-NaOH pH 7.4).

(e) Wash the membrane in 2 × SSPE, dry in air, wrap in u.v. transparent plastic film and place, DNA side down, on a u.v. transilluminator. (The duration of u.v. exposure required for maximal binding of hybridizable target DNA should be determined for each transilluminator and type of plastic wrap (usually 2–5 min).)

hybridize to few restriction fragments in the optimal size range and therefore few RFLPs are revealed.

2 There is no need to purify the cloned bacterial DNA from the vector. The vector sequences become labelled and will hybridize to restriction fragments of bacteriophage λ used as molecular weight standards and electrophoresed in tracks adjacent to the restriction fragments derived from strains of interest. When used in this way these sequences act as controls for the efficiency of the probe labelling, vacuum-blotting, hybridization and detection procedures.

3 Purification of phage DNA in high yield and of suitable purity is accomplished easily.

4 Screening of large numbers of plaques is easier than screening bacterial colonies.

Labelling of probe and hybridization

Recombinant λ phage DNA is purified from bacteriophage particles prepared by the pelleting method as detailed in Table 3. This method, which is modified from Maniatis *et al.* (1982), is recommended since high yields of good quality DNA are obtained rapidly. Lambda phage DNA is labelled with biotin by random priming for 2 h at 37°C. The 100 μl reaction mixture contains:

1 μg dsDNA (heat denatured at 95°C for 10 min in TE buffer).
50 μg BSA.
100 μg salmon sperm primers.
25 μM dATP.
25 μM dCTP.
25 μM dGTP.
5 μM dTTP.
20 μM biotin-11-dUTP
50 mM Tris–HCl (pH 8.0).
2 mM DTT.
5 mM $MgCl_2$.
40 mM KCl.
10 units Klenow fragment.

Labelled probe is used directly or stored frozen at −40°C until required. Membranes (15 × 18 cm) are placed into heat sealable polythene bags and prehybridized for 1 h at 42°C in hybridization buffer (20 ml) containing 45% (v/v) formamide, 5 × SSPE, 5 × Denhardt's solution, 0.1% (w/v) SDS. Following prehybridization, the bag is opened and the buffer is drained off thoroughly but the membrane is not allowed to dry. Biotinylated-probe DNA (100–200 ng) is denatured by heating in a boiling water bath for 10 min, cooled on ice, mixed with hybridization buffer (1 ml) and added to the bag.

TABLE 3. *Preparation of λ phage DNA*

(a) Treat the lysed culture with DNase and RNase (1 μg/ml of each) for 30 min at room temperature.

(b) Add sodium chloride to a concentration of 1 M and incubate on ice for 1 h.

(c) Remove the cellular debris by centrifugation at 10 000 g_{max} for 10 min at 4°C. Make the supernatant 10% (w/v) in polyethylene glycol 8000 (Sigma) and incubate on ice for 1 h.

(d) Collect the precipitate by centrifugation at 10 000 g_{max} for 10 min at 4°C, resuspend in 25 ml of SM buffer containing 0.1 M NaCl, 10 mM $MgSO_4$, 50 mM Tris–HCl pH 7.5 and 0.01% gelatin and extract with an equal volume of chloroform.

(e) Sediment bacteriophages in the aqueous phase by centrifugation at 120 000 g_{max} for 2 h at 4°C.

(f) Gently resuspend the pellet in 1 ml of SM buffer and disrupt the phages by the sequential addition of 0.5 M EDTA pH 8.0 (40 μl), proteinase-K (100 μg), 20% (w/v) SDS (25 μl). Incubate at 65°C for 1 h.

(g) Purify the phage DNA by extraction against equal volumes of (i) phenol solution (phenol containing 0.8% hydroxyquinolone, equilibrated with 0.1 M Tris–HCl pH 8.0 buffer); (ii) phenol solution : chloroform : isoamylalcohol (25:24:1 v/v/v) and (iii) chloroform : isoamylalcohol (24:1 v/v).

(h) Precipitate bacteriophage DNA in the final aqueous phase by addition of 2 volumes of ethanol and then collect it by centrifugation (10 000 g_{max}, 5 min) and dry *in vacuo*.

(i) Remove residual RNA and protein contamination as follows: dissolve the DNA pellet in SET buffer (0.15 M NaCl, 15 mM EDTA, 50 mM Tris–HCl pH 8.0, 1% SDS and DNase free RNase A (50 μg/ml)) and incubate at 37°C for 15 min. Add proteinase-K (100 μg/ml) and continue incubation for 30 min at 65°C. Add 1/2 volume of 7.5 M ammonium acetate and extract with an equal volume for chloroform : isoamylalcohol (24:1 v/v). Precipitate DNA in the aqueous phase by the addition of propan-2-ol (0.54 volumes).

(j) Wash the precipitate twice in 80% (v/v) ethanol, dry *in vacuo* and finally dissolve in TE buffer (10 mM Tris pH 7.0, 1 mM EDTA).

Bubbles should be carefully removed from the bag, which is then resealed and placed in a waterbath. Hybridization is carried out for 16 h at 42°C (a larger volume (10–50 ml) of hybridization solution may be employed if the membrane is agitated and this may increase the rate of hybridization and sensitivity of detection). Membranes are washed under non-stringent conditions (30°C below estimated T_m). Non-stringent conditions are used so that the homologous sequences from divergent strains will hybridize to the probe. In practice, with the probes used to date, this does not result in staining of large numbers of background bands. However, problems of non-specific hybridization might be anticipated in some instances and in such cases the stringency of the washing conditions could be increased.

Probe detection

Biotin-labelled probe/target sequence duplexes are detected by a system based on a streptavidin–poly(alkaline phosphatase) conjugate (BRL). A critical step in the procedure, summarized in Table 4, is the blocking of non-specific conjugate binding sites. It is important to ensure that the filter is entirely immersed in the blocking buffer and that bubbles are excluded from contact with the filter. For reasons which remain unclear, the efficiency of blocking is also dependent upon the source of the fraction V bovine serum albumin. Of those types tested, Sigma (cat. no. A-9647) has given the most consistent results. Following colour development, membranes can be kept stored in the dark for several years with no noticeable colour deterioration.

Interpretation of results

The quality of the probed restriction fragment blot must be assessed in order to check that the patterns observed are likely to be accurate. The λ phage restriction fragment markers are particularly useful indicators of blot quality. The blot should show bands in their correct relative positions corresponding

TABLE 4. *Biotin detection*

(a) Immediately following the posthybridization washes at the required stringency, immerse the membrane in a 3% (w/v) solution of BSA in buffer A containing 0.1 M Tris–HCl pH 7.5 and 0.15 M NaCl (1 ml buffer/10 cm^2 filter) in a sealed plastic box or bag and incubate at 65°C for 1 h in a water bath.

(b) Place the membrane in a plastic tray and drain off excess blocking buffer. Dilute the conjugate to working strength in buffer A (a 10^3 dilution of the BRL streptavidin–polyalkaline phosphatase conjugate stock) and apply to the membrane (1 ml diluted conjugate/20 cm^2 filter) with constant agitation at room temperature for 10 min.

(c) Remove excess conjugate from the membrane by washing, with constant agitation, in two changes (15 min each) of buffer A (1.5 ml buffer/cm^2 filter) followed by 10 min in buffer B containing 0.1 M cm^2 Tris–HCl pH 9.5, 0.1 M NaCl and 50 mM $MgCl_2$.

(d) Develop by application of a solution (1 ml dye solution/20 cm^2 filter) containing 330 mg/litre nitroblue tetrazolium (NBT) and 150 mg/litre 5-bromo-4-chloro-3-indolyl phosphate (BCIP) disodium salt in buffer B. (The dye solution is made up immediately before application from stock solutions of NBT (75 mg/ml in 70% (v/v) dimethylformamide, stored at 4°C) and BCIP (50 mg/ml in water, stored frozen at −20°C).)

(e) Leave bands to develop to a satisfactory level at room temperature under subdued lighting conditions in a sealed plastic bag. (Most colour development occurs within 2 h but some darkening of the bands (and background) is seen for several days.)

(f) Wash thoroughly under running tap water, dry in air and store in a plastic bag.

to each of the phage DNA fragments. In addition, the intensity of most of the bands should be proportional to the size of these DNA fragments. If these criteria are met then it may be concluded that all steps in the restriction fragment analysis from probe labelling and gel electrophoresis through to colour development were successful. If the anticipated pattern of marker bands is achieved but with no signal from the test DNA then it may be concluded that the DNA samples applied to the gel were defective or that no sequences complementary to the probe were present within them. A preliminary inspection of the patterns should also reveal whether partial or non-specific digestion of the test DNA has occurred. Partial digests are characterized by the presence of multiple high molecular weight bands with the low molecular size fragments appearing at reduced intensity. If the DNA is partially degraded due to DNase activity, then the intensity of the larger fragments is reduced and the background corresponding to fragments of low molecular size is increased.

The reproducibility of the restriction fragment patterns observed should be checked thoroughly by the appropriate criteria as described by Saunders *et al.* (1990). An essential first step in assessing the reproducibility of the restriction fragment patterns is to compare analyses of different cultures of the same strain on different blots. This should be done for a number of strains exhibiting different restriction fragment types.

Both the relative intensities and mobilities of the bands should be reproducible. It is however difficult to achieve a consistent level of band colour intensity between blots. For this reason differences between the minor bands (those having an intensity of less than approximately 10% of that of the strongest band) should not be relied upon to distinguish between strains.

In the author's experience it is relatively easy to compare the restriction fragment patterns revealed by probes visually because they are designed to consist of a small number of well spaced bands. Comparison is facilitated by drawing a representation of the bands seen against a molecular weight scale derived from the λ phage standards applied. A solid line should be drawn for strong bands and a dotted line for weak bands (less than approximately 10% of intensity of the most intense band). When drawn in this way (Fig. 1) the restriction fragments patterns of strains analysed on different gels can be compared conveniently and rapidly. In some instances a subtle difference between the patterns of two strains analysed on the same gel may not allow a reliable identification to be made when the strains are analysed on different occasions and compared as described. In such cases the strains should be subjected to repeated restriction fragment analysis in adjacent tracks of the same gel or, alternatively, minor differences may be ignored and the strains classed as being indistinguishable from one another.

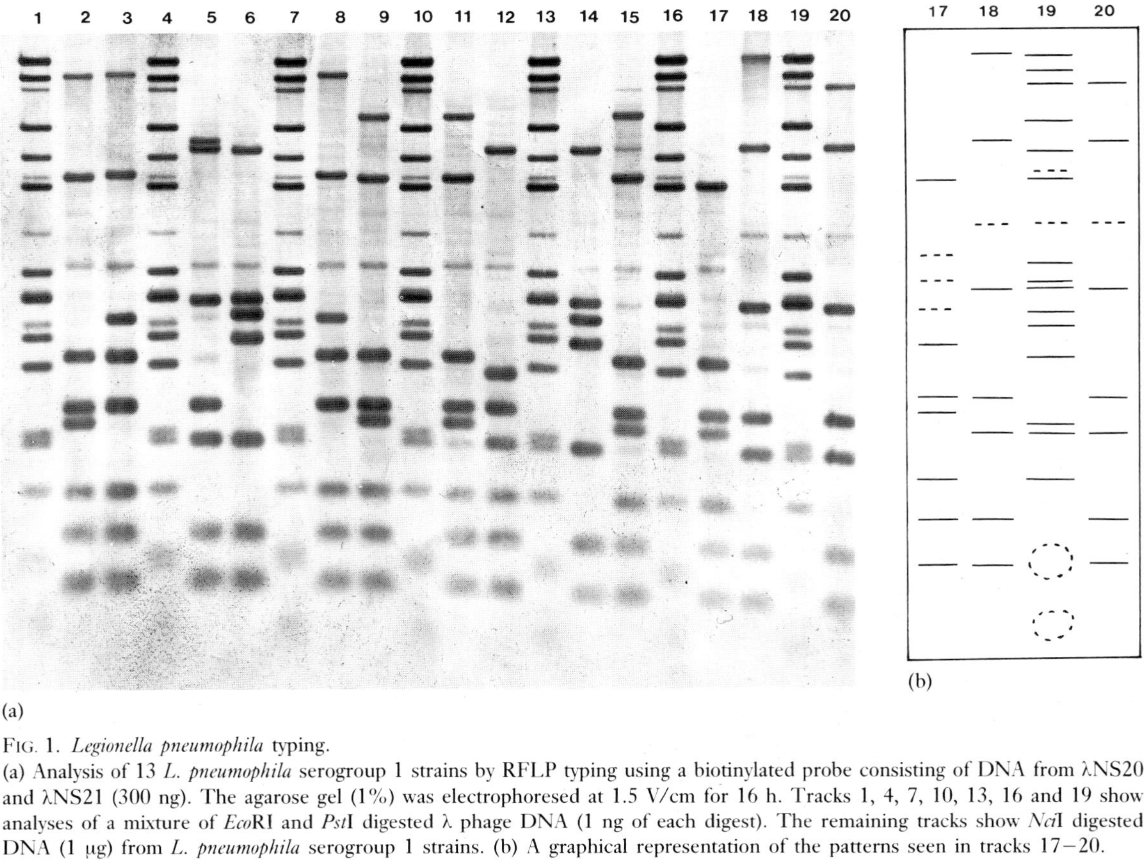

Fig. 1. *Legionella pneumophila* typing.
(a) Analysis of 13 *L. pneumophila* serogroup 1 strains by RFLP typing using a biotinylated probe consisting of DNA from λNS20 and λNS21 (300 ng). The agarose gel (1%) was electrophoresed at 1.5 V/cm for 16 h. Tracks 1, 4, 7, 10, 13, 16 and 19 show analyses of a mixture of *Eco*RI and *Pst*I digested λ phage DNA (1 ng of each digest). The remaining tracks show *Nci*I digested DNA (1 μg) from *L. pneumophila* serogroup 1 strains. (b) A graphical representation of the patterns seen in tracks 17–20.

Examples of Results

Legionella pneumophila *typing*

Figure 1 shows the results of analysis of the restriction endonuclease *Nci*I digested DNAs of 13 typical strains of *L. pneumophila* serogroup 1. The probe employed was a mixture of DNAs (labelled by biotinylation) from two λ phage clones which were selected from a library of *L. pneumophila* genomic fragments in the vector λgtWES.λ.B. (Saunders *et al.* 1988b). Nine distinct patterns can be seen in tracks 2, 3, 5, 6, 9, 12, 17, 18 and 20. The patterns shown in tracks 8, 11, 14 and 15 are indistinguishable from those in tracks 3, 9, 6 and 9 respectively. It can be seen that the strain analysed in track 15 has some additional faintly staining bands not present in the patterns of the other examples of this type (9 and 11). This was subsequently found to be due to contamination with a strain of the restriction fragment type seen in track 20. The use of this probe for typing *L. pneumophila* serogroup 1 isolates has been evaluated for discrimination and reproducibility (Saunders *et al.* 1989a). The discrimination of the method was assessed using a panel of 100 unrelated strains from clinical ($n = 65$) and environmental ($n = 33$) sources (2 unknown). Thirty-four distinct RFLP types were revealed. Reproducibility was established by a variety of criteria and was found to be excellent. The same probes can also be used to type strains of other serogroups of *L. pneumophila*, for example, 18 unrelated serogroup 6 strains could be divided into 9 RFLP types.

Listeria monocytogenes *typing*

Figure 2 shows the restriction fragment patterns of 12 strains of *L. monocytogenes* serogroup 1/2. The DNA was digested with the restriction endonuclease *Nci*I and the probe used was a mixture of DNAs isolated from two clones selected from a library of *L. monocytogenes* genomic fragments in the vector λgtWES.λB. (Saunders *et al.* 1989). Seven different restriction fragment patterns can be seen in tracks 2, 3, 5, 8, 14, 15 and 17. The strains in tracks 6, 9, 11 and 18 are indistinguishable from those in tracks 3, 8, 8 and 8 respectively. Track 12 shows the plethora of high molecular weight bands resulting from the analysis of partially digested DNA from one strain. Studies on this probe-based typing method (Saunders *et al.* 1989) have shown that its discrimination is serogroup-dependent with strains of serovar 4 b being predominantly of a single RFLP type whilst the serogroup 1/2 strains are distributed among many types.

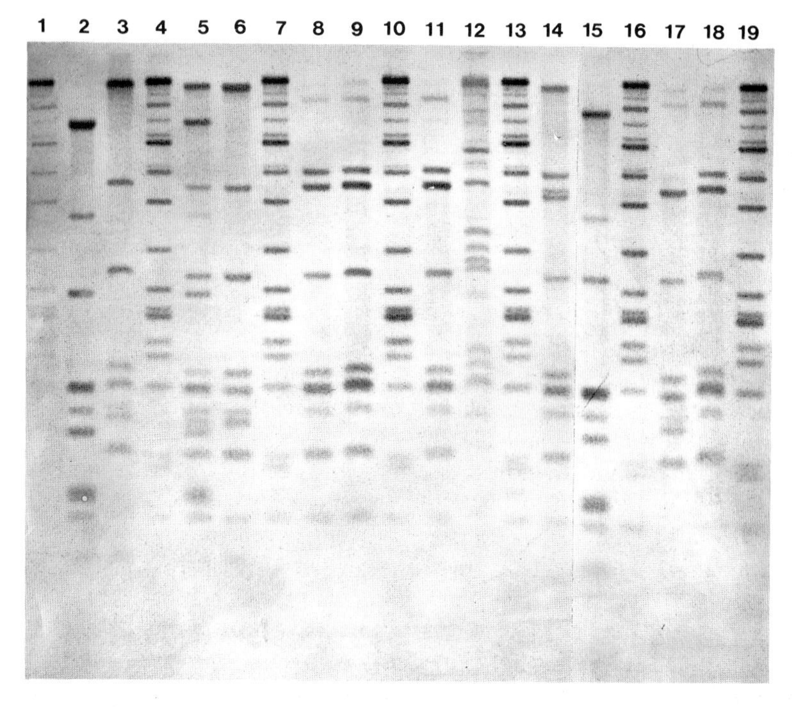

FIG. 2. *Listeria monocytogenes* typing.
Analysis of 12 *L. monocytogenes* serogroup 1/2 strains by RFLP typing using a biotinylated probe consisting of DNA from λNS32 and λNS35 (300 ng). The agarose gel (1%) was electrophoresed at 1.5 V/cm for 16 h. Tracks 1, 4, 7, 10, 13, 16 and 19 show analyses of a mixture of *Eco*RI and *Pst*I digested λ phage DNA (1 ng of each digest). The remaining tracks show *Nci*I digested DNA (1 μg) from *L. monocytogenes* serogroup 1/2 strains.

The use of a rRNA probe for Yersinia *species analysis*

Figure 3 shows the result of analysis of *Nci*I fragments of DNAs from strains of *Yersinia* species and other Enterobacteriaceae with the probe λ142, a cloned rRNA operon from *Legionella pneumophila* (Saunders *et al.* 1988a). The restriction fragment patterns of the *Y. enterocolitica* strains (tracks 14, 16 and 17), although distinct, share many bands. When the patterns obtained from all *Yersinia* strains are compared it can be seen that few bands are common to all isolates and if the strains of other genera are included in the comparison (tracks 2, 3, 4 and 5) even greater pattern divergence is evident. Thus the results correlate well with the published DNA/DNA hybridization data for these species which shows the *Yersinia* species to be 35–75%, and the other Enterobacteriaceae to be approximately 20%, related to strains of *Yersinia*

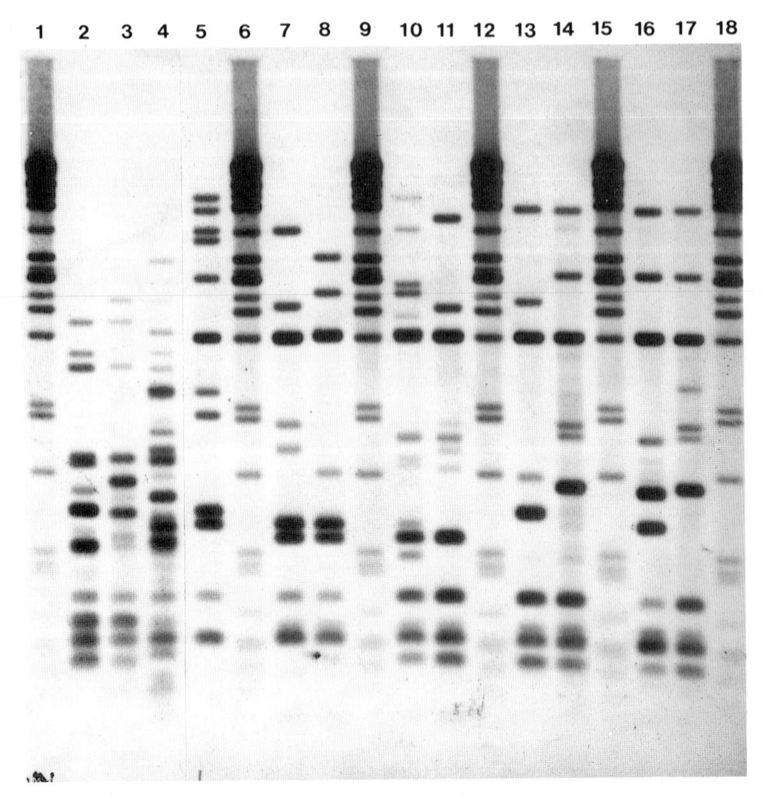

FIG. 3. *Yersinia* species analysis.
Analysis of 12 Enterobacteriaceae using a biotinylated probe consisting of DNA from λ142 (150 ng). The agarose gel (1%) was electrophoresed at 6 V/cm for 3 h. Tracks 1, 6, 9, 12, 15 and 18 show analyses of a mixture of *Eco*RI and *Pst*I digested λ phage DNA (1 ng of each digest). The remaining tracks show *Nci*I digested DNAs (1 μg) from strains as follows: (2), *Citrobacter freundii*; (3), *Salmonella typhimurium*; (4), *Escherichia coli*; (5), *Havnia alvei*; (7), *Yersinia ruckeri* (NCTC 10476); (8), *Yersinia pseudotuberculosis* (NCTC 10275); (10), *Yersinia kristensenii* (NCTC 11471); (11), *Yersinia intermedia* (NCTC 11469); (13), *Yersinia frederiksenii* (NCTC 11470); (14), *Yersinia enterocolitica*; (16), *Yersinia enterocolitica* and (17), *Yersinia enterocolitica* (NCTC 11177).

enterocolitica (Brenner *et al.* 1980). However, these preliminary data indicate that it would not be possible to identify strains of *Yersinia* species reliably on the basis of their restriction fragment patterns which share many interspecies features.

In contrast, the use of λ142 for the analysis of the *Nci*I digested DNAs from species of the genus *Legionella* usually allows accurate identification of the strain to the species level (Saunders *et al.*, 1988 a). This is possible because legionellae are distantly related to species of other genera and in most instances the homology between strains of different *Legionella* species is low (Brenner 1986).

Concluding Comments

Comparison of bacterial isolates by the analysis of RFLPs by means of probes to detect restriction endonuclease fragments derived from a limited portion of the genome is a powerful method for the epidemiological typing of closely related strains and can also be used for species identification.

The main disadvantage of RFLP analysis is that it is a multistep procedure, which must be adhered to closely and it is therefore relatively time-consuming. For this reason it is not usually an appropriate method to apply to large numbers of bacterial isolates. Despite this drawback, the major advantage of the technique is that the results are based on reliable genotypic characteristics rather than on the phenotype and this is sufficient to ensure that it will continue to be of value in various applications.

References

BRENNER, D.J. 1986. Classification of *Legionellaceae*. *Israel Journal of Medical Science* **22**, 620–632.

BRENNER, D.J., URSING, J., BERCOVIER, A., STEIGERWALT, A.G., FANNING, G.R., ALONSO, J.M. & MOLLARET, H.H. 1980. Deoxyribonucleic acid relatedness in *Yersinia enterocolitica* and *Yersinia enterocolitica*-like organisms. *Current Microbiology* **4**, 195–200.

EISENACH, K.D., CRAWFORD, J.T. & BATES, J.H. 1986. Genetic relatedness among strains of the *Mycobacterium tuberculosis* complex. *American Review of Respiratory Disease* **133**, 1065–1068.

GRIMONT, F. & GRIMONT, P.A.D. 1986. Ribosomal ribonucleic acid gene restriction patterns as potential taxonomic tools. *Annals de l'Institut Pasteur/Microbiologie* **137 b**, 165–175.

IRINO, K., GRIMONT, F., CASIN, I., GRIMONT, P.A.D. & the Brazilian Purpuric Fever Study Group. 1988. rRNA gene restriction patterns of *Haemophilus influenza* biogroup aegyptius strains associated with Brazilian Purpuric Fever. *Journal of Clinical Microbiology* **26**, 1535–1538.

JEFFREYS, A.J., WILSON, V. & THEIN, J.L. 1985. Individual specific 'fingerprints' of human DNA. *Nature* **316**, 76–79.

KAISER, K. & MURRAY, N.E. 1985. The use of phage lambda vectors in the construction of representative genomic DNA libraries. In *DNA Cloning*, Vol. I. ed. Glover, D. M. pp. 1–47. Oxford: IRL Press.

KARN, J., BRENNER, S. & BARNETT, L. 1983. New bacteriophage lambda vectors with positive selection for cloned inserts. In *Methods of Enzymology*. Vol. 101, eds Wu, R., Grossman, L. & Moldave, K. p. 3–19. London: Academic Press.

LIN, R.J., CAPAGE, M. & HILL, C.W. 1984. A repetitive DNA sequence, *rhs*, responsible for duplications within the *Escherichia coli* K-12 chromosome. *Journal of Molecular Biology* **177**, 1–18.

MANIATIS, T., FRITSCH, E.F. & SAMBROOK, J. 1982. *Molecular Cloning: A Laboratory Manual.* Cold Spring Harbor, New York: Cold Spring Harbor Laboratory.

MODARSKI, M. 1985. Detection of ribosomal nucleic acid homologies. In: *The Society for Applied Bacteriology Technical series* No. 20. eds Goodfellow, M. & Minnikin, D.E. pp. 41–66. London: Academic Press.

MUSSER, J.M., KROLL, J.S., MOXON, E.R. & SELANDER, R.K. 1988. Clonal population structure of encapsulated *Haemophilis influenzae*. *Infection and Immunity* **56**, 1837–1845.

OWEN, R.J. & PITCHER, D.G. 1985. Current methods for estimating DNA base composition and levels of DNA-DNA hybridization. In *The Society for Applied Bacteriology. Technical Series* No. 20. eds Goodfellow, M. & Minnikin, D.E. pp. 67–93. London: Academic Press.

PITCHER, D.G., SAUNDERS, N.A. & OWEN, R.J. 1989. Rapid extraction of bacterial genomic DNA with guanidium thiocyanate. *Letters in Applied Microbiology* **8**, 151–156.

SAUNDERS, N.A., HARRISON, T.G., KACHWALLA, N. & TAYLOR, A.G. 1988 a. Identification of species of the genus *Legionella* using a rRNA gene from *Legionella pneumophila*. *Journal of General Microbiology* **134**, 2363–2374.

SAUNDERS, N.A., KACHWALLA, N., HARRISON, T.G. & TAYLOR, A.G. 1988b. Cloned nucleic acid probes for detection and identification of legionellae. *Analytical Proceedings* **25**, 128–129.

SAUNDERS, N.A., HARRISON, T.G., HATHTHOTUWA, A., KACHWALLA, N. & TAYLOR, A.G. 1990. A method for typing strains of *Legionella pneumophila* serogroup 1 by analysis of restriction fragment length polymorphisms. *Journal of Medical Microbiology* **31**, 45–55.

SAUNDERS, N.A., RIDLEY, A.M. & TAYLOR, A.G. 1989. Typing of *Listeria monocytogenes* for epidemiological studies using DNA probes. *Acta Microbiologica Hungarica* **36**, 205–209.

SOUTHERN, E.M. 1975. Detection of specific sequences among DNA fragments separated by gel eletrophoresis. *Journal of Molecular Biology* **98**, 503–517.

TOMPKINS, L.S., TROUP, N., LABIGNE-ROUSSEL, A. & COHEN, M.L. 1986. Cloned, random chromosomal sequences as probes to identify *Salmonella* species. *Journal of Infectious Diseases* **154**, 156–162.

The Use of DNA Probes to Detect 'Selected Bacteria' After Their Introduction into a Waste Water Treatment Plant

GABY GAYER-HERKERT AND H.J. KUTZNER
Institut fur Microbiologie, Technische Hochschule Darmstadt, Schnittpahnstrasse 10, 6100 Darmstadt, Germany

Several commercial companies advocate cultures of 'selected bacteria' for the improvement of the waste water treatment process. The strain composition of these preparations, however, remains unknown to the user—as do the mechanisms responsible for the higher efficiency of these bacteria as compared with the autochthonous microflora of the activated sludge.

One important aspect of the application of 'selected bacteria' to such a complex ecological system as a waste water treatment plant is the question as to whether or not they can compete with the autochthonous microflora and establish themselves in this habitat. One way of investigating this point is to isolate and identify numerous strains but, although this is nowadays facilitated by miniaturized tests and computerized identification systems (Kappesser *et al.* 1989) it is still a very laborious task and, since a species identification does not detect strain differences, it may not be sensitive enough.

This chapter describes the use of DNA probes to detect the 'selected bacteria'.

Materials and Methods

Donor strain for DNA probe

The donor strain was an isolate from DBC A_1 (*d*ried *b*acterial *c*ulture) from Flow laboratories GmbH (Meckenheim, Germany); this culture is recommended for the improvement of communal waste water. Studies by Kappesser *et al.* (1989) have shown that DBC A_1 contains a cluster of *Streptococcus* strains which was not found in samples from activated sludge (Table 1). Strain no. 885 was selected for the preparation of a DNA probe.

Genetic Manipulation
0-632-02926-9

TABLE 1. *Hybridization of* Streptococcus-*isolates and DSM reference strains with two DNA-probes from strain No. 885*

		Hybridization with:	
Strain no.	Identified as	Chromosomal DNA	4, 5 kb/ fragment
909	*S. gallinarum*	+	+
922	*S. faecium*	+	+
959	*S. faecium/faecalis*	+	+
882	*S. faecium/faecalis*	+	+
883	*S. faecium/faecalis*	+	+
884	*S. faecium/faecalis*	+	+
885	*S. faecium/faecalis*	+	+
DSM 2146	*S. faecium*	+	+
DSM 20382	*S. faecium*	+	–
DSM 20371	*S. faecalis*	+	–
DSM 20376	*S. faecalis*	+	–
DSM 20628	*S. gallinarum*	+	–
DSM 20063	*S. avium*	+	–
DSM 2072	*S. pyogenes*	+	–
37 Representative strains of many different species		–	–

DSM = Deutsche Sammlung von Mikro organismen
(German Collection of Micro organisms)

Construction of the DNA probe

Two kinds of probes were used in this study: the chromosomal DNA from strain no. 885 and a 4.5 kb fragment from the same strain cloned in the *E. coli* plasmid pUC19. Chromosomal DNA was isolated by a 'rapid method' (Hopwood *et al.* 1985). The 4.5 kb fragment was isolated from *Streptococcus* DNA after digestion with *Pst*I and gel electrophoresis. It was then cloned in pUC19 as depicted in Fig. 1 which also shows the restriction map of the 4.5 kb fragment.

Non-radioactive hybridization

Generally, the BluGENE(TM) reaction kit from Gibco BRL was used for the nick translation and the subsequent staining steps. The method itself was adapted from that described by Schneider & Muller (1988) which is based on established techniques (De Ley 1970; Leary *et al.* 1983; Meinkoth & Geoffrey 1984; Southern 1975). The DNA to be hybridized with the probes came from three sources:

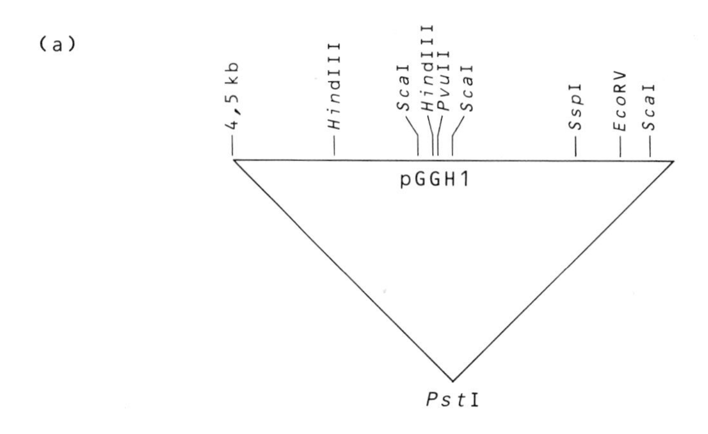

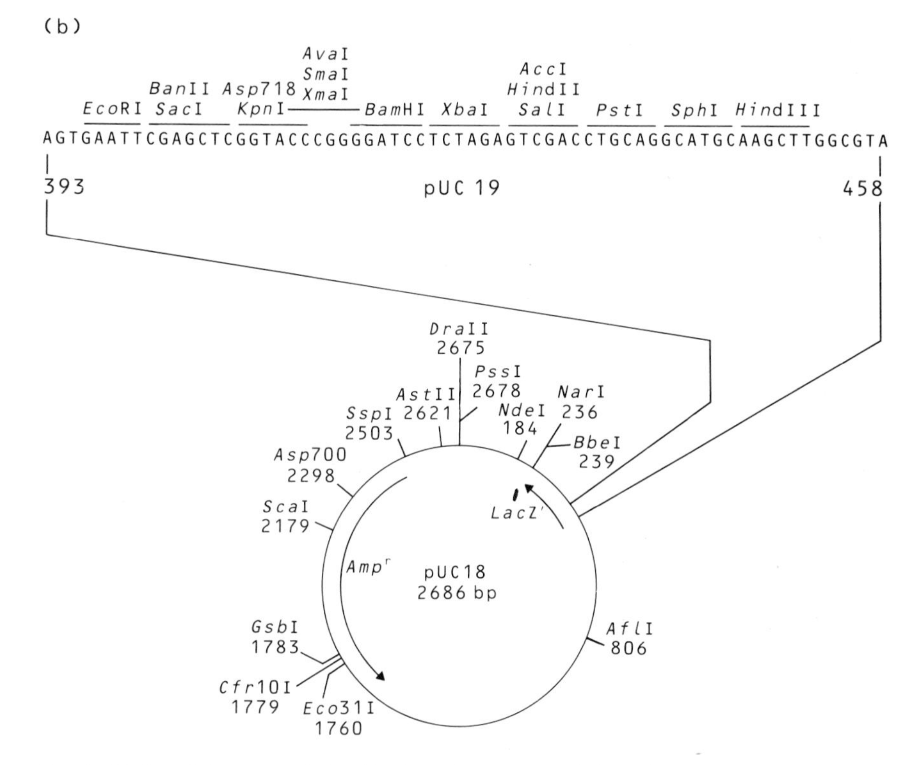

FIG. 1. (a) Restriction map of the 4.5 kb probe constructed from the chromosomal DNA of *Streptococcus* no. 8. (pGGH1); (b) plasmid pUC19 used for cloning and propagation of the 4.5 kb probe into the *Pst*I site of its polylinker.

1 Strains of the *Streptococcus* cluster isolated from DBC A_1.
2 Activated sludge taken before, during and after the addition of DBC A_1.
3 Reference strains (Deutsche Sammlung von Micro organismen—DSM) of the *Streptococcus* cluster.

Hybridization was carried out by two methods: Southern hybridization and dot blot hybridization

Results and Discussion

Hybridization with DBC A_1 isolates and reference strains

As shown in Table 1 and Fig. 2, the two DNA probes prepared from strain no. 885 hybridized with all seven strains of the *Streptococcus* cluster tested (Southern blot hybridization under very stringent conditions gave 95–100% homology). Six of the seven strains from this cluster were identified as *Streptococcus faecium/faecalis*, the other one as *Streptococcus gallinarum*. This result indicates that the identification of this one strain should be reconsidered. In addition, the chromosomal DNA probe of strain no. 885 hybridized with all DSM reference strains belonging to the genus *Streptococcus* (dot blot hybridization). In contrast, the 4.5 kb DNA probe (cloned fragment from the chromosomal DNA of strain no. 885) gave a positive signal only with DSM 2146 (*Streptococcus faecium*). None of the 37 representatives of numerous other genera isolated from DBC A_1 gave positive signals with either of the two probes (random selection shown in Fig. 2). This result shows that the *Streptococcus* strain selected for this study yielded one probe (chromosomal DNA) specific for the genus *Streptococcus* and another (cloned 4.5 kb DNA fragment) specific for one strain of *Streptococcus faecium*.

Hybridization with activated sludge samples

When both probes, i.e. whole DNA and the 4.5 kb fragment respectively, were tested with DNA from activated sludge only samples taken from a plant inoculated with DBC A_1 reacted positively (Fig. 3). Surprisingly the streptococci of the 'selected bacterial culture' could be detected even several weeks after the inoculation had been terminated, proving that they had established themselves in this 'new' habitat.

Conclusions

DNA probes able to detect specifically a certain bacterium among a host of other bacteria, e.g. the microflora of soil, waste water and plant material, have been applied by various authors (see Hazen & Jiminez 1988). In the authors'

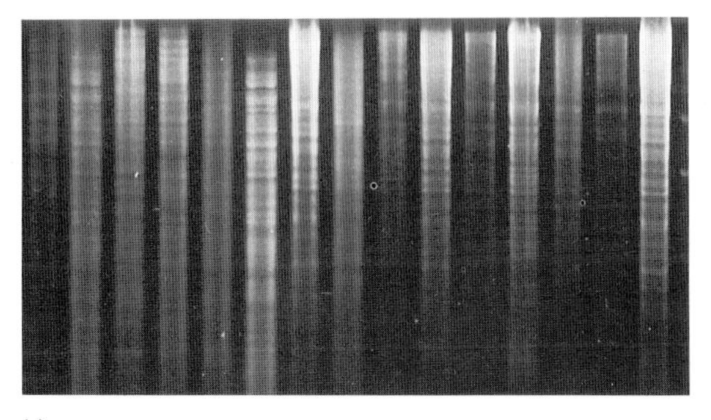

(a)

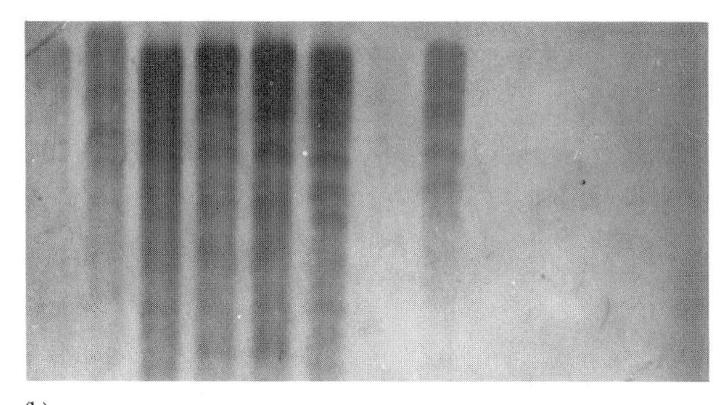

(b)

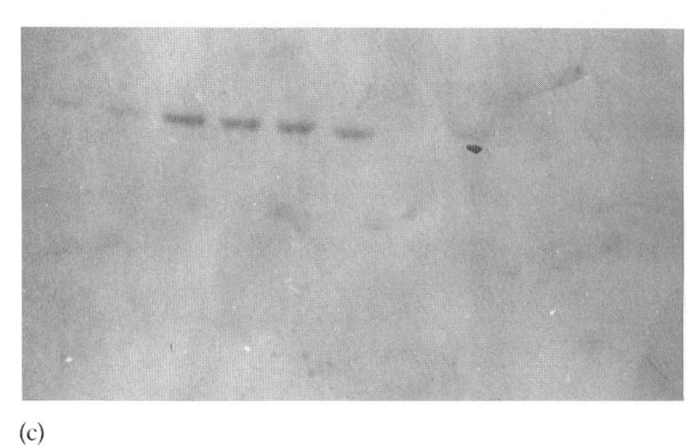

(c)

FIG. 2. Southern hybridization of *Streptococcus* DNA with representative strains from the *Streptococcus* cluster, left side, and a random selection of control strains from DBC A_1, right side. The strain no. 908 is not a member of the *Streptococcus* cluster: it was later identified as a lactobacillus. (a) Agarose gel electrophoresis of *Pst*I digest of DNA; (b) application of the total DNA probe from strain no. 885; (c) application of the probe prepared from the cloned 4.5 kb DNA fragment from strain no. 885.

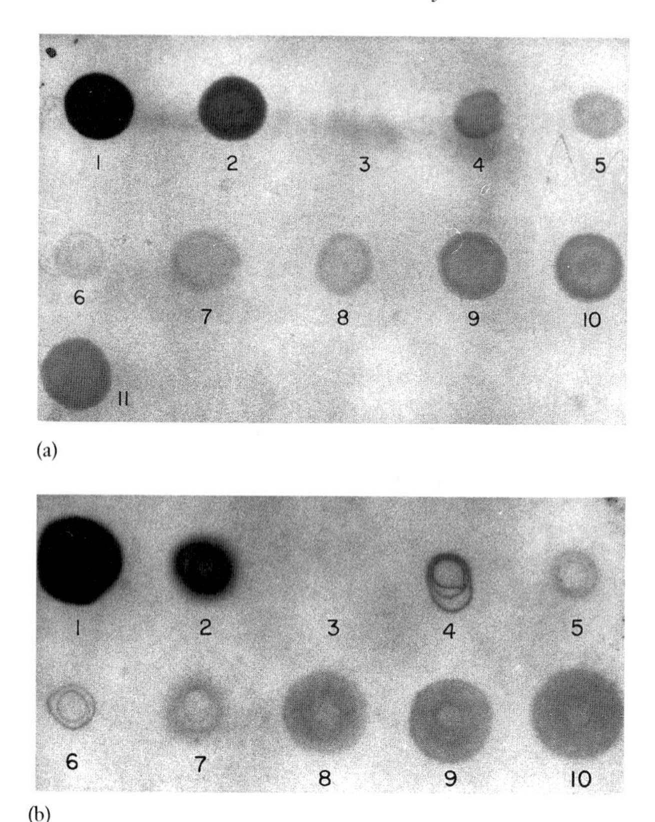

FIG. 3. Dot blot hybridization. The DNA was isolated from activated sludge of a waste water treatment plant before and after addition of the 'special cultures' on 27th September 1988. The last day of addition of the culture was 22nd November 1988. (a) Dot blot hybridizations with whole DNA from strain no. 885; dot no. 1 = positive control (probe). dot no. 2 = DNA from the 'special culture', dot no. 3 = negative control (DNA from activated sludge before the inoculation of the 'special culture'), dots nos. 4–11 = activated sludge probed during and after the application DBC A_1. The last dots were prepared several weeks after terminating the inoculation; (b) dot blot hybridizations with the 4.5 kb fragment prepared from strain no. 885 after cloning in *E. coli* plasmid pUC19 (dots 1–10 as in Fig. 3a).

study the method has been successfully employed to trace members of a 'selected bacterial culture' after its inoculation into a waste water treatment plant. The construction of probes for other bacteria of this system is planned, and their use will without doubt contribute to a better understanding of the population dynamics within an activated sludge plant and the application of the so-called 'selected bacterial cultures' to this process.

Acknowledgements

This work is supported by the grant 02 WA 87263 from the Bundesministerium für Forschung und Technologie (BMFT), Projektrager KFZ-Karlsruhe and a doctoral fellowship from Frankfurter Allgemeine (FAZ).

References

De Ley, J. 1970. Reexamination of the association between melting point, buoyant density, and chemical base composition of deoxyribonucleic acid. *Journal of Bacteriology* **101**, 738–754.

Hazen, T.C. & Jiminez, L. 1988. Enumeration and identification of bacteria from environmental samples using nucleic acid probes. *Microbiological Sciences* **5**, 340–343.

Hopwood, D.A., Bibb, M.J., Chater, K.F., Kieser, T., Bruton, C.J., Kieser, H.M., Lydiate, D.J., Smith, C.P., Ward, J.M. & Schrempf, H. 1985. *Genetic Manipulation of Streptomyces, A Laboratory Manual.* Norwich: The John Innes Foundation.

Leary, J.J., Brigati, D.J. & Ward, D.C. 1983. Rapid and sensitive colorimetric method for visualizing biotin-labelled DNA probes hybridized to DNA or RNA immobilized on nitrocellulose Bio-blots. *Proceedings of the National Academy of Sciences of the United States of America* **80**, 4045–4049.

Meinkoth, J. & Geoffrey, W. 1984. Hybridization of nucleic acids immobilized on solid supports. *Analytical Biochemistry* **158**, 267–284.

Schneider, J. & Muller, L. 1988. DNA-DNA-Hybridisierung mit biotinylierten DNA-Sonden. *Forum Mikrobiologie* **11**, 254–262.

Southern, E.M. 1975. Detection of specific sequences among DNA fragments separated by gel electrophoresis. *Journal of Molecular Biology* **98**, 503–517.

Kappesser, S., Rude, E. & Kutzner, H.J. 1989. Microbiological studies of 'selected bacterial cultures' for the aerobic treatment of waste water. Dechema: Biotechnology Conferences, No. 3. 1989.

The Use of Genetic Manipulation to Study Bacterial Pathogens

R. Wilson and Karen E. Webb
Host Defence Unit, Department of Thoracic Medicine, National Heart and Lung Institute, Dovehouse Street, London SW3 6LY, UK

The pathogenic potential of a bacterium is usually a complex function which is dependent on multiple determinants. This is particularly true of bacteria that normally have a commensal relationship with their host, and those that cause opportunistic infections. This chapter describes methods that use genetic manipulation to study the pathogenicity of bacteria that are associated with acute and chronic bronchial respiratory infection. First, it is necessary to understand that most of them are not virulent in the usual sense of the word. They have become adapted to survive either in the environment, or in a commensal relationship with their host, and disease when it occurs can be thought of as an accidental consequence of the microbial factors which are advantageous for bacterial survival, often together with a change in the host defences.

The bacteria that commonly cause acute tracheobronchitis, infectious exacerbations of chronic bronchitis, or are associated with chronic respiratory infection, such as *Streptococcus pneumoniae* and *Haemophilus influenzae*, are frequently found as commensals in the upper respiratory tract of normal healthy individuals. Therefore their pathogenicity should be investigated in the wider context of conditions which allow their perpetuation on mucosal surfaces. The commensal microbe is highly adapted to its host, and deploys strategies to ensure its survival, proliferation and dissemination. Under normal circumstances, this may result in no discernable damage to the host and is known as the carrier state. However, on occasions, the organism may multiply and spread, e.g. to the lower respiratory tract, resulting in disease of varying severity. The carrier state can be thought of as a balance which is altered during disease, usually by a change in the host state rather than by a change in bacterial properties. For this reason, bacterial infections of the bronchial tree are commonly preceded by a virus infection, which creates a deficiency in the host defences that bacteria can exploit.

Genetic Manipulation
0-632-02926-9

Pseudomonas aeruginosa is a Gram-negative bacterium that inhabits moist environments as diverse as mountain streams and the laboratory sink. The bacterium is recognized by the production of a blue phenazine pigment called pyocyanin. This is a secondary metabolite which is the end-product of a complicated biochemical pathway. In nature its production could possibly protect the organism from u.v. radiation, and have a role in acquisition of iron for growth. *P. aeruginosa* is not a commensal and does not infect normal individuals, but is an opportunist pathogen. In cystic fibrosis patients particularly, and in other forms of chronic chest sepsis, it infects the bronchial tree and is difficult to eradicate, leading to a deterioration in respiratory function. Pyocyanin (Fig. 1) is a redox compound and has been shown to have antibiotic properties against other bacterial species (which could help during competition for an ecological niche), to adversely affect the cells of the respiratory epithelium, and to affect lymphocyte and neutrophil function *in vitro*. Pyocyanin may therefore play a part in the pathogenesis of respiratory infection by *P. aeuruginosa*, although in evolutionary terms it probably had a very different role.

The potential of a bacterium to compete successfully for essential nutrients, to evade host defences, to propagate from one host to another, and to cause tissue damage involves a complex phenotype governed by many genes. Until recently, the analysis of bacterial pathogenicity depended upon identifying phenotypic characteristics that correlated with pathogenicity. A well-known example is the use of the coagulase test to identify virulent strains of staphy-

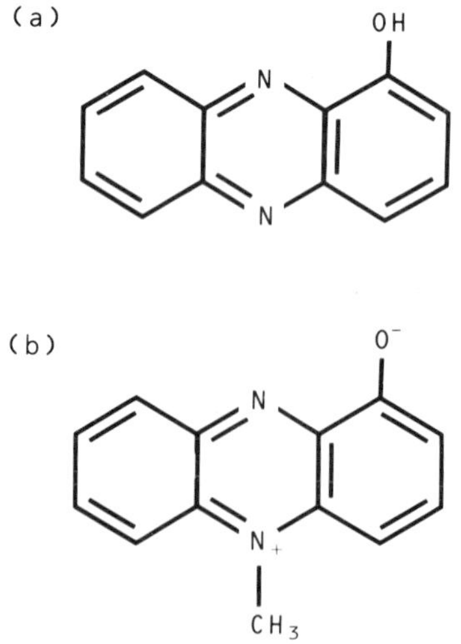

(a) OH N N

(b) O^- N N_+ CH_3

FIG. 1. The molecular structure of 1-hydroxyphenazine (a); and pyocyanin (b).

lococci. However, a specific property can correlate with virulence potential but lack any direct relevance itself to the mechanisms of pathogenicity.

The ability to analyse the so-called pathogenic personality of microorganisms at a molecular level has been greatly facilitated through the application of genetic and recombinant DNA techniques. One highly successful approach has been the comparison of isogenic strains. These are strains that differ only in being sufficient or deficient in one or more genes which are required for the expression of a particular virulence determinant. It should be emphasized that studies of virulence require biologically relevant model systems so that results can be related to human disease. *In vitro* experiments and animal models of human disease are of necessity divorced from the *in vivo* situation in man. Nevertheless, the cautious interpretation of results comparing the performance of isogenic strains in experimental infection of animals, organ cultures, and isolated cell systems, has been very instructive.

A Genetic Analysis of Pneumolysin (a virulence determinant of *Streptococcus pneumoniae*)

Recombinant DNA techniques offer many other opportunities for the study of bacterial pathogenicity. For example, genes responsible for bacterial toxin production can be cloned into *E. coli* expression vectors and when they are transcribed they produce protein at levels sufficient to facilitate purification of recombinant toxin. Oligonucleotide-mediated site-directed mutagenesis is a technique that utilizes the cloned gene to engineer single amino acid substitutes in the protein (Kunkel 1985). Comparisons can then be made between the function of native and mutant toxins.

S. pneumoniae is often responsible for community-acquired pneumonia and infectious exacerbations of chronic bronchitis. Although the polysaccharide capsule of the bacterium has long been held to be the major virulence determinant, recent work has emphasized the contribution of protein toxins produced by the bacterium. Pneumolysin is a sulphydryl-activated toxin which causes lysis of red blood cells. The toxin binds to cholesterol in cell membranes and then oligomerizes with other toxin molecules to form transmembrane channels that lead to cell lysis. At sublytic concentrations pneumolysin affects the function of polymorphonuclear leucocytes and monocytes and, at higher concentrations, it activates the classical pathway of complement.

Cloning of the pneumolysin gene was accomplished by constructing genomic libraries of *S. pneumoniae* DNA sequences in the λ phage insertion vector λgt10 (Walker *et al.* 1987). Overlaying recombinant phage plaques with sheep erythrocytes revealed the presence of haemolytic recombinants. One clone λPLY003 contained a 5.5 kb insert containing the pneumolysin gene. A 2.5 kb *Tth*111 fragment was subcloned into pUC8 which, when transformed into

E. coli, directed production of pneumolysin. This provided enough pneumolysin to study its activity in a number of biological models.

The effect of pneumolysin on human respiratory epithelium *in vitro* was to slow ciliary beating, and at higher concentrations to disrupt epithelial surface integrity by causing cell extrusion. By using the technique of oligonucleotide mediated site-directed mutagenesis to produce single amino acid substitutions in the toxin it has been possible to create mutant toxins with loss of either haemolytic or complement activating activity (Saunders *et al.* 1989). Mutant toxins with reduced haemolytic activity had less effect on respiratory epithelium, while a change in the complement-activating properties of the toxin did not change its final effect on epithelium. Another mutant toxin had normal haemolytic and normal complement-activating activity, yet had a delayed onset of its effect on respiratory epithelium, presumably due to altered binding of the toxin to receptors or altered oligomerization of toxin molecules in the cell membrane.

The Effect of *Pseudomonas aeruginosa* on Respiratory Epithelium

A cell biology approach

The mucociliary system (Fig. 2) forms a first-line defence mechanism of the bronchial tree against inhaled particles, including bacteria. For efficient mucus transport, each individual cilium must not only beat in a co-ordinated manner itself but must do so in concert with all the other cilia on the same and adjacent cells. Some bacteria produce factors which disturb this sytem by both slowing and disorganizing ciliary beating (Wilson 1988). It has been suggested that by perturbing mucociliary clearance the bacterium can create time to multiply in the respiratory tract, and that by slowing ciliary beating it removes a physical barrier which prevents the bacterium adhering to the respiratory epithelium.

We first noted that 18 h culture filtrates of *P. aeruginosa* slowed and disorganized human ciliary beating *in vitro* (Wilson *et al.* 1987). Prolonged incubation caused ciliary stasis and epithelial disruption. Assays were developed to measure the amount of known virulence factors in the filtrates and to correlate these with ciliary slowing activity. Only the phenazine pigment content of the filtrates correlated. Gel filtration yielded only one peak of ciliary slowing activity which co-eluted with the pigments, and the accumulation of pigment during bacterial culture correlated with an increase of ciliary slowing activity.

Two phenazines, pyocyanin and 1-hydroxyphenazine, were extracted from cultures and purified by high performance liquid chromatography. 1-Hydroxyphenazine caused immediate onset of ciliary slowing and dyskinesia (disorganized beating) which was not associated with epithelial disruption.

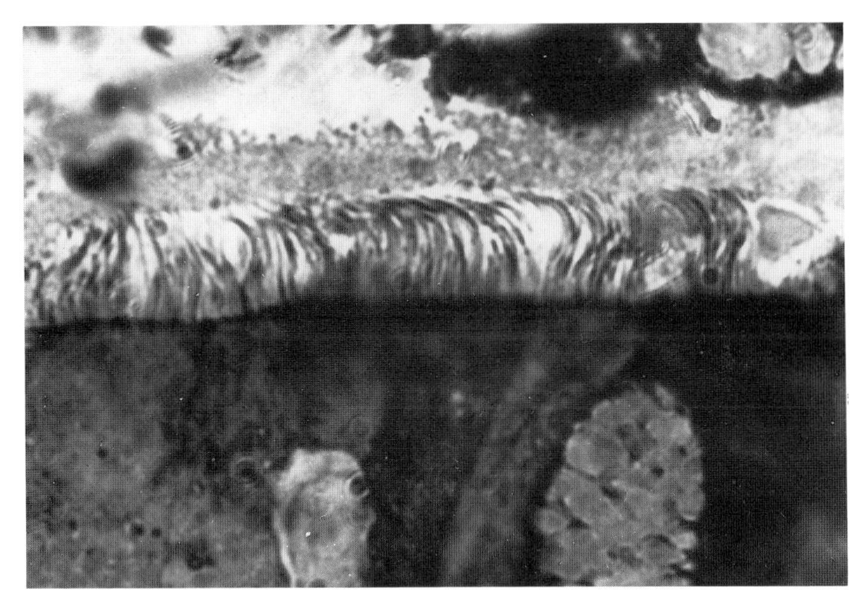

FIG. 2. Human ciliated respiratory epithelium. Inhaled particles including bacteria stick onto the overlying mucus layer. The mucus is transported out of the respiratory tract by the beating of the underlying cilia within the watery periciliary fluid.

Pyocyanin caused a gradual slowing of ciliary beating associated with epithelial disruption. Both these compounds have been extracted from the sputum of patients infected by *P. aeruginosa* at concentrations similar to those required to produce an effect *in vitro*, and both slowed mucociliary clearance in the guinea pig *in vivo*. A bolus dose of l-hydroxyphenazine slowed mucociliary clearance immediately, although it subsequently gradually recovered, while a bolus dose of pyocyanin had no immediate effect but later clearance fell without any recovery.

Therefore, when taken with the other studies mentioned above, a reasonable case has been made for the phenazine pigments contributing to the pathogenesis of *P. aeruginosa* respiratory infection. However, the bacterium produces many other toxins, some of which also slow mucociliary clearance. In order to investigate the contribution of phenazines to pseudomonas pathogenesis we have sought to create isogenic phenazine-producing and non-producing strains.

A genetic approach

P. aeruginosa differs from other Gram-negative bacteria such as *E. coli* and *Salmonella typhimurium* in that genes for biosynthetic pathways are not clustered on the chromosome. Therefore it was unlikely that all the genes necessary for phenazine production could be represented in a single clone in a gene library. We have chosen to use transposon mutagenesis (Fig. 3) to create mutant strains deficient in phenazine production which are otherwise identical to the wild type. Other approaches could have been adopted and some of these will be described in the context of our investigation. The techniques outlined are also applicable to other bacterial species, although the detailed methodology may need modification.

Transposons can be thought of as 'jumping genes'. They are usually 5–15 kb in size and carry genes for transposase activity and further functions, usually determining resistance to antibiotics. They have the ability to replicate and insert one copy at a new location in the genome. When a transposon inserts into a gene, the function of that gene is lost because the genetic code for transcription is interrupted and the protein encoded can no longer be formed. If the gene into which the insertion had occurred was a vital one, e.g. controlling cell division, this would be lethal to the bacterium, but if the gene is non-essential then the transposon can provide a positive selection marker (because it codes for antibiotic resistance) of the mutant phenotype.

Transposition in nature is a rare event controlled by feedback regulation. It confers an advantage by promoting exchanges and rearrangements of genes, and it accounts for much of the spread of antibiotic resistance between pathogenic bacteria. Experimentally, transposons can be used as mutagens. An ideal transposon in this context inserts into target DNA at (almost) random sites. The transposon is carried on a plasmid which is replicated stably in the donor cell but not in the recipient cell. The plasmid is transferred from the donor to the recipient, but as it cannot replicate in the recipient, the plasmid is lost. The transposon can only survive if it jumps into the recipient's chromosome. This event occurs with a probability of 10^{-4} to 10^{-6} per cell and can be selected for by the antibiotic marker. Once an interesting mutant

FIG. 3. (a) Transposon mutagenesis of *Pseudomonas aeruginosa* by TN5–751 carrying genes for trimethoprim (Tp) and kanamycin (K) resistance on the plasmid pME305 (pME9) which itself carries genes for carbenicillin (C) and tetracycline (Tc) resistance: (i) transfer of the plasmid pME9 from *E. coli* (streptomycin sensitive) to *P. aeruginosa* (streptomycin resistant) by conjugation; (ii) screen on antibiotic containing agar to select for successful transfer of the plasmid; (iii) growth at 43°C leads to elimination of the temperature-sensitive plasmid. Only bacteria in which the transposon has survived by jumping into the pseudomonas genome survive. (b) Use of transposon mutagenesis to create a probe to locate a cloned virulence gene within a pseudomonas genomic library: (i) as described above, the plasmid bearing the transposon is transferred from *E. coli* to *P. aeruginosa*; (ii) the transposon hops randomly into the pseudomonas genome and the plasmid is eliminated. By chance, the virulence gene of interest is interrupted by the transposon insertion. The transposon is cloned, carrying with it flanking regions of pseudomonas DNA; (iii) a pseudomonas gene library is probed with the cloned transposon. The homology of the flanking DNA allows selection of the cloned virulence gene.

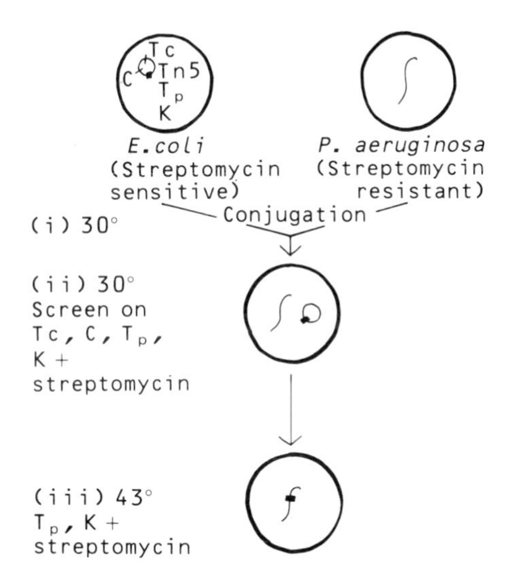
(a)
Tc
C
Tn5
Tp
K
E.coli
(Streptomycin
sensitive)
P. aeruginosa
(Streptomycin
resistant)
Conjugation
(i) 30°
(ii) 30°
Screen on
Tc, C, Tp,
K +
streptomycin
(iii) 43°
Tp, K +
streptomycin

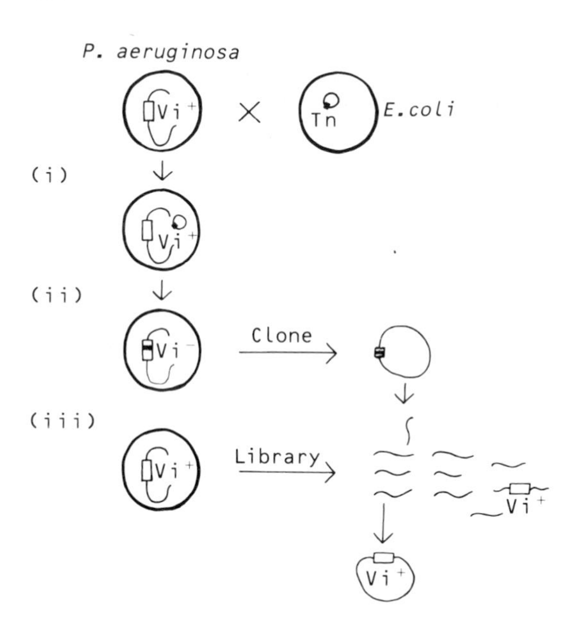
(b)
P. aeruginosa
Vi+
E.coli
Tn
(i)
Vi+
(ii)
Vi−
Clone
(iii)
Vi+
Library
Vi+
Vi+

has been identified, the region of DNA containing the transposon can be cloned, selecting for the antibiotic resistance markers on the transposon. The cloned DNA fragment (the transposon flanked by pseudomonas DNA) can then itself be used as a probe for the identification of the wild-type gene in a *P. aeruginosa* gene library (Fig. 3).

Genetic material can be moved between bacteria by conjugation, transformation and transduction. These important techniques are described below. The experimental methods have been applied to the study of pyocyanin production by *P. aeruginosa*.

Methods

Conjugation

This describes mating between two bacterial cells when part of the genetic material from one cell is transferred to the other. For example, the first step in transposon mutagenesis is the transfer of a plasmid containing the transposon from the donor cell (usually *E. coli*) to the recipient cell (*Pseudomonas* in our experiments). The methodology below describes the transfer of plasmid pME9 (carrying the genes for resistance to tetracycline and carbenicillin and the transposon Tn5-751 with genes for resistance to kanamycin and trimethoprim) from the streptomycin-sensitive *E. coli* ED8654 to the streptomycin-resistant *P. aeruginosa* PA06049.

1 Grow PA06049 in 10 ml nutrient broth containing streptomycin 1 000 μg/ml at 43°C overnight without aeration, and ED8654pME9 in 10 ml nutrient broth containing tetracycline 25 μg/ml and kanamycin 25 μg/ml at 30°C. Growing pseudomonas at 43°C without aeration allows replication but reduces its restriction enzyme systems and therefore increases successful conjugation. pME9 is a temperature-sensitive plasmid which does not replicate at 43°C.

2 Inoculate 10 ml of fresh nutrient broth with 1 ml of overnight cultures and grow under the above conditions for 4 h.

3 Centrifuge at 3 000 rpm for 5 min.

4 Decant supernatant and resuspend pellets in 500 μl broth.

5 Mix 100 μl of PA06049 and ED8654pME9 and drop on to a 4 cm^3 patch on nutrient agar and allow to dry. Both bacteria alone should be used as controls. Mating efficiency may be increased by dropping the mixture on to a 0.45 μm filter laying on top of the agar.

6 Incubate for 4 h at 30°C.

7 Scrape cells into 1 ml of phosphate buffered saline and resuspend.

8 Plate out for single colonies on nutrient agar containing streptomycin 1 000 μg/ml, kanamycin 300 μg/ml, trimethoprim 500 μg/ml, carbenicillin 500 μg/ml and tetracycline 125 μg/ml. Incubate overnight at 30°C. The only surviving bacteria will be PA06049pME9. No bacteria should grow in control experiments.

Transposon mutagenesis

Transposon mutagenesis with Tn5-751 on the temperature-sensitive plasmid pME9 (Rella *et al.* 1985) has been used to create phenazine minus isogenic strains. The culture conditions have been adjusted so that only isolates in which the transposon has 'jumped' into pseudomonas DNA survive.

1 Pick single colonies of PA06049pME9 on to nutrient agar containing kanamycin 300 μg/ml, trimethoprim 500 μg/ml and streptomycin 1 000 μg/ml. Culture at 43°C.

2 Pick single colonies on to pigment-enhancing (King's) agar to screen for changes in phenazine pigment production and on to nutrient agar containing carbenicillin 500 μg/ml and tetracycline 125 μg/ml to demonstrate loss of plasmid pME9 (no growth).

3 The presence of the transposon in the pseudomonas genome is confirmed by DNA hybridization. Restriction endonuclease fragments of mutant pseudomonas genomic DNA are separated by agarose gel electrophoresis and then transferred to nitrocellulose by Southern blotting. The DNA is fixed to the nitrocellulose by baking. DNA–DNA hybridization is carried out with radiolabelled Tn5-751 DNA.

Transduction

This relies on a bacteriophage accidentally packaging cellular DNA sequences and transferring them to the next bacterium it infects (Stapleton *et al.* 1984). In this way, mutations in the phenazine biosynthetic pathway can be transferred to different *P. aeruginosa* strains. Bacteriophages which propagate on *P. aeruginosa* are very common. For example phage F116L transduces markers at a frequency of 1×10^{-7} to 5×10^{-7} per plaque forming unit.

1 Subculture propagating strain (e.g. PAO6049 phenazine mutant) into 5 ml nutrient broth, incubate at 37°C overnight.

2 Make a 1 in 20 dilution of overnight broth in molten soft agar (0.3% agar) maintained at 50°C.

3 Add sufficient bacteriophage to give equivalent of 10 × RTD (routine test dilution—the highest dilution of stock phage just giving confluent lysis) in soft agar mixture.

4 Pour immediately onto surface of cold nutrient agar plate (9 cm diameter).

5 Allow to set, incubate (lid uppermost) at 30°C overnight.

6 Harvest phage: to each plate add 3 ml nutrient broth, gently break up soft agar layer with a sterile bent glass rod, leave for 15 min.

7 Pipette fluid into a sterile container, wash plate surface with a further 2 ml nutrient broth and add to above.
8 If phage is chloroform-resistant, add 0.5 ml chloroform for each plate harvested and centrifuge at 3 000 rpm for 30 min or, if phage is chloroform-sensitive, centrifuge at 10 000 rpm for 30 min and filter the supernatant through a 0.22 μm filter.
9 Titrate phage (see below).
10 Culture recipient cells (another pseudomonas strain to which you wish to transfer the phenazine mutation) to late log phase in nutrient broth.
11 Mix phage with these cells at a multiplicity of infection of one and plate in 0.7% soft agar over nutrient plates containing kanamycin anamycin and trimethoprim (transposon antibiotics). Drug-resistant colonies are screened on King's agar for change in pigment production.

Titration of stock phage

1 Take 1.0 ml of nutrient broth containing stock phage (10^{-1} dilution).
2 Distribute 0.9 ml nutrient broth into each of seven tubes.
3 With a sterile pipette, mix 10^{-1} dilution and pipette 0.1 ml into tube one (= 10^{-2} dilution). Discard pipette.
4 With a sterile pipette, mix 10^{-2} dilution and pipette 0.1 ml into tube 2 (= 10^{-3} dilution). Discard pipette.
5 Repeat step 4 for remaining dilutions up to 10^{-8}.
6 Flood nutrient agar plate with a 4 h broth culture of the propagating strain and remove excess. Allow the plate to dry (without lid) for 15 min.
7 Starting at the 10^{-8} dilution, apply one 20 μl drop of each dilution to the plate. Incubate at 30°C overnight.
8 Note the dilution with the highest number of countable (distinct) plaques. Let this dilution be 10^{-x} and the number of plaques be T. The phage titre is $T \times 50 \times 10^{x}$. The routine test dilution is the highest dilution just failing to give confluent lysis.

Transformation

Transduction and conjugation share the common feature that the transferred DNA is protected from the environment, either by the bacteriophage capsid or the sex pilus, whereas transformation occurs with exposed DNA which the bacterium takes up from the environment. Some bacteria, e.g. *H. influenzae* (Talmadge & Herriot 1960), have developed highly efficient genetically determined systems to do this, whereas others, such as *E. coli*, must be subjected to special conditions before they will take up DNA. It has to be said that transformation of *P. aeruginosa* is difficult and efficiency varies between strains (Bagdasarian & Timmis 1982).

Transformation can be used to demonstrate that the change in phenazine production relates to the insertion of the transposon. For example, once DNA around the site of the transposon insertion has been cloned into a plasmid (Fig. 3b), this plasmid can be transformed into the pigment-negative mutant strain. Complementation of genetic information then leads to the return of pigment production.

Method

1 Grow the bacteria (e.g. a phenazine minus mutant of PAO6049) in nutrient broth to exponential phase (about 2×10^8 cells/ml) and harvest by centrifugation at +4°C.
2 Wash the bacteria in an equal volume of cold buffer I (10 mM 3-*N*-morpholino-propanesulphonic acid (MOPS, Sigma Chemicals) pH 7.0, 10 mM rubidium Cl, 100 mM MgCl) by resuspension and centrifugation.
3 Resuspend bacteria and hold for 30 min at 0°C in an equal volume of cold buffer II (100 mM MOPS pH 6.5, 10 mM rubidium Cl, 100 mM $CaCl_2$).
4 Centrifuge bacteria and resuspend in 1/10 volume of cold buffer II and incubate 0.2 ml portions of this suspension with the purified DNA (e.g. plasmid-containing cloned pseudomonas DNA involved in the phenazine biosynthetic pathway) for transformation (0.2–1.0 μg) at 0°C for 45 min and then at 42°C for 1 min.
5 Dilute the bacterial suspension by addition of 3 ml nutrient broth and incubate at 30°C for 90 min with shaking, prior to dilution and plating on nutrient agar. Successful transformation with plasmid DNA will result in the return of pigment production due to complementation of genetic information.

References

BAGDASARIAN, M. & TIMMIS, K.N. 1982. Host: vector systems for gene cloning in pseudomonas. In *Current Topics in Microbiology and Immunology* Vol. 96. eds Henle, W., Hofschneider, P.H., Koprowski, H., Melchers, F., Rott, R., Schweiger, H.G. & Vogt, P.K. Berlin & Heidelberg: Springer-Verlag.

KUNKEL, T.A. 1985. Rapid and efficient site-specific mutagenesis without phenotypic selection. *Proceedings of the National Academy of Sciences of the United States of America* **82**, 488–492.

RELLA, M., MERCENIER, A. & HAAS, D. 1985. Transposon insertion mutagenesis of *Pseudomonas aeruginosa* with a Tn5 derivative: application to physical mapping of the ARC gene cluster. *Gene* **33**, 293–303.

SAUNDERS, F.K., MITCHELL, T.J., WALKER, J.A., ANDREW, P.W. & BOULNOIS, G.J. 1989. Pneumolysin, the thiol-activated toxin of *Streptococcus pneumoniae*, does not require a thiol group for *in vitro* activity. *Infection and Immunity* **57**, 2547–2552.

STAPLETON, M.J., JAGGER, K.S. & WARREN, R.L. 1984. Transposon mutagenesis of *Pseudomonas aeruginosa* exoprotease genes. *Journal of Bacteriology* **157**, 7–12.

TALMADGE, M.B. & HERRIOT, R.M. 1960. A chemically defined medium for growth, transformation, and isolation of nutritional mutants of *Haemophilus influenzae*. *Biochemical and Biophysical Research Communications* **2**, 203–206.

WALKER, J.A., ALLEN, R.L., FALMAGNE, P., JOHNSON, M.K. & BOULNOIS, G.J. 1987. Molecular cloning, characterisation and complete nucleotide sequence of the gene for pneumolysin, the sulphydryl activated toxin of *S. pneumoniae*. *Infection and Immunity* **55**, 1184–1189.

WILSON, R. 1988. Secondary ciliary dysfunction. *Clinical Science* **75**, 113–120.

WILSON, R., PITT, T., TAYLOR, G., WATSON, D., MACDERMOT, J., SYKES, D., ROBERTS, D. & COLE, P.J. 1987. Pyocyanin and 1-hydroxyphenazine produced by *Pseudomonas aeruginosa* inhibit the beating of human respiratory cilia *in vitro*. *Journal of Clinical Investigation* **79**, 221–229.

RNA Detection Methods in Basic Cardiac Research

P.J.R BARTON,[1] W.J. VALLINS,[1] G.E. LYONS,[2] N.J. BRAND[1] AND M. YACOUB[1]

[1]*Department of Cardiothoracic Surgery, National Heart and Lung Institute, Dovehouse Street, London SW3 6LY, UK;* [2]*Department of Molecular Biology, Institut Pasteur, 28 Rue du Dr Roux, 75724 Paris Cedex 15, France*

The technique of molecular cloning provides a powerful and versatile tool for the analysis of cardiac development, function and disease. By means of cloned probes it is possible to define the pattern of expression of genes encoding isoforms of proteins present in the heart. This is of use both in defining transitions in normal development, and in describing abnormalities in gene expression associated with disease. In addition, recent advances in our understanding of gene regulation allow us to define the molecular mechanisms underlying many of these processes.

The shift towards viewing cardiac function in terms of molecular biology (i.e. regulation by altered gene expression) has been heralded by some as a new paradigm in cardiac research (Katz 1988). This is likened to the earlier paradigms of Starling's law (regulation by changing end-diastolic fibre length; Starling 1918), and the theory of excitation–contraction coupling (regulation by biochemical changes in the myocardial cell—see Katz 1955), both of which resulted in a major shift in our understanding of cardiac function. Whatever the overall impact of this new technology, it is clear that many of the adaptive changes seen during development, and in disease, are directly linked to alterations in gene activity. Molecular biology provides a new approach towards our understanding of these processes and, ultimately, to the treatment of disease.

The ability to fractionate nucleic acids and to detect specific sequences by molecular hybridization provide powerful tools in molecular biology. The techniques discussed here demonstrate methods of detecting specific mRNA sequences, either in isolated RNA, or *in situ* with sections of cardiac or fetal tissue. Both techniques are based on the use of hybridization as a reliable and highly specific method for monitoring mRNA accumulation and thereby the activity of specific genes. A large variety of cloned gene probes are currently available and a comprehensive list of cloned human sequences has

0-632-02926-9

recently been compiled (Schmidtke & Cooper 1990). By means of these data it is possible to generate probes for direct use in monitoring gene activity. For genes which have not already been cloned, oligonucleotide probes can be generated by predicting the mRNA sequence from amino acid sequence data. For genes where no amino acid sequence data are available, cloning strategies can be devised based on functional domains predicted to be present in the protein (e.g. consensus sequences for calcium binding domains, active enzyme sites, phosphorylation sites, DNA binding domains, etc.). The latter approach has been greatly facilitated by the development of the polymerase chain reaction (PCR) (see the chapter by Brand *et al.*, this volume).

This chapter describes some basic methods for mRNA detection, and demonstrates their application to aspects of cardiac research.

Isolation of RNA and mRNA Detection by Northern Blotting

RNA preparation

All RNA work must be carried out in a ribonuclease (RNase)-free environment using RNase-free water and glassware. For general procedures relating to RNase decontamination see Blumberg (1987). RNA can be quickly and conveniently isolated from cardiac muscle and other tissue by homogenization in guanidium isothiocyanate and centrifugation through a CsCl cushion. This method provides high molecular weight RNA free of contaminating DNA and protein. Guanidium isothiocyanate is a powerful protein denaturant and is extremely effective as an RNase inhibitor, and problems associated with endogenous RNase are therefore eliminated. The method routinely used by the authors is adapted from Chirgwin *et al.* (1979) as follows.

1 Homogenize tissue (either fresh, or frozen in liquid nitrogen), e.g. in a Waring blender, in 5 M guanidium isothiocyanate, 5 mM Na acetate, 0.5% sarcosyl, 1% β-mercaptoethanol, 1% antifoam. (This should be made fresh from concentrated stocks.)

2 Layer the homogenate onto a 5.7 M CsCl cushion (6 ml homogenate per 2 ml cushion) in sterile polyallomer ultracentifuge tubes, and centrifuge overnight at 35 000 rpm at 20°C in a Sorvall TH641 rotor or similar.

4 Following centrifugation, remove and discard the supernatant and most of the CsCl cushion using a pipette. Pour off the remaining CsCl and invert tube to drain.

5 Resuspend the RNA pellet (quantities of less than 1 mg may be invisible), in 3 × 100 μl of DEPC-treated sterile water.

6 As an optional precaution at this stage, RNA can be extracted with an equal volume of phenol:chloroform:isoamylalcohol, and reprecipitated with 0.3 M Na acetate and 2.5 volumes of ethanol.

The yield of RNA can be determined by optical density at 260 nm (1 OD unit = 40 μg RNA), and the quality can be checked on a simple 1% agarose mini-gel. Yields may vary considerably depending on the source and condition of the tissue used. Biopsies from old, fibrous or diseased cardiac muscle may yield as little as 20 μg RNA per 100 mg tissue. Fresh cardiac muscle from young animals or fetal heart can give up to 100 μg RNA per 100 mg tissue. RNA should be stored in DEPC-treated water at −20°C or, for long-term storage, as an ethanol precipitate.

Northern blotting

1 Denature 5–20 μg of RNA (per track) in 20 μl of 50% formamide, 6% formaldehyde, 10 mM MOPS pH 7.4; 0.5 mM EDTA, 0.1% SDS, 10% glycerol, 0.02% bromphenol blue and xylene cyanole, at 65°C for 10 min. Then load onto 1–2% agarose gel made in 10 mM MOPS pH 7.4, 0.5 mM EDTA and run in 10 mM MOPS pH 7.4, 0.5 mM EDTA. RNA stained with ethidium bromide transfers less efficiently than unstained RNA and marker tracks should be loaded and stained separately.
2 For blotting, nitrocellulose gives excellent results but the nylon based membranes provide greater durability. We routinely use Hybond N^+ (Amersham International PLC). RNA is transfered from agarose gels using a system as shown in Fig. 1. The gel is placed on Whatman 3MM filter paper which is dipped into 10 × SSC; a prewetted filter is placed on the gel, followed by two sheets of filter paper and paper towels. Transfer occurs by capillary action overnight. Care should be taken to avoid trapping air bubbles between the gel and membrane.
3 After transfer, the filter is baked at 80°C for 1 h, and can be stored at room temperature prior to hybridization.

Probes for Hybridization Experiments

Radiolabelled probes can be produced from cloned DNA by a variety of means such as nick translation, random priming, and end-labelling (for general methodology see Sambrook *et al.* 1989). Here we describe the use of oligonucleotides, which are particularly useful in generating probes from published sequence data, and of RNA probes derived from transcription vectors, which are particularly well suited to *in situ* hybridization.

Oligonucleotide probes

Oligonucleotide probes of 17–30 base-pairs are suitable for use in Northern blot hybridization. These may be derived from published gene or mRNA

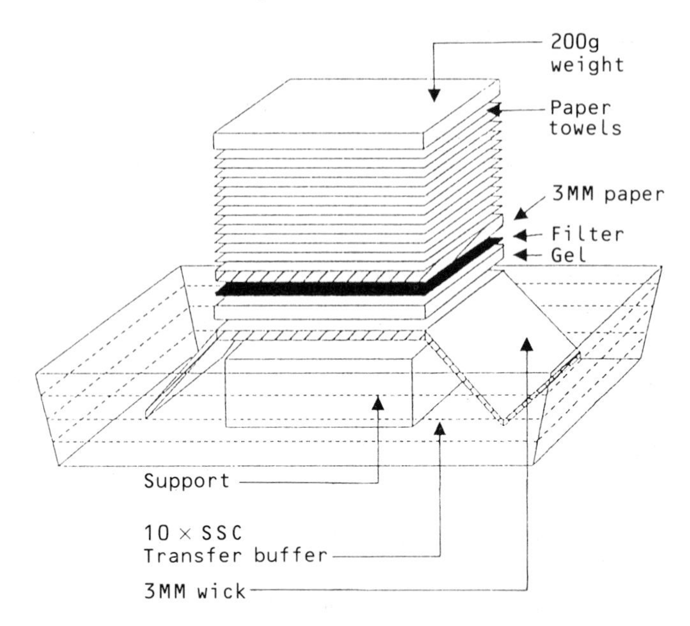

FIG. 1. Northern blotting system. RNA is transferred from the gel to the filter by capillary action (see text).

sequences, and may also be used where protein sequence data alone is available. In this case the mRNA coding sequence is predicted from the protein sequence. Because of the redundancy of the genetic code it is necessary to generate mixed sets of oligonucleotides to cover all possible coding sequences. The number of oligonucleotides required can be limited by taking into account codon bias, and by using third position options which allow pairing with more than one base (for a fuller description on the design of oligonucleotides see Lathe 1985).

The stability of hybridization is closely linked to the G/C content of the oligonucleotide and for best results the hybridization conditions should be adjusted accordingly. For short oligonucleotides (up to 20 residues) the thermal melting point can be calculated from:

$$T_m(°C) = 2 \times (A + T) + 4 \times (G + C)$$

(see Wahl *et al.* 1987)

Hybridization is best done at 5°C below the T_m. Oligonucleotides can be and labelled with (γ-^{32}P)ATP and polynucleotide kinase as follows:

1 Add 50–100 μCi (γ-^{32}P)ATP (3000 Ci/mmol) to 5–10 pmol of oligonucleotide in 20 μl of 67 mM Tris–HCl (pH 8.3), 10 mM $MgCl_2$, 10 mM DTT.

2 Incubate at 37°C for 30 min with 5 units of polynucleotide kinase.
3 The reaction can be stopped by heating to 95°C for 2 min, and the labelled oligonucleotide purified on a 20% acrylamide gel.
4 Alternatively, labelled oligonucleotides can be quickly and conveniently separated from unincorporated nucleotides by precipitation twice in 2 M ammonium acetate + 2.5 volumes of ethanol at −20°C.

Hybridizing Northern blots with oligonucleotides

Using the calculations given above, labelled oligonucleotides should be hybridized to filters as follows:
1 Filters are prehybridized with 5 × SSC, 0.1% SDS, 1 × Denhardt's solution (0.02% Ficoll, 0.02% PVP, 0.02% BSA), 10 μg/ml heat denatured salmon sperm DNA, at 60°C for 1−4 h.
2 Hybridization is in the same buffer containing 10^6 cpm/ml of oligonucleotide, at T_m −5°C overnight.
3 Filters are washed in 5 × SSC 0.1% SDS at room temperature and exposed to X-ray film at −70°C.

Fig. 2 shows a typical Northern blot hybridized with a (^{32}P)-labelled oligonucleotide (5′−CCGTCATCCTGACTGGAAGGTAGATGG−3′) complementary to human α-cardiac actin mRNA. The specificity of hybridization is demonstrated by the lack of hybridization to skeletal muscle RNA, and the presence of a single hybridizing band in atrial muscle RNA.

Riboprobes

RNA probes derived from transcription vectors are rapidly becoming the probe of choice for most hybridization experiments. They are produced by specific transcription from one of the prokaryotic promoters T3, T7 or Sp6 located next to the cloned sequence in a suitable cloning vector. The authors routinely use pBluescript vectors (Stratagene Inc., USA), which have both T3 and T7 promoters located one either side of a multiple cloning site polylinker (see Fig. 3). In order to generate probes specific for the cloned DNA fragment, the vector is first linearized with a suitable restriction endonuclease. In this way 'run-off' transcripts of a discrete size are generated, and vector sequences (other than the small region between the promoter and the cloned insert) are not transcribed. In addition, by making use of restriction sites within the cloned fragment, it is possible to generate probes from different parts of the sequence.

Radiolabelled riboprobes can be produced by incorporating [α−^{32}P] or [α−^{35}S] labelled ribonucleotides in the transcription reaction as follows:
1 For [α-^{35}S]-labelled probes add in the following order:
5 μl transcription buffer (200 mM Tris, pH 8.0, 40 mM $MgCl_2$,

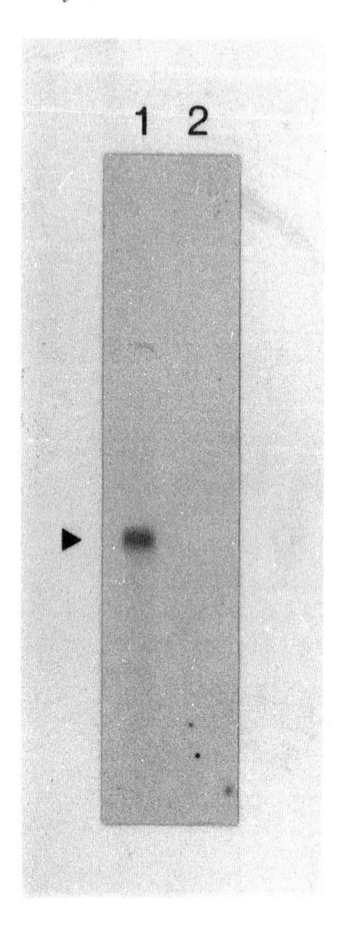

FIG. 2. Northern blot hybridization with oligonucleotide probes. Ten μg of RNA from human atrial muscle (track 1) and human quadriceps (track 2) were run on a 1.5% agarose gel, transferred to Hybond N+ and hybridized with [^{32}P]-labelled oligonucleotide, complementary to α-cardiac actin mRNA, as described in the text. The filter was washed in 5 × SSC at room temperature and exposed for 3 days.

10 mM spermidine 250 mM NaCl);
0.5 μg DNA template;
1 μl each of 10 mM rATP, rGTP, rCTP;
1 μl 100 mM DTT;
1 μl RNase inhibitor;
5 μl [$\alpha-^{35}$S]-rUTP: > 1000 Ci/mmol, 10 mCi/ml;
10 U of T7 or T3 RNA or SP6 polymerase;
DEPC-treated water to 25 μl.

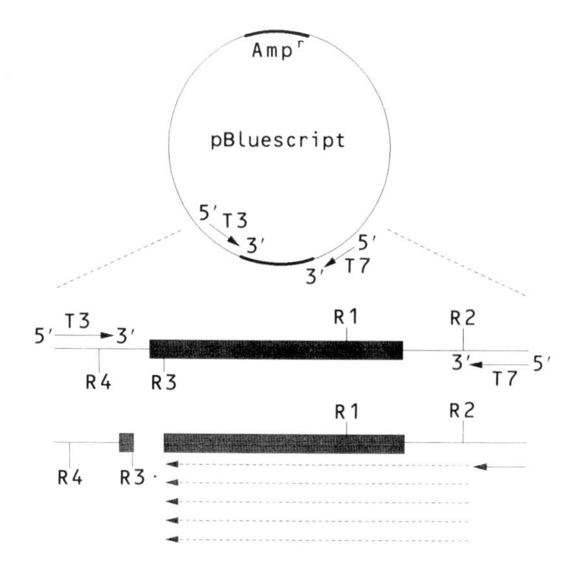

FIG. 3. RNA probes derived from transcription vectors. Cloned DNA inserted into the multiple cloning site of pBluescript (Stratagene Inc., USA) can be transcribed into RNA using either T3 or T7 RNA polymerase to produce copies of the upper or lower DNA strands respectively. In order to produce transcripts of a discrete size, and corresponding to inserted DNA only, the plasmid is cut with a suitable restriction site either external to (R2 and R4), or internal to (R1 and R3) the DNA insert. In this example, the insert (bold line) is cut internally at site R3 for transcription from the T7 promoter. In this way probes corresponding to different regions of the cloned sequence can be generated. Amp^r: ampicillin resistance gene, T3 and T7: sites of initiation of polymerase transcription, arrows indicate T3 and T7 primers.

2 Incubate at 37°C for 1–2 h.
3 Add 5–10 units of RNase-free DNase, and incubate at 37°C for 15 min to destory DNA template.
4 Add 10 μg total yeast RNA as carrier, extract three times with phenol/chloroform, and precipitate twice in 0.3 M Na acetate and 2.5 volumes of ethanol.
5 Unincorporated nucleotides can be removed by passage through an RNase-free Sephadex G-50 column in 10 mM Tris, 5 mM EDTA, 0.1% SDS, 10 mM DTT. Add 500 μg total yeast RNA to column before running sample.

Riboprobes produced in this way are suitable for hybridization to Northern and Southern blots, for use in ribonuclease protection assays, and are particularly well suited to *in situ* hybridization.

In situ Hybridization

In situ hybridization is a powerful technique of great value for determining both the spatial distribution and temporal pattern of gene expression during development. It can be applied to any tissue sample provided that it is fresh or was correctly fixed or frozen (−70°C) when fresh. *In situ* hybridization offers many advantages over the use of isolated RNA and Northern blotting. It provides detailed information on mRNA accumulation at the cellular level, of particular importance in mixed cell populations, and serial sections can be hybridized with different probes to build a complete picture of gene expression within a tissue or organ.

The exact conditions for *in situ* hybridization required for a particular experiment vary both according to the tissue itself (primarily affecting fixation steps) and the relative abundance of the mRNA (this may affect the choice of probes and method of detection used—for example, abundant mRNAs or viral sequences may be successfully detected by one of the non-radioactive methods). For work with cardiac muscle, when looking at the expression of contractile protein genes, the following protocol is suitable.

Preparation of tissue

Fresh tissue should be dissected in cold phosphate buffered saline (PBS) solution and transferred to 4% paraformaldehyde in PBS. After overnight fixation, tissue is rinsed in PBS at 4°C (30 min), 0.85% saline solution at 4°C (30 min), then at room temperature in 50% ethanol; 0.4% saline (2 × 15 min), 70% ethanol (30 min), 85% ethanol (30 min), 95% ethanol (30 min), 100% ethanol (30 min), xylene (2 × 30 min), and 1:1 xylene and melted paraplast at 60°C (45 min). Tissue is then transferred to paraplast at 60°C for three changes for 20 min each, oriented in plastic moulds and cooled. Tissue blocks are stored at 4°C until cut.

If tissue is to be frozen, orientate the sample in a mounting medium (e.g. OCS) and freeze in melting isopentane for 5 min, then transfer to liquid nitrogen storage.

Preparing sections

Paraffin embedded sections are cut at 5−7 μm on a standard paraffin microtome and placed on a drop of water on subbed slides at 40−50°C where they are allowed to relax to their original conformation. Slides are drained by the capillary action of a corner of blotting paper and the section is pressed flat on the slide using moist blotting paper. Sections are dried flat overnight at room temperature and stored at 4°C until used. Before hybridization, paraffin is removed from the sections in xylene (2 × 10 min), rehydrated through an ethanol series (100−30%), rinsed in 0.85% saline (5 min), then in 1 × PBS (5 min) and fixed in freshly prepared 4% paraformaldehyde in 1 × PBS (20 min). Following rinsing in PBS (2 × 5 min) sections are treated with proteinase-K (20 μg/ml) in 50 mM Tris–HCl, 5 mM EDTA, pH 7.2) for 7 min. Sections are then rinsed in PBS and fixed in 4% paraformaldehyde for 5 min, rinsed in water and acetylated in 0.1 M triethanolamine + 0.25% acetic anhydride (10 min). Finally, sections are rinsed in PBS (5 min), 0.85% saline (5 min) and dehydrated through an alcohol series (30, 50, 70, 85, 95, 100% × 2) and air-dried.

In situ *hybridization*

A useful and efficient probe for *in situ* hybridization is [α-^{35}S]-labelled RNA derived from a suitable cloning vector into which has been inserted the gene sequence concerned (see previous section). Probes should ideally be short (50–100 nucleotides) and if the cloned fragment is longer, the transcripts should be partially hydrolysed by alkali with the average size of fragments checked by gel analysis.

High specific activity [^{35}S]-labelled RNA probes synthesized with [α-^{35}S]-UTP (specific activity >1000 Ci/mmol) are applied directly to the sections (75 000 cpm per μl) in 10–30 μl of hybridization buffer (50% deionized formamide, 0.3 M NaCl, 20 mM Tris–HCl pH 7.4, 5 mM EDTA, 10 mM $NaPO_4$, 10% dextran sulphate, 1 × Denhardt's solution, 50 μg/ml yeast RNA) covered with a siliconized cover slip (22 × 22 mm) and hybridized for 16 h at 50°C in a humid chamber. Coverslips are floated off in 5 × SSC, 10 mM DTT, at 50°C, and sections are then washed at 65°C in: 50% formamide, 2 × SSC, 0.1 M DTT, for 20 min, then rinsed in washing buffer (400 mM NaCl, 10 mM Tris–HCl, 5 mM EDTA pH 7.5) at 37°C, followed by washing buffer containing RNase A (20 μg/ml) and RNase T1 (2 u/ml) for 45 min at 37°C. Slides are then rinsed in 2 × SSC (15 min) and 0.1 × SSC (15 min) at 37°C and dehydrated through an alcohol series containing 0.3 M ammonium acetate.

Following hybridization and washing, slides are dipped in photographic emulsion (Kodak NTB-2 nuclear track emulsion) and exposed. Photographic development is carried out in Kodak D-19 developer (3.5 min at 16°C), and fixed in Kodak X-ray fixer undiluted, and lightly stained in 0.2% toluidine blue. Slides are dehydrated through a series of alcohols and coverslipped with CytoSeal 60 after dipping 2 × 5 min in xylene. Analysis is best carried out by both light and dark field microscopy on a Zeiss Axiophot microscope.

Cardiac development analysed by in situ *hybridization*

One area of research particularly well suited to analysis by *in situ* hybridization is that of cardiac development. Hybridization of [^{35}S]-labelled gene specific probes allows the detailed analysis of the timing and spatial distribution of

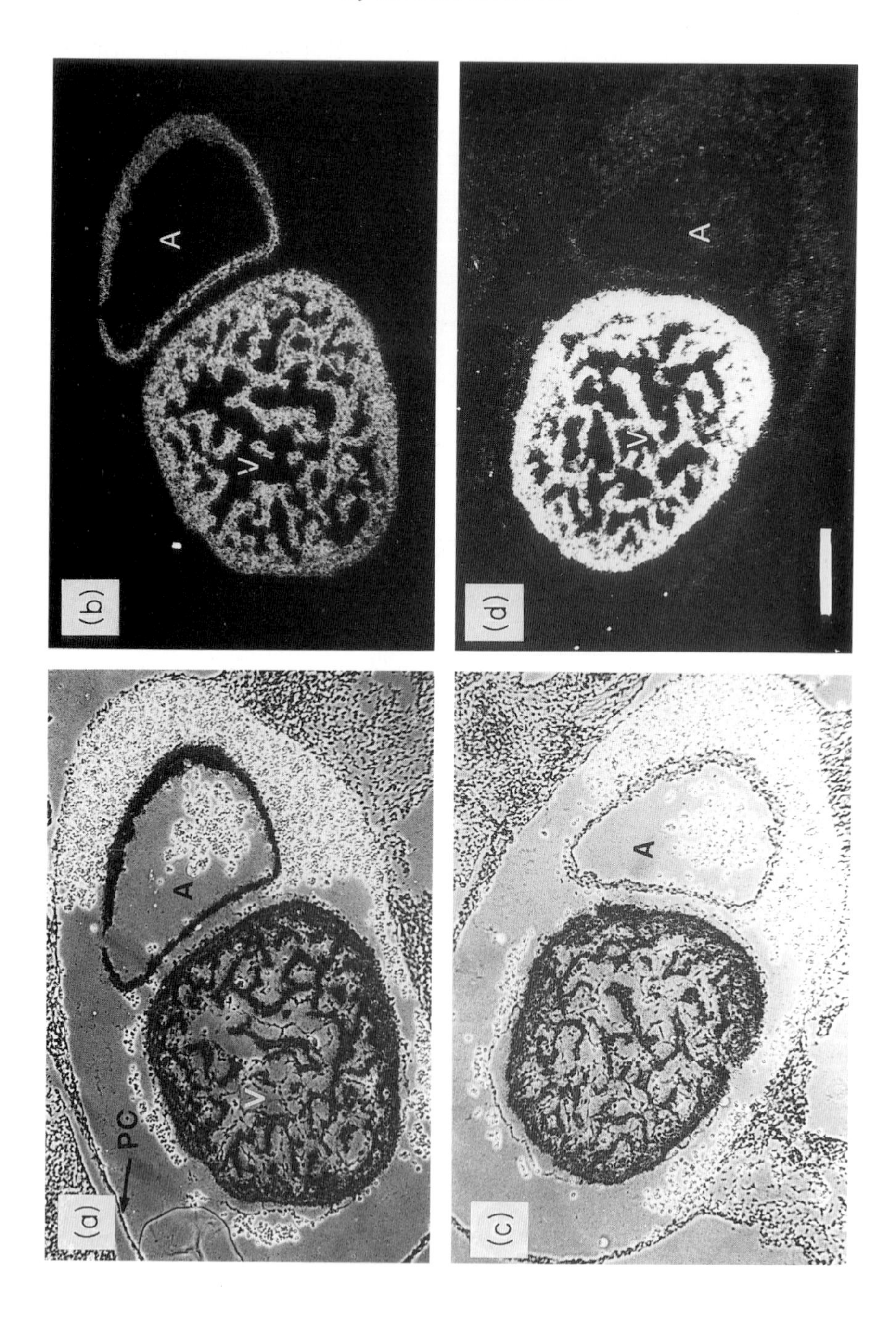
(a)
PC
A
V
(b)
A
V
(c)
A
(d)
A
V

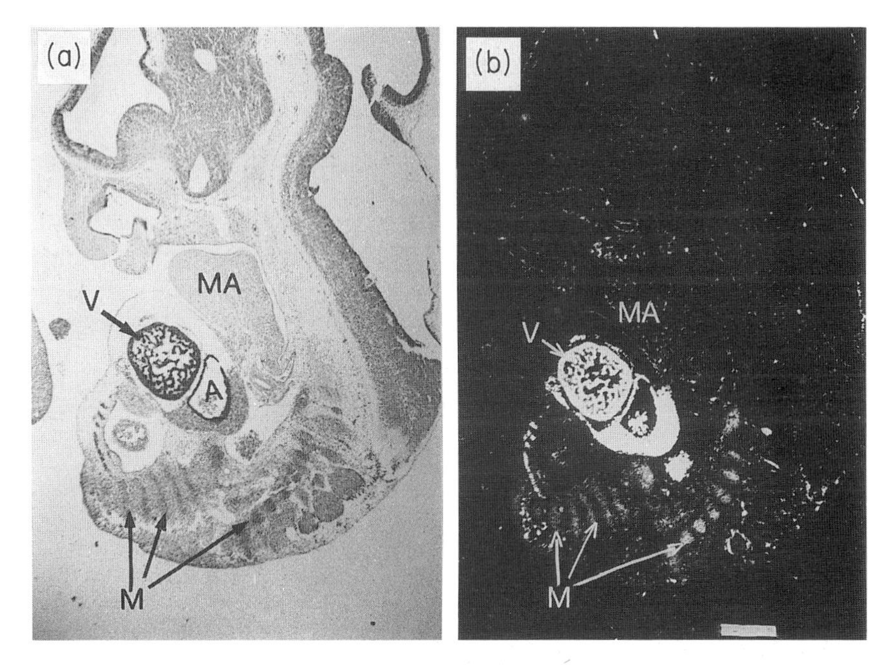

FIG. 5. Gene expression throughout the whole embryo. (a) Five μm parasagittal section of an 11 day fetal mouse hybridized with myosin light chain $MLC1_A$ specific probe; (b) dark field image showing distribution of hybridization in ventricular (V), atrial (A) and developing skeletal muscle (M). MA = mandible. Scale bar = 500 μm. NB: blood cells in and around the atrium are highly refractile in dark field image.

mRNA accumulation throughout the heart. In smaller animals, analysis of mRNA accumulation throughout the whole embryo is possible. One advantage of *in situ* hybridization with cloned gene probes is that the non-coding sequences of mRNA encoding very similar proteins can be used. In this way it is possible to avoid problems associated with cross-hybridization and to analyse the specific expression of proteins too similar to generate specific antibody probes. An example of such an approach is the analysis of the expression of α-cardiac and α-skeletal actin genes in the developing embryo (Sassoon *et al.* 1988). α-cardiac and α-skeletal actin differ in only 4 out of 375 amino acids

FIG. 4. Gene expression in the developing heart. (^{32}S)-labelled riboprobes specific for mRNAs encoding the myosin light chain $MLC1_A$ or β-myosin heavy chain (β-MHC) were hybridized to serial sections of 11 day mouse fetal heart. (a) and (c): Parasagittal sections showing developing ventricular, V and atrial, A muscle; (b) and (d): dark field images of the same sections as in (a) and (c) hybridized with (b) $MLC1_A$ specific probe; (d) β-MHC specific probe. Scale bar = 200 μm. PC = pericardium.

and isoform specific antibodies cannot be produced. The 5′, 3′ non-coding sequences of α-cardiac and α-skeletal muscle mRNAs are, however, very different, and can therefore be used as gene specific probes.

The authors have analysed atrial myosin alkali light chain ($MLC1_A$) and β-myosin heavy chain (β-MHC) expression in fetal heart. Eleven day mouse embryos were paraffin embedded as described, and serial 5 μm parasagittal sections were hybridized with [^{35}S]-labelled RNA probes produced by the pBluescript system (Figs 4 and 5). The probes correspond to the 3′ non-coding sequences of $MLC1_A$ and β-MHC mRNAs. Following autoradiography, the sections were photographed under both light and dark field optics. Under dark field optics, the silver grains found in the photographic emulsion refract light and therefore appear white. Figure 4 shows serial sections hybridized with $MLC1_A$ (A and B) or β-MHC (C and D) probes. $MLC1_A$, which is expressed only in atrial muscle in the adult heart (see Barton & Buckingham 1985), is clearly present throughout the developing myocardium. β-MHC mRNA, on the other hand, is restricted primarily to the ventricular muscle at this stage. $MLC1_A$ is known to be expressed also in fetal skeletal muscle and Fig. 5 shows the pattern of mRNA accumulation throughout the 11 day mouse embryo where hybridization is evident over the forming skeletal muscle.

References

BARTON, P.J.R. & BUCKINGHAM, M.E. 1985. The myosin alkali light chain proteins and their genes. *Biochemical Journal* **231**, 249–261.

BLUMBERG, D.D. 1987. Creating a ribonuclease-free environment. In *Guide to Molecular Cloning Techniques*, eds Berger, S.L. & Kimmel, A.R. *Methods in Enzymology* **152**, pp. 20–24. New York: Academic Press.

CHIRGWIN, J.M., PRZYBYLA, A.E., MACDDONALD, R.J. & RUTTER, W.J. 1979. Isolation of biologically active ribonucleic acid from sources enriched in ribonuclease. *Biochemistry* **18**, 5294–5299.

KATZ, A.M. 1988. Molecular biology in cardiology, a paradigmatic shift. *Journal of Molecular and Cellular Cardiology* **20**, 355–366.

KATZ, L.N. ed. 1955. Symposium on the regulation of the performance of the heart. *Physiological Reviews* **35**, 90–168.

LATHE, R. 1985. Synthetic oligonucleotide probes deduced from amino acid sequence data. Theoretical and practical considerations. *Journal of Molecular Biology* **183**, 1–12.

SAMBROOK, J., FRITSCH, E.F. & MANIATIS, T. 1989. *Molecular Cloning—A Laboratory Manual.* 2nd edn. Cold Spring Harbor, New York: Cold Spring Harbor Laboratory.

SASSOON, D.A., GARNER, I. & BUCKINGHAM, M. 1990. Transcripts of α-cardiac and α-skeletal actins are early markers for myogenesis in the mouse embryo. *Development* **104**, 155–164.

SCHMIDTKE, J. & COOPER, D.N. 1989. A comprehensive list of cloned human DNA sequences. *Nucleic Acids Research* **18** Sequences supplement, 2413–2548.

STARLING, E.H. 1918. *The Linacre Lecture on the Law of the Heart.* London: Longmans, Green and Co.

WAHL, G.M., BERGER, S.L. & KIMMEL, A.R. 1987. Molecular hybridization of immobilized nucleic acids: Theoretical concepts and practical considerations. In *Guide to Molecular Cloning Techniques*, eds Berger, S.L. & Kimmel, A.R. *Methods in Enzymology* **152**, pp. 399–406. New York: Academic Press.

The Polymerase Chain Reaction and its Application to Basic Research in Molecular Biology

N.J. Brand, W.J. Vallins, M. Yacoub and P.J.R. Barton
Department of Cardiothoracic Surgery, National Heart and Lung Institute, Dovehouse Street, London SW3 6LY, UK

The Polymerase Chain Reaction

The polymerase chain reaction (PCR) represents a powerful new technique for the molecular biologist, equally applicable to routine diagnostic procedures and to basic research. Fundamentally, the method involves the *in vitro* enzymatic amplification of a specific DNA fragment of known sequence by successive rounds of DNA synthesis, often yielding microgram quantities from a minute amount of starting material (for example, genomic DNA from just a few hundred cells). A double-stranded DNA target is denatured by heating to 94°C in the presence of deoxynucleotide *tri*phosphates (dNTPs) and molar excesses of each of two oligonucleotide primers which are specific for opposite strands in the DNA duplex, and which flank the region to be amplified (Fig. 1a). The temperature is reduced so that the oligonucleotide probes selectively hybridize to their complementary sequences and the temperature is then raised to 72°C, stimulating polymerization by a thermostable DNA polymerase (*Taq* polymerase) which uses the bound oligonucleotides as primers for extension (Fig. 1b). One cycle of denaturation, annealing and extension results in a doubling of the amount of the fragment targeted for amplification, two rounds quadruples the amount (Fig. 1c), and so on in an exponential manner, allowing a theoretical amplification of the specified target by about 10^7-fold in 23 cycles. The DNA products are then analysed by electrophoresis in agarose or polyacrylamide gels and, if necessary, they may be transferred to nylon or nitrocellulose membranes by Southern blotting for hybridization with a specific radiolabelled probe. For a fuller description of the method and variations upon it, and key references, the reader is referred to the recent review article by White *et al.* (1989).

The key to the success of PCR as a routine laboratory technique is the thermostable polymerase (*Taq* polymerase), which was isolated from the hot

Genetic Manipulation
0–632–02926–9

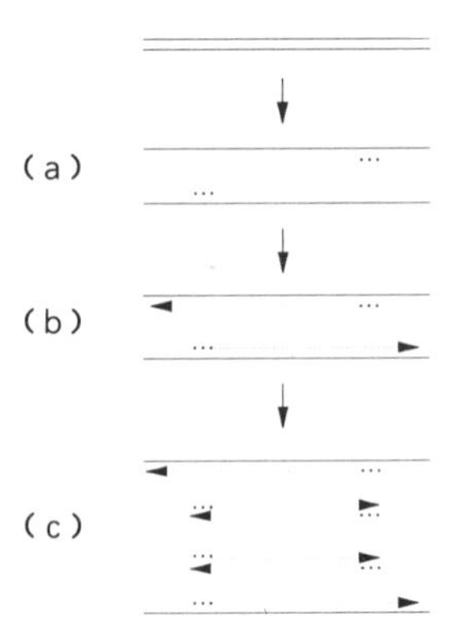

FIG. 1. The PCR reaction: (a) a double-stranded DNA molecule is denatured by heating, producing two single DNA strands. Sequence-specific oligonucleotide primers (dotted lines) flanking the region to be amplified are hybridized selectively to either the sense (top) DNA strand or the antisense DNA strand by reducing the temperature (see text); (b) in the presence of dNTPs and molar excess of the oligonucleotide primers, *Taq* polymerase synthesizes complementary strands (fine dotted lines) by extending the hybridized primers, thereby doubling the amount of the target DNA. This represents one complete cycle of PCR; (c) at the end of successive rounds of PCR DNA strands of defined length are synthesized (centre) which will, at the end of the reaction, represent the major PCR product.

spring bacterium *Thermus aquaticus*. This enzyme, commercially available from many molecular biology companies, has an optimal temperature for DNA polymerization of around 72°C and can withstand heating to temperatures as high as 95°C for short periods of time. Of no lesser importance has been the development of sophisticated heating devices, allowing the programming of a cycle of several incubations at selected temperatures and for set durations (equivalent to the denaturation, annealing and extension steps of the reaction) and the repetition of that cycle, thus driving the DNA amplification. Various thermal cyclers are available at present, including refridgerator-cooled models such as that marketed by Perkin-Elmer Cetus, and water-cooled devices such as those sold by Techne and BioExcellence.

The method has found many applications in its basic form, requiring DNA sequence information at two points flanking the amplification target so that oligonucleotide primers can be synthesized. For example, DNA fingerprinting, often from minute quantities of starting material, is becoming increasingly popular in forensic science; PCR is applicable to population genetics for RFLP mapping and in the diagnosis of genetic disorders, disease susceptibility and oncogene expression. But it is largely in the field of fundamental research that a number of important variations have been made upon the basic method. We will describe some of these in the following sections, giving detailed protocols, and discuss how we are using these to study gene expression in the heart.

Methods

PCR from a double-stranded DNA template

This basic PCR protocol, requiring the DNA sequences for two oligonucleotides flanking the amplification target, is applicable to both genomic DNA and cloned DNA (for example, a complementary DNA (cDNA) cloned into a plasmid vector). The method and suggested incubation conditions are given below.

Method 1

1 The following components are mixed together in a standard 0.5 ml Eppendorf microcentrifuge tube or equivalent.
1 μg genomic DNA or 10 ng linearized plasmid DNA;
5 μl 10 × *Taq* buffer (see Appendix);
50 pmol of each oligonucleotide primer (1 μM final concentration);
2 μl of 4 mM dNTP mix, pH 7–8 (see Appendix) (1 mM of each dNTP, giving final concentration of 40 μM each dNTP);
Water to 50 μl;
1 U (typically 0.5–1 μl) of *Taq* polymerase (Cetus or New England Biolabs enzymes are preferred).
2 Overlay with a few drops of mineral oil to prevent evaporation of the reaction mixture during the cycles.
3 Suggested PCR cycle:
Denaturation: 94°C/1.5 min;
Annealing: 50°C/2 min;
Extension: 72°C/2 min.
Between 25 and 50 such cycles may be performed sequentially. At the end of the cycle, an extra extension step of 6 min at 72°C is performed to make sure that all the amplified double-stranded fragments are full-length.
4 Load 5–10% of the reaction mixture on a 2% agarose gel, separate the PCR products from the oligonucleotides by electrophoresis, stain the gel with ethidium bromide (EtBr) solution and visualize the DNA on a u.v. transilluminator. Alternatively, run the products out on a 10% acrylamide gel containing EtBr (useful for fragments of 100 bp or shorter).

Subcloning PCR products for DNA sequencing

The oligonucleotide primers can be modified by adding 5′ tails which are not complementary to the template but which contain specific recognition sites for restriction enzymes. The amplified PCR product is then digested with those enzymes and the fragment subcloned into a suitable vector linearized with the same enzymes. It is best to include a three or four nucleotide overhang 5′ to the restriction site as this improves the efficacy of binding of the restriction enzyme to DNA and hence digestion by the enzyme. Also, because *Taq* polymerase leaves 'ragged' DNA ends, the PCR products should be treated with Klenow polymerase in the presence of 250 μM dNTPs to fill in the ends prior to subcloning (see Appendix).

This form of PCR is useful in subcloning particular regions of a cDNA when suitable flanking restriction sites are not available. For example, to express part of a protein in order to assess its function, the corresponding region of the cDNA could be amplified by PCR with appropriate restriction sites built into the flanking oligonucleotides. Mammalian proteins, or parts of them, are often over-expressed as fusion proteins in *E. coli* by cloning a cDNA fragment into a suitable vector such that the cDNA is downstream of, and in frame with, the *E. coli lacZ* gene. The fusion protein contains amino acids from β-galactosidase at its N-terminus, allowing subsequent purification of the protein from a crude cell lysate by immunoprecipitation with antibodies directed against part of the recombinant protein. Recently, cDNAs representing specific regions of the human cardiac β-myosin heavy chain and ventricular light chain 1 were amplified by PCR and expressed as β-galactosidase fusion proteins in *E. coli* (Eldin *et al.* 1989). These fusion proteins were immuno-adsorbed to antibodies which had been raised against the native cardiac proteins and which were known to interfere, when bound, with various functions attributed to the proteins such as ATPase activity and the interaction between actin and myosin. In this way, it was possible to attribute these functions to particular regions of the myosin heavy chain or myosin light chain polypeptides.

Applications to Basic Research

Undoubtedly, one of the major applications of PCR is in cDNA cloning, with RNA as the starting material. RNA is prepared from the tissue in which the target gene is known or suspected to be expressed, either by preparing total RNA by standard protocols or, more simply, by boiling some of the tissue for 5 min in 50 μl of water containing 0.1% diethyl pyrocarbonate (DEPC), then spinning out the cell debris and taking the supernatant for PCR (Sherman *et al.* 1989). The first step is to synthesize a single-stranded cDNA template of the appropriate mRNA, using reverse transcriptase and with the anti-sense oligonucleotide (complementary to the mRNA) acting as primer. *Taq* polymerase is added and the PCR cycle started as normal.

PCR from an RNA template

Reverse transcriptase works efficiently in *Taq* polymerase buffer, allowing first-strand cDNA synthesis and PCR cycling to be done in the same reaction mixture.

Method 2

1 Mix together the following:
2 μl 10 × *Taq* buffer;
0.1–1 μg total or polyA + RNA;
50 pmol antisense primer;
DEPC-treated water to 20 μl.
Heat the mixture to 65°C for 5 min in order to melt out any secondary structure, and allow to cool to 50°C (or 5°C below the minimum T_m of the primers). Then chill on ice.
2 Add the following components in the order shown:
3 μl 10 × *Taq* buffer;
2 μl 4 mM dNTPs, pH 7–8;
1 μl 100 mM DTT;
0.5 μl (10–15 U) placental RNase inhibitor (e.g. RNAsin; Amersham International PLC);
DEPC-treated water to 50 μl;
15 U (typically 0.5–1 μl) of AMV reverse transcriptase (e.g. Pharmacia).
Incubate at 42°C for 60 min to synthesize the first cDNA strand (the antisense oligonucleotide is used as primer).
3 Add 50 pmol of the sense primer plus 1 U *Taq* polymerase, overlay with mineral oil and start the PCR reaction; the reverse transcriptase is inactivated by the first denaturation cycle. Analyse the products by electrophoresis (see *Method 1*, step 4).

Figure 2 shows the results of an experiment in which a fragment of human β-myosin heavy chain mRNA was amplified from less than 1 μg of human ventricular muscle RNA under the conditions described above (lane 1). The size of the product (indicated by the closed triangle) agrees well with the predicted size of 104 base pairs including the restriction site tails. The unincorporated oligonucleotides (open triangle) run as a smear ahead of the PCR product.

An advantage of this method is that one can clone both abundant and rare mRNA species directly from the tissue in which they are expressed, or from a small number of cells. Additionally, the method can be used to study differential mRNA splicing in a particular gene, using specific oligonucleotide primers flanking the splice junction to amplify the region spanning the junction. (Note that PCR is especially useful in this regard as the major PCR products will be amplified from the mRNA and not from any contaminating genomic DNA, introns generally being too large to be amplified.) By careful design of the

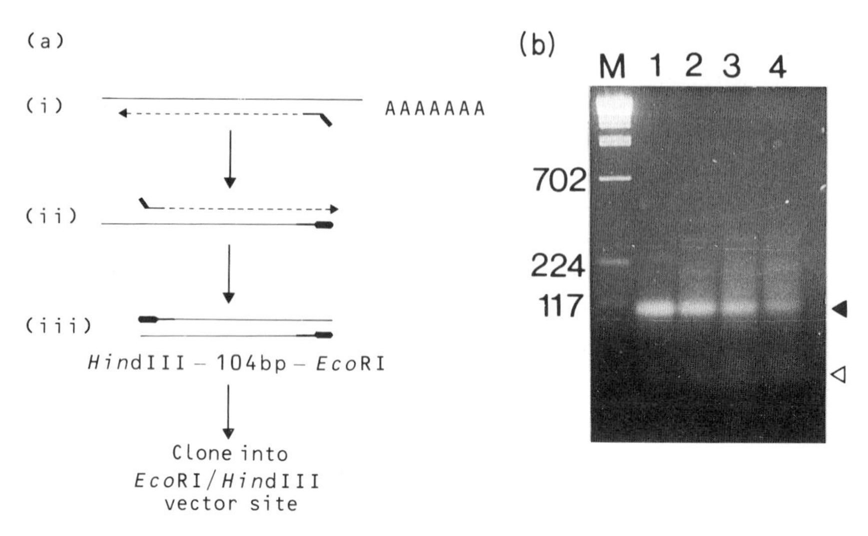

FIG. 2. Amplification of β-myosin heavy chain sequences. (a) Two oligonucleotide primers were generated from published sequence data for the human β-myosin heavy chain (β-MHC) mRNA; (i) the first primer, complementary to part of the 3′ non-coding sequence with a 5′ tail containing the *Eco*RI recognition site, was used to prime a reverse transcriptase reaction from human ventricular muscle RNA; (ii) using the purified cDNA transcript from (i), and a mixture of the initial primer and a second (mRNA-strand) primer carrying a *Hind*III recognition site tail, a PCR reaction was initiated; (iii) the amplified product is cut with both *Eco*RI and *Hind*III to generate a fragment suitable for cloning. (b) Gel analysis of the PCR product described above. The 104 bp product is indicated by the closed triangle, and oligonucleotides by the open triangle. The PCR reaction was carried out in increasing Mg^{2+} concentration. M = Marker DNA, track 1 PCR product using 1.3 mM Mg^{2+}. Track 2 = 2.5 mM Mg^{2+}. Track 3 = 4 mM Mg^{2+}. Track 4 = 8 mM Mg^{2+}. Marker DNA sizes are shown in base-pairs.

oligonucleotides, complex splicing patterns can be studied in a variety of tissues.

Strategies for cDNA cloning with PCR

PCR from an RNA template offers a quick and convenient way of isolating new cDNAs by relatedness to known gene families. This circumvents more laborious cloning methods involving screening recombinant cDNA libraries with radiolabelled probes and then subcloning potential new cDNA inserts into plasmid or M13 vectors for DNA sequencing. Two approaches are described below.

Members of the steroid hormone receptor family of transcription factors show a high level of homology at the amino acid level, particularly, in certain functionally conserved regions of the receptor. For example, parts of the hormone-binding domain and the DNA-binding region of these receptors are highly conserved at the amino acid level amongst most members of the family identified to date (Petkovich *et al.* 1987). In contrast, an intervening hydrophilic region of 46 residues is poorly conserved and thus is diagnostic for each type of receptor (Petkovich *et al.* 1987; Brand *et al.* 1988). Oligonucleotides specific for highly conserved sequences in the flanking hormone and DNA-binding regions can be used to amplify cDNA sequences for the hydrophilic region, using mRNA as a starting material. The products can then be sequenced in order to identify them by reference to the hydrophilic region. The PCR fragment can be radiolabelled later for use as a specific probe for isolating a full-length cDNA from a library.

Though an amino acid sequence may be completely conserved between related proteins, the genetic code is degenerate. Furthermore, in some cases one may want to clone members of gene families that are not well conserved, and thus any degeneracy at the amino acid level must be taken into account when designing the oligonucleotide primers. Consequently, degenerate sets of antisense oligonucleotides are used to prime cDNA synthesis starting from an RNA template. Because of the power of PCR, it is possible to use oligonucleotide mixtures in which only one primer out of more than a hundred is complementary to the target RNA species. For example, this approach has recently been applied successfully to cloning cDNAs representing new members of the POU gene family of transcription factors, using fully degenerate sense and antisense primer sets flanking the diagnostic POU and homeo-specific domains of these factors. The primers used covered sequences encoding nine codons each (He *et al.* 1989).

Anchored PCR

In some instances, only limited DNA sequence information is available for constructing an oligonucleotide probe. For example, members of a gene family might be related solely by a single small region of homology or there may be only very limited amino acid sequence data available. The anchored method of PCR (Loh *et al.* 1989), also known as the 'rapid amplification of rare cDNA ends' or RACE (Frohman *et al.* 1988), requires a DNA sequence from only one region of a gene. In principle, an antisense primer for this region is used to synthesize the first cDNA strand and a homopolymer tail is then attached to the free 3′ end. A complementary homopolymer oligonucleotide is then annealed to the tail, forming an 'anchor' which defines the unknown end of the cDNA, and is used to prime the second-strand synthesis with *Taq* polymerase. Subsequent amplification by PCR thus uses one specific oligonucleotide and the homopolymer anchor primer. The method can be used to clone rare messages and is adaptable for cloning both 5′, 3′ sequences;

```
MyoD1      S N P N Q R L P K V E I L R N A I R Y I E G
Myogenin   L N P N Q R L P K V E I L R S A I Q Y I E R
N-myc      L V K N E K A A K V V I L K K A T E Y V H S
L-myc      L A S C S K A P K V V I L S K A L E Y L Q A
c-myc      L E N N E K A P K V V I L K K A T A Y I L S

Consensus                  P K V E I L R N A
                                 V     K K
```

FIG. 3. Amino acid comparison of a region conserved between the myogenic regulatory factors MyoD1 and myogenin with *myc* oncogenes. Blocks of homology with MyoD1/myogenin are indicated. The consensus sequence used to synthesize a degenerate antisense oligonucleotide set for anchored PCR cloning is also shown (see text).

Frohman *et al.* (1988) used this method to amplify 5′ and 3′ sequences of cDNAs representing rare differential splicing products of the mouse *int*2 gene.

A method for anchored PCR is given below, based on that of Frohman *et al.* (1988) which we have adapted in an attempt to isolate cDNAs encoding cardiac regulatory factors with homology to *myc* oncogenes. Recently, cDNAs for several regulatory factors implicated in skeletal muscle development have been identified. The amino acid sequences of *Myo*D1 (Tapscott *et al.* 1988), and myogenin (Wright *et al.* 1989; Edmonson & Olson 1989), predicted from the cloned cDNA sequences, show a small region of homology with the *myc* family of oncogenes (Fig. 3). The authors have constructed an antisense oligonucleotide set against this *myc* region and are currently analysing cDNA clones isolated by anchored PCR from cardiac RNA.

Anchored PCR from an RNA template

An antisense oligonucleotide set is used to prime cDNA synthesis from an RNA template by reverse transcriptase. The cDNA is then tailed with poly-C and a sense poly-G oligonucleotide (the anchor primer) is annealed. Second-strand and amplification are carried out using *Taq* polymerase, often yielding several DNA species. The amplified material is then gel-purified and each specific DNA product is re-amplified to increase both the purity and the yield of the fragment.

Method 3

1 *First-strand synthesis*: mix together the following reagents;

2 μl 10 × *Taq* buffer;

0.1–1 μg of total RNA;

50 pmol of antisense oligonucleotide (i.e. complementary to the mRNA);

Add DEPC-treated water to 20 μl.

Anneal the oligonucleotide set to the RNA as described in *Method 2*, step 1. Then add the following reagents to the annealing mixture:

3 μl 10 × *Taq* buffer;
2 μl 4 mM dNTPs, pH 7–8;
1 μl 100 mM DTT;
0.5 μl placental RNase inhibitor;
0.5 μl of a [^{32}P]-labelled dNTP (e.g. of specific activity 800 Ci/mmol, 10 mCi/ml);
DEPC-treated water to 50 μl;
15 U AMV reverse transcriptase.

Incubate at 42°C for 1 h, then add 150 μl sterile TE buffer pH 8 (see Appendix) and keep on ice. In general, the addition of [^{32}P]-dNTP allows sufficient incorporation to act as a tracer in the following purification step.

2 *cDNA purification*: It is important to separate the cDNA from the unincorporated label and free oligonucleotides as contaminating primers act as efficient templates for tailing by terminal transferase if they are not removed from the cDNA. This may be done by applying the cDNA sample to a 2 ml Bio-Gel A-5 m (Bio-Rad) column equilibrated with TE buffer and collecting fractions representing the first radioactive peak containing the cDNA products (Frohman *et al.* 1988). As an alternative, we use a 2 ml plastic disposable syringe, plugged with a glass fibre filter, and containing 2.5 ml of Sephadex G-100 (Pharmacia) equilibrated with sterile TE buffer pH 8. The cDNA mixture is applied to the top of the column. Fractions (100 μl) are collected and the radioactivity determined by Cherenkov counting. The fractions containing RNA and labelled cDNA (approximately 300 μl) are then pooled and dried by centrifugation under vacuum.

3 *Homopolymer tailing*: terminal transferase, isolated from calf thymus, is an enzyme which efficiently adds nucleotides to the 3′ end of single-stranded DNA molecules. We routinely add a poly(C) tail. To add the homopolymer tail, resuspend the precipitated cDNA in 22 μl water. Add these to the tube:

6 μl 5 × terminal transferase buffer (BRL recipe; supplied with enzyme).
1 μl 6 mM dCTP.
1 μl (approx. 14 U) terminal transferase (BRL).

Incubate at 37°C for 10 min only, then heat-inactivate the enzyme at 65°C for 15 min. Dilute the reaction mixture to 500 μl with TE buffer, pH 8. This procedure should add 15–30 nucleotides to the 3′ end of the cDNA.

4 *First PCR reaction*: to 1–10 μl of poly-C tailed cDNA, add the following:

5 μl 10 × *Taq* buffer:
2 μl 4 mM dNTPs, pH 7–8.
50 pmol antisense oligonucleotide.

50 pmol poly-G anchor (sense) primer.

DEPC-treated water to 50 μl.

Perform the standard PCR reaction for at least 30 cycles and separate the DNA products on a 2% low-gelling temperature agarose gel (available from BRL).

Figure 4 shows the results of a first anchored PCR reaction using mouse cardiac RNA as a template and a degenerate antisense oligonucleotide set representing the *myc* homology region. Typically, the first reaction (lane 1) produces one or sometimes several bands seen against a background of non-specific DNA. The products can be further amplified by a second PCR reaction (same figure, lane 2).

5 *Second PCR reaction*: excise a small plug of agarose containing a first-round cDNA fragment from the gel (a Pasteur pipette is ideal for this purpose) and add to 500 μl of TE buffer, pH 8, in an Eppendorf tube. Elute the DNA by

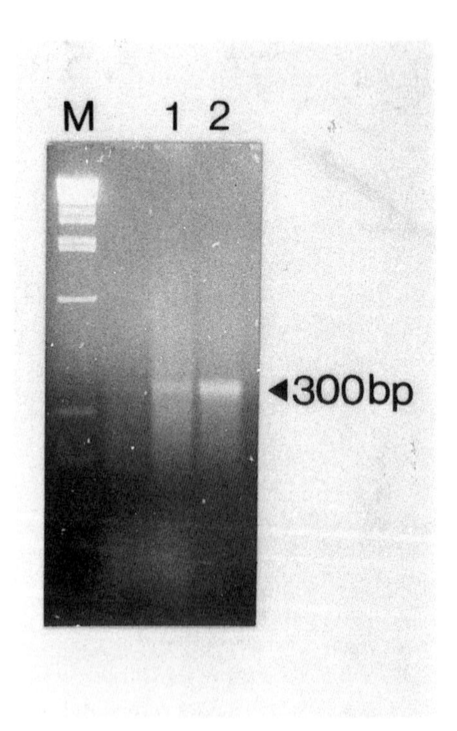

FIG. 4. Example of anchored PCR, using an oligonucleotide set directed against the *myc* region of homology (see Fig. 3). A 300 bp product was amplified in this case; the first-round product (lane 1) and second-round re-amplified product (lane 2) are shown.

heating at 70°C for 10 min to melt the agarose, then dilute the DNA 100-fold in TE buffer, as a 10 000-fold dilution of the first-round cDNA is recommended for a second-round of PCR (Haqqi *et al.* 1988). Take 5 μl for the second PCR reaction and repeat the PCR as in the previous section. When separated on a gel, the products should be present in higher yield and greater purity. The purity may be improved further by using a 'nested' primer within the 5′ tail of the anchor oligonucleotide (Haqqi *et al.* 1988). The anchor primer that we use has the sequence:

*Pst*I
5′ GAG**GTTTTCCCAGTCACGA***CTGCAG*GGGGGGGGGGGGGG 3′

This incorporates a G_{16} anchor tail at the 3′ side of the primer, a 5′ tail containing the sequence of the commercially available M13 (−40) sequencing primer (shown in bold type), and a *Pst*I restriction site (in italics) which may be utilized to remove the M13 sequence and facilitate subcloning of the PCR products into M13 or plasmid vectors. The M13 primer may be used as the nested primer in the second PCR reaction. In Fig. 4, the yield of the anchor product re-amplified by means of the nested primer (lane 2) is approximately five times greater than that of the original PCR fragment (Fig. 4, lane 1).

Conclusions and Perspectives

The PCR method is being applied to many biological systems both as a diagnostic and research tool. This chapter has largely concentrated on the application of PCR to the isolation of specific cDNA clones. PCR offers several advantages over conventional cloning methods. It is possible to obtain cDNA fragments within a period of days, in sharp contrast to more laborious (and hazardous) conventional cloning methods involving the screening of lambda or cosmid libraries by filter hybridization with [^{32}P]-labelled DNA probes. It also provides an extremely sensitive approach for cloning rare mRNA species or for cloning from small numbers of cells (Rappolee *et al.* 1988). Products of PCR obtained in this way, which contain specific flanking restriction enzyme sequences, may be inserted into DNA sequencing vectors for rapid clone identification, or may be used directly for double-stranded DNA sequencing (Higuchi *et al.* 1988). It should be noted, however, that the kinetics for the fidelity of replication by *Taq* polymerase are not clearly defined as yet, and there is evidence for an inherent error rate of about 0.1% (see Saiki *et al.* 1988). Cloning by PCR should ideally be used for rapidly identifying sequences of interest, but sequence determinations of new mRNAs should not be based solely on clones obtained in this way. Products of PCR can of course be easily used as homologous probes for isolating a *bona fide* cDNA or gene copies from an appropriate library.

References

Brand, N., Petkovich, M., Krust, A., Chambon, P., de The, H., Marchio, A., Tiollais, P. & Dejean, A. 1988. Identification of a second human retinoic acid receptor. *Nature* **332**, 850–853.

Edmondson, D.G. & Olson, E.N. 1989. A gene with homology to the *myc* similarity region of MyoD1 is expressed during myogenesis and is sufficient to activate the muscle differentiation program. *Genes and Development* 3, 628–640.

Eldin, P., Cornillon, B., Cathiard, A.M., le Cunff, M., Anoal, M., Leger, J.O.C., Mornet, D., Vosberg, H.P. & Leger, J.J. 1989. Actin and myosin light chain binding sites on human cardiac myosin heavy chain fragments synthesized by *E. coli*. *Journal of Molecular and Cellular Cardiology* **21**, Suppl. III, s8.

Frohman, M.A., Dush, M.K. & Martin, G.R. 1988. Rapid production of full-length cDNAs from rare transcripts: amplification using a single gene-specific oligonucleotide primer. *Proceedings of the National Academy of Sciences of the United States of America* **85**, 8998–9002.

Haqqi, M.T., Sarkar, G., David, C.S. & Sommer, S.S. 1988. Specific amplification with PCR of a refractory segment of genomic DNA. *Nucleic Acids Research* **16**, 11844.

He, X., Treacy, M.N., Simmons, D.M., Ingraham, H.A., Swanson, L.W. & Rosenfield, M.G. 1989. Expression of a large family of POU-domain regulatory genes in mammalian brain development. *Nature* **340**, 35–42.

Higuchi, R., Von Beroldingen, C.H., Sensabaugh, G.F. & Erlich, H.A. 1988. DNA typing from single hairs. *Nature* **332**, 543–546.

Lathe, R. 1985. Synthetic oligonucleotide probes deduced from amino acid sequence data. Theoretical and practical considerations. *Journal of Molecular Biology* **183**, 1–12.

Loh, E.Y., Elliott, J.F., Cwirla, S., Lanier, L.L. & Davis, M.M. 1989. Polymerase chain reaction with single-sided specificity: analysis of T cell receptor δ chain. *Science* **243**, 217–228.

Petkovich, M., Brand, N.J., Krust, A. & Chambon, P. 1987. A human retinoic acid receptor which belongs to the family of nuclear receptors. *Nature* **330**, 444–450.

Rappolee, D.A., Brenner, C.A., Schultz, R., Mark, D. & Webb, Z. 1988. Developmental expression of PDGF, TGH-α, and TGF-β genes in preimplantation mouse embryos. *Science* **241**, 1823–1825.

Saiki, R.K., Gelfand, D.H., Stoffel, S., Scharf, S.J., Higuchi, R., Horn, G.T., Mullis, K.B. & Erlich, H.A. 1988. Primer-directed enzymatic amplification of DNA with a thermostable DNA polymerase. *Science* **239**, 487–494.

Sherman, D.R., Geliehter, J. & Cross, G.A.M. 1989. Rapid and simple amplification of a specific RNA sequence by the polymerase chain reaction. *Trends in Genetics* **5**, 137.

Tapscott, S.J., Davis, R.L., Thayer, M.J., Cheng, P.-F., Weintraub, H. & Lassar, A.B. 1988. MyoD1: A nuclear phosphoprotein requiring a Myc homology region to convert fibroblasts to myoblasts. *Science* **242**, 405–411.

White, T.J., Arnheim, N. & Erlich, H.A. 1989. The polymerase chain reaction. *Trends in Genetics* **5**, 185–189.

Wright, W.E., Sassoon, D.A. & Lin, V.K. 1989. Myogenin, a factor regulating myogenesis, has a domain homologous to MyoD. *Cell* **56**, 607–617.

Appendix

Buffer composition

10 × Taq *polymerase buffer (Cetus recipe)*

100 mM Tris–HCl, pH 8.4.
500 mM KCl.
15 mM $MgCl_2$.
1 mg/ml gelatin or BSA.
Filter-sterilize and keep frozen in aliquots.

dNTPs mixture

Make 50 mM stocks of each dNTP, neutralized to pH 7–8 with NaOH. Prepare 4 mM mixture (1 mM each dNTP) from these stocks and store frozen in aliquots.

TE buffer, pH 8

10 mM Tris–HCl, pH 8.
0.1 mM EDTA.

Filling in DNA ends with Klenow polymerase

To 'fill-in' the ends of a DNA molecule, incubate with 1 U of the Klenow fragment of *E. coli* DNA polymerase-I in the presence of 250 μM of each dNTP for 30 min at 37°C. The reaction is carried out in 10 mM Tris–HCl pH 7.5 and 10 mM $MgCl_2$. Then inactivate the enzyme by heating to 68°C for 10 min. Extract with phenol and chloroform, precipitate from ethanol, and finally resuspend the DNA in a small amount of TE pH 8 buffer.

Problems and Troubleshooting

The reader should check key references for discussions of the consequences of varying certain parameters (Saiki *et al.* 1988; Frohman *et al.* 1988; Haqqi *et al.* 1988). The most important are briefly described below.

Divalent cation concentration

Taq polymerase requires magnesium as a source of divalent cations for efficient DNA replication. The buffer described above contains $MgCl_2$ at a final concentration of 1.5 mM. The optimal concentration of Mg^{2+} for a PCR reaction should ideally be titrated at the start of each project as concentrations of *free* $MgCl_2$ within the range of 1–10 mM may significantly alter the yield of the PCR-amplified products (possibly by binding to the negatively-charged phosphate DNA backbone, affecting the annealing efficiency of primers to target sequences). In the protocols described here, the final concentration of dNTPs in the reaction is about 160 μM total, which will sequester an equimolar amount of Mg^{2+}, leaving a free Mg^{2+} concentration of about 1.3 mM. Figure 2 b shows the effect of titrating the $MgCl_2$ regarding the amplification of a β-MHC cDNA fragment. A major product of 104 bp is seen with the standard conditions of 1.3 mM free Mg^{2+} (lane 1), but as the Mg^{2+} concentration increases (same figure, lanes 2–4), the product yield is dramatically reduced, accompanied by the appearance of aberrant higher molecular weight products.

dNTP concentration

As with the Mg^{2+} concentration, this should be kept low, certainly no higher than 200 μM final concentration of each dNTP. High dNTP concentrations may be associated with poor fidelity of replication and chelation of primers and Mg^{2+} ions. Figure 5 shows a PCR experiment in which a specific region of the human retinoic acid receptor α (Petkovich *et al.* 1987; Brand *et al.* 1988) was amplified in the presence of two different concentrations of dNTPs. Under the same conditions of template, primer and enzyme concentrations,

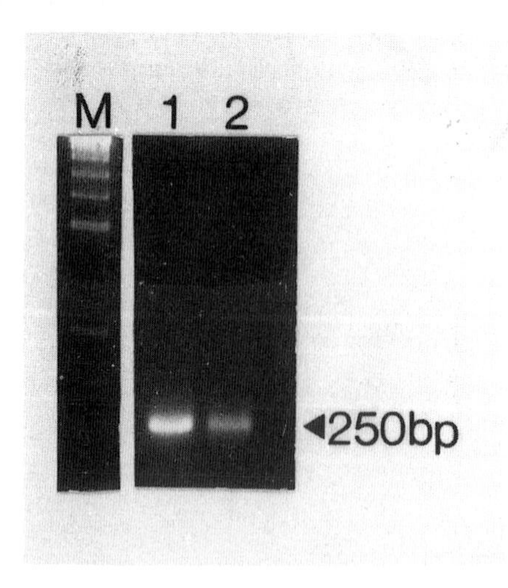

FIG. 5. A region of the retinoic acid receptor α was amplified by PCR in the presence of either 80 μM (lane 1) or 160 μM (lane 2) of each dNTP. The expected PCR product is indicated.

the level of amplification was at least 10 times more efficient when carried out in the presence of 80 μM each dNTP (Fig. 5, lane 1) than with 160 μM each dNTP (same figure, lane 2).

Oligonucleotide design

The oligonucleotide can be 16–30 nucleotides long (excluding restriction site, or other 5′-tail if appropriate). The base composition should be roughly 50% G:C-rich, avoiding poly(A:T) or poly(G:C) tracts which could result in hybridization to non-target sequences. The 3′ end should contain a few G or C residues, ensuring that this end is tightly bound to the template in order for efficient extension by the polymerase to occur. Check that no secondary structure formation is possible within a primer and that two primers in the same reaction will not cross-hybridize, thereby forming 'primer dimers' as the major amplification product. Degenerate primer sets should not be too short (i.e. less than 20 nucleotides) to take account of possible primer:template mismatches and reduced efficiency of annealing.

Cycle conditions

These can be varied so as to determine the optimal conditions for amplification. Of particular importance are the number of cycles, the annealing temperature chosen (which for shorter primers can be calculated from the theoretical T_m for the oligonucleotide) and the period of time allowed for the drop from denaturation temperature to the annealing temperature. The latter two points can be used to optimize the specificity of binding of the primer for its correct target. See Lathe (1985) for details on calculating T_m values and general oligonucleotide design.

The Effect of Nutrient Limitation on the Stability and Expression of Recombinant Plasmids

A. WARNES AND J.R. STEPHENSON
Division of Biologics, Centre for Applied Microbiology and Research, Porton Down, Salisbury, Wiltshire SP4 0JG, UK

Genetically engineered plasmids promise to be one of the most valuable tools the biotechnology industry possesses. The adaptation of laboratory-generated plasmids to the fermentation conditions applicable in industry can be plagued with many problems. Potentially one of the most severe problems is the inability of genetically engineered plasmids to be maintained in large-scale cultures. It is the purpose of this chapter to review the field of plasmid stability, to describe the methods available to detect genetically-engineered proteins and to explain how to monitor the stability of a plasmid in a culture. There are two main areas of plasmid instability, segregational instability which is due to defective partitioning and structural instability which may arise by deletion, insertion or rearrangement of the plasmid DNA.

Structural Instability

Structural instability may be caused by deletion, addition or rearrangement of genetic material. There are numerous examples in the literature of structural instability of recombinant plasmids (Cohen *et al.* 1977; Timmis *et al.* 1978; Imamaka *et al.* 1980) but few reports of these changes occurring in continuous culture (Goodwin & Slater 1979). Noack *et al.* (1981), investigating the stability of pBR plasmids, noticed that prior to the loss of stability of pBR325 in carbon limited medium, the tetracycline-resistance determinant was lost. Later work by Chew *et al.* (1986) indicated that transposition of IS1 from the host chromosome of *E. coli* HB101 into the tetracycline resistance gene of pAT153 resulted in the loss of expression and the development of a more stable host/vector system. Work by Brownlie & Cole (personal communications) and Chew *et al.* (1988) has shown that the tetracycline resistance gene in pAT153 when grown in *E. coli* HB101 under a number of nutrient limitations and at various dilution rates is prone to mutational deletions and insertions.

0–632–02926–9

Segregational Instability

Genetic control of stability

Segregational instability can occur due to the lack of a *par* function, a reduction in copy number or toxicity of a gene product. *Par* functions are thought to be able to help in the segregation of plasmid DNA by binding to the outer membrane of the host as demonstrated by Gustafsson *et al.* (1983). In addition Wahle & Kornberg (1988) have shown that *par* is a specific binding site for DNA gyrase, although it is not thought to affect plasmid stability through its supercoiling activity or by an influence on DNA replication. Thus in naturally occurring plasmids with low copy numbers, e.g. IncFII, (Timmis *et al.* 1981) and pSC101 (Meacock & Cohen 1980) the presence of a *par* function usually ensures that they are actively segregated at cell division, thus avoiding segregational instability. The higher copy number plasmids (e.g. ColEI) and many other genetically engineered plasmids do not contain a *par* function and are segregated randomly. It has been suggested that random partitioning may be sufficient to ensure the stability of multicopy vectors and that active partitioning may not be essential (Summers & Sherratt 1984). Obviously a reduction in the copy number of a plasmid without *par* could lead to segregational instability. A range of factors can influence the copy number of a plasmid including nutrient limitation, an increase in the host's growth rate, an increase in the genetic load imposed upon the host's replication system, an increase in plasmid multimers or by the toxic effects of the protein expressed by the plasmid.

Plasmids without a partitioning function are therefore segregated randomly and the probability $P(0)$ of either daughter cell failing to inherit a plasmid is given by a binomial distribution:

$$P(0) = 2(1/2)^c$$

Where c is the number of plasmid molecules per cell (copy number) at division (but see discussions in methods). In fact Cooper *et al.* (1987) have taken the mathematical modelling a stage further to produce a formula which will indicate whether the instability of a plasmid is due either to the loss of the plasmid on segregation or to differences in the growth rates between cells with or without the plasmid. This may be of limited value in predicting the area of instability of particularly unstable plasmids, although far greater efforts may be needed to cure instability. Although the copy number of widely used derivatives of ColEI, such as pBR322, is known to be high, approximately 50 per host chromosome in *E. coli* HB101 (Hashimoto-Gotoh & Timmis 1981), there is a probability of plasmid-free cells arising under certain conditions such as nutrient limitations or during rapid cell growth (Jones *et al.* 1980b; Nugent *et al.* 1983). This problem can be prevented by continually selecting for phenotypic traits such as antibiotic resistance: but this may not be a desirable solution for large-scale cultures because of the costs of producing a product free from contamination and the problems involved in waste disposal. Partitioning regions have been cloned into plasmids such as pBR322 (Primrose *et al.* 1983) thus enhancing the stability of these plasmids. Although partitioning functions are often used to overcome problems with instability, the extent of curing may be variable (Nilsson & Skogman 1986). Even though *par* is commonly used in plasmid constructions its mode of action is still unknown. It does not code for any proteins or contain any transcription or translation start signals. There are areas of the *par* locus which are capable of forming regions of intrastrand secondary structures and these may play some role in partitioning (Miller *et al.* 1983). *Par* has also been shown to bind the outer-membrane of the host (Gustafsson *et al.* 1983) and bind DNA gyrase (Wahle & Kornberg 1988).

The replication of stringent plasmids is of necessity linked to chromosomal replication, hence their low copy number. Low copy number plasmids usually contain the *tra* and *mob* genes for conjugation while the plasmid has a *bom* site. The *tra* genes promote growth of pili for the transfer of linear plasmid DNA (Williams & Skurray 1980), while the *mob* genes are involved in the nicking of the plasmid DNA producing linear forms which can then be transfered to recipient hosts (Willetts & Crowthers 1981; Bagdasarian *et al.* 1981). The *bom* site is an area on the plasmid which allows the DNA to be nicked and linearized. The loss of the *bom* site or the *mob* genes from plasmids would therefore result in the loss of transfer as nicking cannot occur. The loss of the *tra* genes from plasmids may not have this effect as recipient hosts may be able to form conjugation tubules allowing the transfer of plasmid DNA (Young & Poulis 1978). Most new vectors developed by genetic manipulation are now non-mobile to satisfy legislation on containment.

The effect of copy number on stability

Plasmid stability can be affected by changes in copy number as shown by the stability of pAT153, which is a derivative of a relatively unstable plasmid pBR322 (Jones & Melling 1984). The reason for this increased stability may be due to the increase in copy number of pAT153 relative to pBR322 (Twigg & Sherratt 1980), or the fact that too few generations had elapsed before segregants could be detected under laboratory conditions. Other workers have also found pAT153 to be stable for 90 generations (Caulcott *et al.* 1985), although Chew *et al.* (1986) found pAT153 to be unstable (30 generations) at low dilution rates under all nutrient limitations, but stably maintained (90 generations) at high dilution rates (0.3–0.5/h) in glucose-limiting media.

Finally, a sequence of DNA was discovered in ColEI called *cer* which causes recombination to occur and reduces plasmid dimers or multimers to monomers (Summers & Sherratt 1984). This is extremely important in plasmids lacking a partitioning function as an increase in multimers effectively reduces the copy number of the plasmid which could then affect plasmid stability. The *cer* region has also been cloned into pBR322 thus increasing stability. The same workers also discovered a region in the plasmid DNA called *Rom*, which codes for a 63 amino acid peptide that negatively controls transcription of RNAII and consequently plasmid replication. The presence of this region leads to reduced copy numbers, which may result in plasmid instability.

The effects of nutrient depletion on plasmid stability

Melling *et al.* (1977) were one of the first groups to investigate the effects of nutrient limitation on plasmid stability while working on the RP1 plasmid in *E. coli* W3110. The results showed the plasmid to be stable at various dilution rates (0.05–1.0/h) under nutrient limitations of carbon, magnesium and phosphate. These results were later confirmed showing there was no loss of antibiotic resistance under prolonged chemostat runs (Jones *et al.* 1980a). As the copy number of RP1 was only 1–2 per cell, a partitioning function was thought to be involved in the plasmids replication (Nordstrom *et al.* 1980). However, Klemperer *et al.* (1979) found that the specific growth rate of cells without the plasmid (R−) was double that of cells containing the plasmid (R+) under low phosphate concentrations in batch culture. Similar results were shown in competition experiments where R+ cells were out-grown by a 1% inoculum of R− cells in phosphate limited chemostats (Melling *et al.* 1977). Converse experiments always showed that R− cells came to predominance under phosphate-limiting conditions, whereas the outcome of other limitation studies relied only on the size of the inoculum of R+ cells.

Further studies showed that plasmid-free segregants of pBR322 (*par* minus) arose under glucose or phosphate limiting studies in continuous culture, which correlated with a drop in copy number after 30 generations (Jones *et al.* 1980b; Wouters *et al.* 1980). An increase in the stability of pBR322 (to 240 generations) was seen with an increase in dilution rate. Since nutrient restriction is less severe at higher dilution rates, it may take many more generations before plasmid-free segregants arise. Similar results were obtained with decreases in temperature, the plasmid being far more stable at 30°C than at 37 or 42°C. In contrast, results by Noack *et al.* (1981) showed no loss of pBR322 from *E. coli* grown in glucose or nitrogen-limiting conditions. These conflicting results could be due to variations in the genetic backgrounds of the host organisms used by the different workers. In fact, variations in strains of bacteria have been observed at the Centre for Applied Microbiology and

Research in the large-scale production of genetically-engineered products. Recently Sayadi *et al.* (1988) have shown an increase in the stability of pBR322 to 240 generations when the host was immobilized. However, this paper fails to note which nutrient was limiting with the medium consisting of tryptone yeast extract and 0.1 M KCl.

The effect of size on stability

Warnes & Stephenson (1986), investigating the effect of various sizes of cytomegalovirus DNA inserted into the tetracycline resistance gene of pAT153, found the plasmid with the smallest insert (2 kb) was stably maintained for 80 generations at varying dilution rates (0.1−0.4 h). The plasmids with the larger inserts of DNA (8 and 21 kb) became increasingly unstable and the maximum specific growth rate of the host organism was also greatly reduced with this increase in the size of the plasmid. This indicates that an increased metabolic load imposed upon the cell could instigate plasmid instability as reported by DaSilva & Bailey (1986). The plasmid CloDF13 was shown to be stably maintained at a copy number of 10 in *E. coli* by a competitive process at the level of transcription, with segregation being influenced by plasmid size and copy number (Hakkaart *et al.* 1985). These authors also showed that the effect of copy number was dominant over plasmid size in causing instability.

Effects of host mutations on plasmid stability

Fermenter adaptation of the host organism by genetic changes within the chromosomal DNA may lead to fluctuations in the stability of the plasmid. These mutations have resulted in the appearance of more competitive bacteria with the changes usually occurring in nutrient limited conditions (Goodwin & Slater 1979; Helling *et al.* 1981; Warnes & Stephenson 1986). The results, in this chapter, show the stability of pAT153 containing a 8 kb insert of cytomegalovirus DNA grown under carbon, magnesium and phosphate-limiting conditions at a number of dilution rates. The re-emergence of the plasmid in the population corresponded to a change in host phenotype from *leu* minus to *leu* plus, resulting in a fitter host strain. Sayadi *et al.* (1988) have shown similar changes occurring in nutrient-rich media over a period of generations.

Experimental Techniques used to Monitor Plasmid Stability

Screening for antibiotic resistance

Screening bacterial colonies is one of the simplest methods for detecting the presence of a plasmid in a culture. As no allowances are made for copy number reductions in each host the results may include false negatives. Screening can be performed by picking one hundred colonies from viable count plates onto L-agar plates and incubating for 16 h at 37°C. The colonies are then replicated, by means of sterile velvet discs or filter paper, onto L-agar plates, both with and without the required antibiotic, and incubated for 16 h at 37°C. The results can then be expressed as the percentage antibiotic resistant cells in the population, which is related to the amount of cells harbouring the plasmid. Obviously a range of antibiotics are often used in plasmids to ensure selection, and the concentration of each in the L-agar plates varies. The antibiotics illustrated use the concentrations listed in Table 1.

Theoretical considerations in copy number determination

A key parameter in the study of plasmid stability is the estimation of copy number of the plasmid in each cell. This is often defined as the ratio of the number of moles of plasmid DNA to the number of moles equivalent of chromosomal DNA. However, unless one is looking at the host's growth at a specific point in the logarithmic growth phase, i.e. in continuous culture (Stüber & Bujard 1982) it may be more meaningful to define copy number as the number of moles of plasmid DNA to the number of host organisms. However, as all the studies mentioned here involve the comparison of copy numbers in bacteria grown under similar, carefully controlled conditions, the definition of copy number as number of plasmids per chromosome is valid. Therefore this definition is used throughout the work described below and is the basis of the method used. A number of methods are available for the estimation of copy number, although none is entirely satisfactory (Projan *et al.* 1983). The methods are all based on one of three analytical principles: physical separation of plasmid from chromosomal DNA, hybridization or measurement of gene dosage. The methods based on physical separation include density or alkaline velocity density centrifugation and these rely on the plasmid being in the covalently closed circular form and do not take into

TABLE 1. *The concentration of antibiotics required to inhibit the growth of organisms without a specific resistance*

Antibiotic	Final concentration (μg/ml)
Ampicillin	50
Chloramphenicol	10
Kanamycin	50
Streptomycin	25
Tetracycline	15

account the amount of plasmid DNA lost from the preparation and are therefore minimum estimates (Thomas 1987). Hybridization methods are difficult to perform and require strict working conditions. Furthermore, steps in the procedures could lead to a loss of plasmid DNA although the form of the DNA is irrelevant (Rush *et al.* 1975; Shepard *et al.* 1979). Methods based on gene dosage rely on the fact that expression is proportional to the number of plasmid copies (Uhlin & Nordstrom 1977). This may be adequate for stringent plasmids with copy numbers under ten, although most genes in relaxed plasmids usually do not follow this pattern. These methods are obviously time-consuming and because of this they are unsuitable for handling large numbers of samples. These problems have been largely overcome by using agarose gel electrophoresis which separates the plasmid from the chromosomal DNA by virtue of its molecular weight. This method was first used by Eckhardt (1978), who stained two bands of chromosomal and plasmid DNA with ethidium bromide. Fluorescence spectrophotometry could then be used to estimate the relative quantities of plasmid and chromosomal DNA. The method used in this work was originally adapted by Twigg & Sherratt (1980), in which photographic negatives were used instead of fluorescent spectrophotometry. The other major addition was the use of internal standards so that differences in copy number could be estimated accurately. Recent advances in instrumentation have made it possible to visualize the DNA bands directly, thus avoiding errors arising from ethidium bromide binding or photography.

Methodology for copy number determination by gel electrophoresis

Direct visualization of the plasmid content was obtained by means of a modification of the method of Twigg & Sherratt (1980):

1 A 0.5 ml sample of culture was centrifuged (12 000 rpm) at room temperature in a minifuge for 5 min and the supernatant removed.

2 The pellet was resuspended in 100 μl of lysis solution (20 mM Tris–HCl pH 8.0; 10 mM EDTA, 100 mM NaCl, 20% w/v sucrose, 2 mg/ml lysozyme and 5 units/ml pancratic RNase) and incubated at 37°C for 30 min.

3 100 μl 2% SDS was then added and the sample vortexed for 1 min before freeze-thawing twice in liquid nitrogen and rewarming to 37°C.

4 10 μl/ml proteinase-K were then added and the mixture was incubated at 37°C for 30 min.

5 An equal amount of sample to loading buffer (50% glycerol, 0.02% xylene cyanol, 0.02% bromophenol blue) was then run on a 1.0% agarose gel (made up as above) for 4 h at 100 mV.

6 The gel was stained and photographed and the negative scanned with a Joyce–Loebl scanning densitometer and the areas under the peaks integrated (Fig. 1).

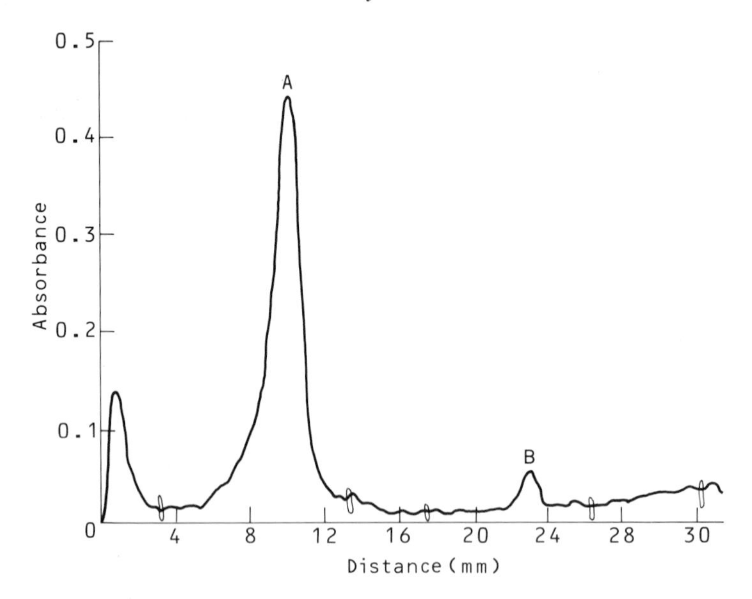

FIG. 1. Scan of a photographic negative taken from a gel stained with ethidium bromide, showing the chromosomal DNA (track A) and plasmid DNA (track B).

7 Using the following formulae, as described by Projan *et al.* (1983), the copy number per host chromosome can be calculated:

$$Cp = \frac{Dp \times I \times Mc}{Dc \times Mp}$$

where

Cp = Plasmid copies per host chromosome;
Dp = Amount of plasmid DNA in the gel;
Dc = Amount of chromosomal DNA in the gel;
Mc = The total chromosomal DNA per cell (3.8×10^6);
Mp = Molecular weight of the plasmid (pBR322 = 50);
I = Difference in the binding of ethidium bromide to chromosomal or protein DNA (1.36).

pBR322 in HB101 can be used as an internal standard when grown in log-phase. A copy number of 50 has been quoted (Hashimoto-Gotoh & Timmis 1981).

Detection of Recombinant Proteins

The detection of recombinant products is dependent on the protein and its level of expression. The most straightforward method available is to visualize the proteins on a SDS-PAGE gel, according to their apparent molecular weights. However, *E. coli* proteins may interfere with the detection of recombinant gene products. This interference may be reduced by Western blotting or radioimmune precipitation which involves probing the gel or the proteins transferred to nitrocellulose with specific labelled antibodies. Other methods available for reducing interference include pulse labelling with ^{35}S-methionine (following expression of the recombinant gene by an inducible promoter), or by using the mini or maxi-cell techniques (Reeve 1979; Sancar *et al.* 1981). The former involves the production of mini-cells which contain plasmid DNA but no chromosomal DNA. This results in translation depletion of normal host proteins leaving only those encoded by the plasmid. The latter uses specific hosts with mutated DNA repair mechanisms which, in conjunction with u.v. radiation, inactivates the chromosomal DNA, leaving the high levels of plasmid DNA relatively intact. Both of these methods reduce the proteins encoded by the chromosomal DNA, resulting in a reduction of background interference.

Although these methods give an indication of size, they cannot detect low levels of expression, or determine whether the protein is in its native form. Immunological techniques are important in confirming the identity of the expressed proteins. Antigens can be detected by using either monoclonal or polyclonal antibodies which can indicate any major changes in structural organization. A variety of methods can be used to monitor expression such as ELISA, radioimmune assays, immunofluorescence, immunoblotting and immunodiffusion. The immunological assays are also employed to screen large numbers of colonies for the expression of recombinant proteins, and are also particularly useful for epitope mapping (Broome & Gilbert 1978). The methods mentioned above may be redundant if a biochemical assay is available to measure the activity of the recombinant product, such as β-lactamase (Sutcliffe 1979). Complementation of mutants in the host chromosome is another method of detecting expressed proteins as shown with β-galactosidase. However, in the detection of recombinant proteins, these last two methods are rarely available.

ELISAs are now the preferred methods for the detection of recombinant proteins, as not only can they be used for screening but they can also give quantitative values. PAGE analysis and Western blotting can then be used in conjunction with the ELISA to determine the specificity and sensitivity of the assay system.

ELISA

This system will be illustrated by describing a capture assay for staphylococcal protein A (SPA), developed by the authors. This assay uses human IgG bound to the plate as the capture antibody to which any SPA in the sample should bind. The concentration of human IgG used in the test was arrived at empirically. Internal standards containing 100 ng/ml SPA (Sigma) were run on every plate. Other samples contained lysate buffer or lysates of *E. coli* or *S. aureus* strains, both those expressing and those not expressing SPA. Known concentrations of SPA were added to these samples to detect interference by bacterial constituents.

The first detection antibody was a commercial rabbit anti-SPA serum; the inclusion of this stage greatly improved the sensitivity of the assay. The second detection antibody was a commercial goat anti-rabbit-IgG peroxidase conjugate. 5-Amino salicylic acid was used as the chromogenic agent. ELISA plates (Dynatech M129A) were coated with 100 μl/well of capture antibody (commercial human IgG (Dako)) at optimal dilution in 15 mM Na_2CO_3, 35 mM $NaHCO_3$, 0.02% NaN_3 (pH 9.6) at room temperature overnight. After each stage of the assay the plates were washed four times in phosphate buffered saline (PBS) containing 0.1% Tween 20 (PBS-T). This solution was found to be adequate to block all non-specific adsorption of protein to the plates. PBS-T was used as the diluent for all stages of the ELISA procedure. Internal standards and other samples were diluted appropriately, and two-fold dilution series were set up in 100 μl volumes in Sterilin M25A plates which had been presoaked in PBS-T for 2 min to block adsorption of protein. Then 90 μl samples of these dilution series were transfered to the ELISA plates and incubated for 90 min at room temperature. Bound SPA was detected by incubation with rabbit anti-SPA serum (Sigma) for 90 min, then with goat (anti-rabbit IgG) IgG–horse radish peroxidase conjugate (Bio-Yeda) (GARP) for 90 min, both at room temperature. The bound conjugate was then incubated for 15 min at room temperature with 0.1% 5-amino salicylic acid, 0.005% H_2O_2 in 50 mM Na_2HPO_4/NaH_2PO_4, pH 6.0 and the optical density of the coloured product was determined at 450 nm, in a Titertek Multiskan MCC automatic plate reader linked to an Acorn BBC B microcomputer running data capture reduction software written in the authors' laboratory. The end-point dilution was defined as the dilution producing a given optical density above background and was transformed into SPA concentration by comparison with internal standards run on every plate.

Polyacrylamide slab gel electrophoresis

The polyacrylamide slab gel electrophoresis (PAGE) buffer systems are usually based on those developed by Laemmli (1970), for use with gradient and standard gels. The gradient gels used in our studies were 50% (w/v) acrylamide and 0.8% *N,N'*-bis-methylene acrylamide stock solution diluted with buffer (final concentration 0.1% SDS, 375 mM Tris–HCl, pH 8.8) and polymerized by the addition of 0.042% tetramethylethylenediamine (TEMED) and 0.7% ammonium persulphate (APS). Gradient gels (7.5–20%) were produced by means of a gradient maker (BRL, Cambridge), with a 99:1 (w/w) proportion of acrylamide to bisacrylamide throughout. The standard gels had a running gel of final concentration 10% (w/v) acrylamide, 0.17% (w/v) *N,N'*-bis-methylene acrylamide, 0.15% SDS, 500 mM Tris–HCl pH 8.8 and polymerized by the addition of 0.1% APS and 0.01% TEMED. The stacking gels both had a final concentration of 5% (w/v) acrylamide and 0.17% (w/v) *N,N'*-bis-methylene acrylamide, 0.1% SDS, 125 mM Tris–HCl pH 6.5 and polymerized by 0.1% APS and 0.01% TEMED. The samples were prepared for PAGE under denaturing conditions in 0.1% (w/v) SDS, 0.1 M Tris–HCl pH 6.8 containing a trace of bromophenol blue and heated to 100°C for 1 min. The gel was then run in running buffer (Tris–HCl 6.32 g, glycine 4.0 g and SDS 1.0 g made up to 1 litre in distilled water) at 10 V/cm. It should, however, be emphasized that PAGE systems frequently have to be modified to suit the particular task at hand.

Western blotting

Polyacrylamide slab gels are first equilibriated in two changes of blotting buffer (7.5 mM $Na_2HPO_4.12\ H_2O$, 17.5 mM $NaH_2PO_4.2H_2O$) for 30 min each at room temperature. A nitrocellulose membrane (Bio-Rad) is soaked in the same buffer for 30 min and proteins are then electrophoretically transferred from the polyacrylamide gel to the nitrocellulose by applying a transverse electric field of 15 V (2.5 V/cm) of gel for 2 h. Portions of the membrane were either blocked overnight at 4°C in ELISA diluent (PBS-T) containing 3% gelatin and 0.02% NaN_3, to prevent non-specific binding, or stained in 0.1% Coomasie blue in 50% methanol, 45% distilled water, 5% acetic acid, and destained in the same solvent. For immunological probing, the membrane was washed in diluent and exposed to antibodies (90 min each) by sealing it in a polythene bag containing 10 ml of serum or conjugate at appropriate dilutions in PBS-T + 1% gelatin. After final washing in diluent, it was washed in substrate buffer (50 mM sodium acetate buffer pH 5.0) and rocked in substrate solution (0.16 mg/ml amino-ethyl carbazole, 0.03% H_2O_2) until developed (usually 5–10 min at room temperature; Fig. 2). The reaction was stopped by washing the membrane in water and leaving it to dry. Problems with excessive background colour could be eliminated by using gold labelled goat anti-rabbit IgG with a silver enhancer (Jensen Products, ICN Biomedicals Ltd, High Wycombe: Auroprobe BLplus GAR and IntenSE II) and used as specified by the manufacturer's instructions.

Assay for β-lactamase

β-lactamase activity can be estimated by the hydrolysis of nitrocefin as described by O'Callaghan *et al.* (1972). A stock solution of nitrocefin was first prepared

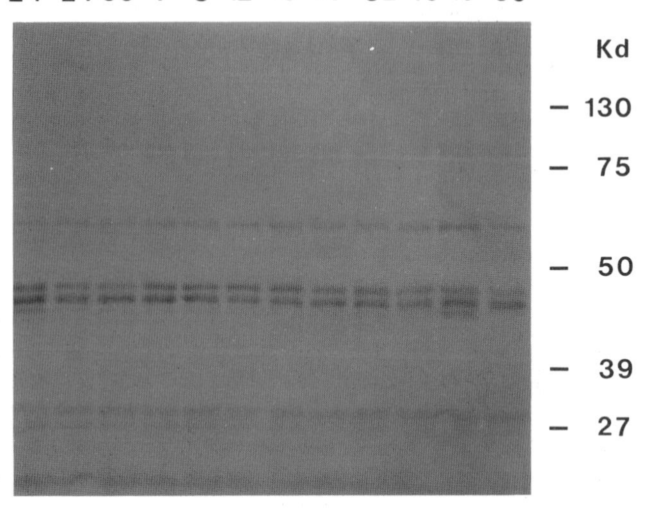

FIG. 2. Western blot analysis of samples taken from a fermenter run with the plasmid pPA16 which is expressing recombinant protein A. The tracks contain samples taken at various time intervals throughout the experiment as depicted at the top of the figure.

by dissolving 5 mg in 0.5 ml demethyl sulphoxide which was then added to 9.5 ml of phosphate buffer (pH 7.0). A 4 ml amount of this solution was added to 36 ml of phosphate buffer (pH 7.0) and used as the substrate for the assay. Dilutions of samples were then mixed with the substrate and monitored for changes in absorbance at 482 nm.

$$\text{Activity } (\mu\text{mol/min}) = \frac{\text{change in OD (482 nm)} \times V \times D}{\text{Time} \times \text{Em}_{\text{M}}}$$

where V = volume of the sample;
D = dilution of the sample;
Em_{M} = 15.9 (mM absorbance of hydrolysed nitrocefin at 482 nm).

Estimation of the Nutritional Requirements of the Host Organism

It is important to estimate the concentrations of each nutrient in the medium in order to ascertain whether that nutrient is limiting or not. The chemically defined medium mentioned in this work is based on that originally developed by Klemperer *et al.* (1979) (Table 2). The host organism was grown in batch

TABLE 2. *A chemically defined medium permitting maximum growth of* E. coli

Nutrient	Concentration (mM)
Ammonium chloride	20.0
Ammonium ferric sulphate	0.005
Glucose	25.0
Magnesium sulphate	0.4
Potassium chloride	0.4
Proline	13.0
Sodium phosphate (buffer, pH 7.0)	6.5
Thiamine	0.1

culture in 100 ml fluted flasks on an orbital incubator shaker at 150 rpm and 37°C. The bacterium was first grown for 16 h in complete chemically defined medium, then washed twice in chemically defined medium without the nutrient under investigation and resuspended to give an optical density of 2.5 at 470 nm. Then 0.25 ml amounts of this cell suspension were used to inoculate 24.75 ml of chemically defined medium with the appropriate concentration of nutrient under investigation. By using various concentrations of each limiting nutrient, growth curves can be produced and, if the end of non-limited growth is then plotted against nutrient concentration, the concentration of each nutrient to produce growth to a particular optical density can be calculated.

Monitoring the Stability of Plasmids Expressing Staphylococcal Protein A in *E. coli*

The authors have used the above methods to monitor the stability and expression of plasmids containing the gene coding for staphylococcal protein A (SPA). Fermenters were used to grow the host *E. coli* JM83 in continuous culture so that the stability of the plasmid (pPA16) could be monitored under defined conditions. The experiment illustrated here (Fig. 3) shows the stability of pPA16 in *E. coli* JM83 when grown under glycerol limiting conditions at a dilution rate of 0.1 per h. The results show that, with regard to the percentage of ampicillin-resistant cells in the population, instability was initiated after 20 generations in continuous culture, resulting in total loss of resistance at 70 generations. Although the results for copy number estimation and SPA production show a similar pattern, plasmid instability was initiated before that seen in the case of ampicillin-resistance. Restriction analyses of the samples (Fig. 4) not only show the loss of the plasmid from the culture but also whether any major structural changes have occurred throughout experiment. Fig. 5 is a gel showing the chromosomal and plasmid bands of DNA which

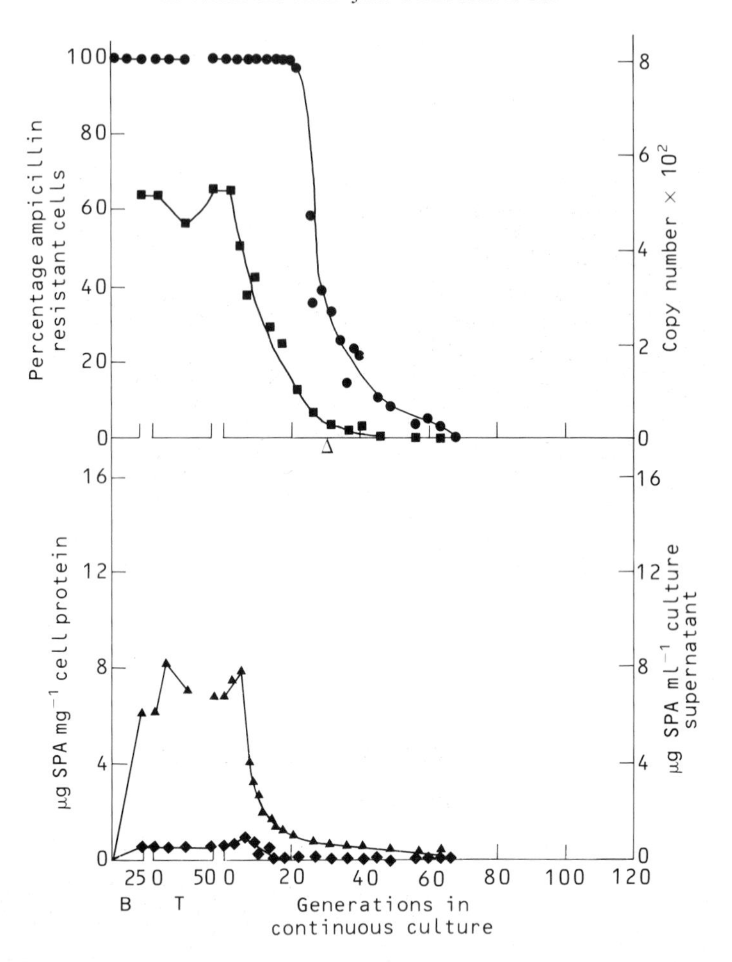

FIG. 3. The stability and expression of pPA16 in *E. coli* JM83 when grown in glycerol-limited medium at a dilution rate of 0.1/h: the *x*-axis has been broken down into the batch phase (B), transitional phase (T) which represents the growth during the time from switching on the nutrient pump to reaching the steady state, and the number of generations in continuous culture. The *y*-axis depicts the percentage of ampicillin-resistant cells in the population (●); the copy number of the plasmid in each host chromosome (■); the amount of recombinant protein A (μg) in 1 mg of cell protein (▲); and the amount of recombinant protein A (μg) in 1 ml of culture supernatant (◆). Changes in the nutrient medium reservoir are denoted by the symbol (△).

are used to estimate the copy number of the samples. Finally, Western blots of the samples show not only the presence of SPA but also the number of forms in which the protein exists (Fig. 6).

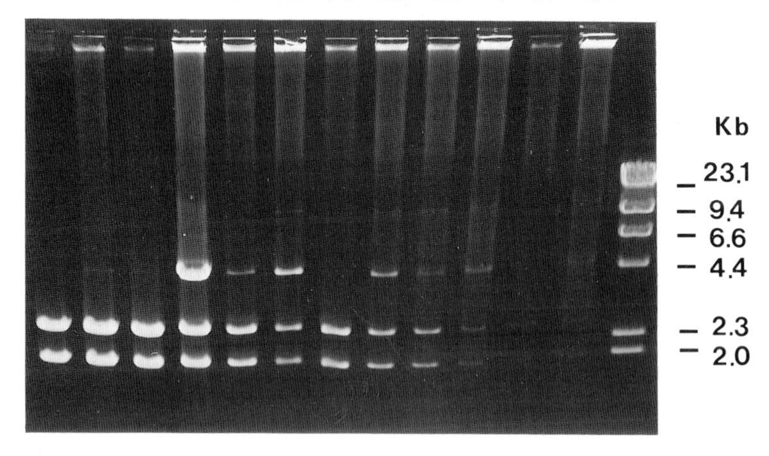

FIG. 4. Restriction analysis of plasmid DNA taken from samples at different time intervals as illustrated in Fig. 3. The tracks are denoted in time as described in Fig. 3. Molecular weight markers are shown on the right of the gel.

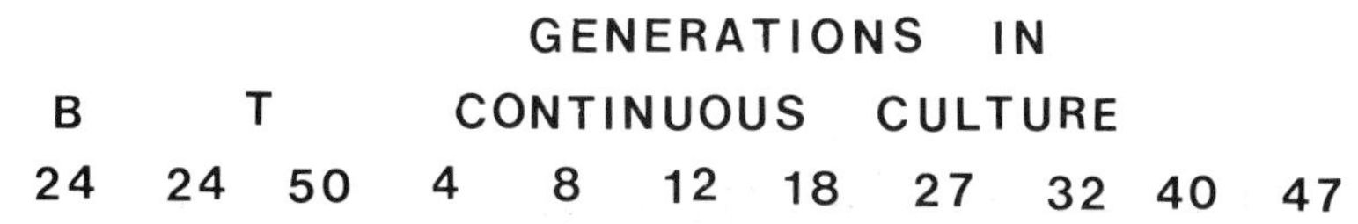

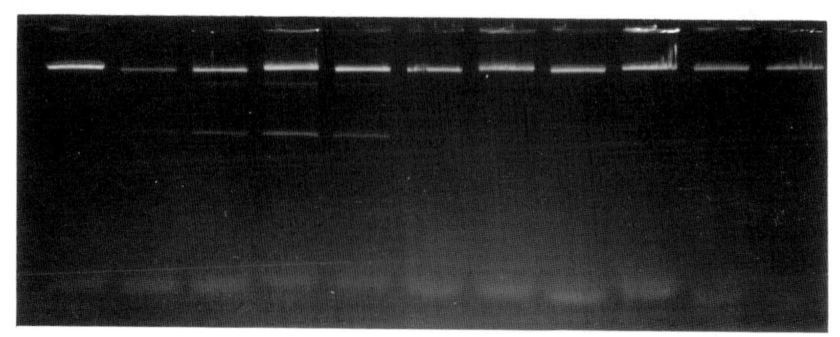

FIG. 5. Copy number analysis of plasmid and chromosomal DNA taken from samples at different time intervals as described in Fig. 3 and designated in Fig. 4.

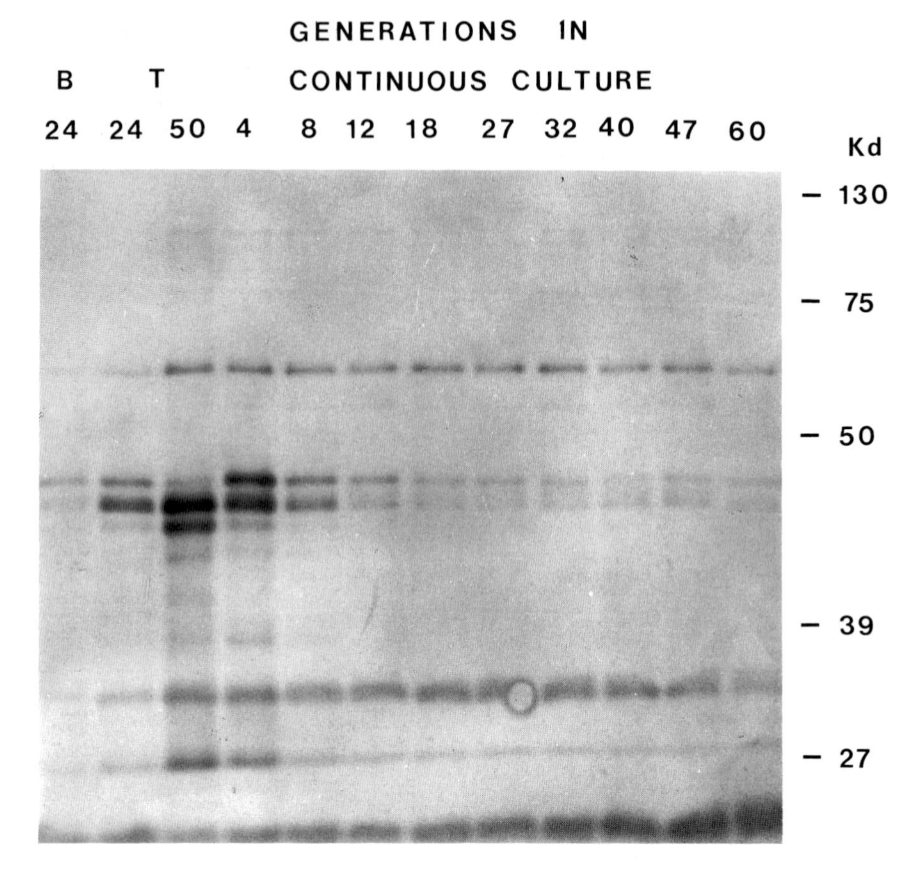

FIG. 6. Western blot analysis of samples taken at different time intervals as illustrated in Fig. 3. The tracks are designated as in Fig. 4. Molecular weight markers are shown on the right of the gel.

Summary

The integrity and stability of genetically-engineered plasmids is a crucial factor in the success of applying recombinant DNA technology to the Biotechnology industry. Several workers have shown that genetically-engineered plasmids can be unstable when grown for several generations in either batch or continuous culture. This instability can be due to a number of factors including genetic instability, segregational problems, host mutation, metabolic load and nutritional limitations. Although instability can be a severe problem with many genetically-engineered plasmids grown under industrial conditions, careful genetic construction can reduce this problem significantly. In addition, careful selection of growth medium and proper monitoring of the plasmid and its products can solve the problem of instability in most cases.

Finally, there are now reliable and sensitive methods for monitoring the genetic status of the host and its plasmids as well as assessing the quantity and quality of the recombinant proteins being produced. It should be possible to adapt most of these processes to on-line technology and thus enable plasmid stability to be maintained.

Acknowledgements

The authors are grateful to Dr G. Wilkinson for his helpful discussions and to A.R. Fooks for his excellent technical assistance.

References

Bagdasarian, M., Lurz, R., Ruckert, B., Franklin, F.C.H., Bagdasarian, M.M., Frey, J. & Timmis, K.N. 1981. Specific purpose plasmid cloning vectors. *Gene* **16**, 237–247.

Broome, S. & Gilbert, W. 1978. Immunological screening method to detect specific translation products. *Proceedings of the National Academy of Sciences of the United States of America* **75**, 2746–2749.

Caulcott, C.A., Lilley, G., Wright, E.M., Robinson, M.K. & Yarranton, G.T. 1985. Investigation of the stability of plasmids directing the expression of Met-prochymosin in *Escherichia coli. Journal of General Microbiology* **131**, 3355–3365.

Chew, L.C.K., Tacon, W.C.A. & Cole, J.A. 1986. Increased stability and maintenance of pAT153 in *Escherichia coli* HB101 due to transposition of IS1 from the chromosome into the tetracycline resistance gene. *FEMS Microbiology Letters* **36**, 275–280.

Chew, L.C.K., Tacon, W.C.A. & Cole, J.A. 1988. Effect of growth conditions on the rate of loss of the plasmid pAT153 from continuous cultures of *Escherichia coli* HB101. *FEMS Microbiology Letters* **56**, 101–104.

Cohen, S.N., Cabello, F., Chang, A.C. & Timmis, K. 1977. DNA cloning as a tool for the study of plasmid biology. In *Recombinant Molecules. Impact on Science and Technology*. eds Beers, R.F. & Basset, E.G. pp. 91–105. New York: New York Press.

Cooper, N.S., Caulcott, C.A., Brown, M.E. & Rhodes, P.M. 1987. The statistical analysis of plasmid stability data. Abstract P.B29–16. *Microbe '86*, Manchester.

DeSilva, N.A. & Bailey, J.E. 1986. Theoretical growth yield estimates for recombinant cells. *Biotechnology and Bioengineering* **28**, 741–746.

Eckhardt, T. 1978. A rapid method for the identification of plasmid deoxyribonucleic acid in bacteria. *Plasmid* **1**, 584–588.

Goodwin, D. & Slater, H. 1979. The influence of the growth environment on the stability of drug resistance plasmids in *Escherichia coli* K12. *Journal of General Microbiology* **111**, 201–210.

Gustafsson, P., Wolf-Watz, H., Lind, L., Johansson, K.E. & Nordstrom, K. 1983. Binding between the *par* region of plasmids R1 and pSC101 and the outer-membrane fraction of the host bacteria. *EMBO Journal* **2**, 27–32.

Hakkaart, M.J.J., Vangelman, B., Veltkamp, E. & Nijkamp, H.J.J. 1985. Maintenance of multicopy plasmid CloDF13: Role of plasmid size and copy number in partitioning. *Molecular and General Genetics* **198**, 364–366.

HASHIMOTO-GOTOH, T. & TIMMIS, K.N. 1981. Incompatibility properties of ColE1 and pMB1 derivative plasmids: Random replication of multicopy replicons. *Cell* **23**, 229–238.

HELLING, R.B., KINNEY, T. & ADAMS, J. 1981. The maintenance of plasmid-containing organisms in populations of *Escherichia coli*. *Journal of General Microbiology* **123**, 129–141.

IMAMAKA, T., TSUNIKAWA, H. & AIBA, S. 1980. Phenotypic stability of *trp* operon recombinant plasmids in *Escherichia coli*. *Journal of General Microbiology* **118**, 235–261.

JONES, S.A. & MELLING, J. 1984. Persistence of pBR322-regulated plasmids in *Escherichia coli* hosts grown in chemostat cultures. *FEMS Microbiology Letters* **22**, 239–243.

JONES, S.A., DEARNEY, K., BENNET, P.M. & MELLING, J. 1980a. The stability of antibiotic resistance plasmids in *Escherichia coli* hosts grown in continuous culture. *Society of General Microbiology Quarterly* **8**, 44.

JONES, I.M., PRIMROSE, S.B., ROBINSON, A. & ELWOOD, D.C. 1980b. Maintenance of some Col E1-type plasmids in chemostat culture. *Molecular and General Genetics* **180**, 579–584.

KLEMPERER, R.M.M., ISMAIL, N.T.A.J. & BROWN, M.R.W. 1979. Effect of plasmid RP1 on the nutritional requirements of *Escherichia coli* in batch culture. *Journal of General Microbiology* **115**, 325–331.

LAEMMLI, U.K. 1970. Cleavage of structural proteins during the assembly of the head of bacteriophage T4. *Nature* **227**, 680–685.

MEACOCK, P. & COHEN, S.N. 1980. Partitioning of bacterial plasmids during cell division. *Cell* **20**, 529–542.

MELLING, J., ELWOOD, D.C. & ROBINSON, A. 1977. Survival of R-factor carrying *Escherichia coli* in mixed culture in the chemostat. *FEMS Microbiology Letters* **21**, 87–89.

MILLER, C.A., TUCKER, W.T., MERACOCK, P.A., GUSTAFSSON, P. & COHEN, S.N. 1983. Nucleotide sequence of the partition locus of *Escherichia coli* plasmid pSC101. *Gene* **24**, 309–315.

NILSSON, J. & SKOGMAN, S.G. 1986. Stabilization of *Escherichia coli* tryptophan-production vectors in continuous cultures: A comparison of three different systems. *Biotechnology* **4**, 901–903.

NOACK, D., ROTH, M., GEUTNER, R., MULLER, G., UNDISZ, K., HOFFMIEIR, C. & GASPAR, S. 1981. Maintenance and genetic stability of vector plasmids pBR322 and pBR325 in *Escherichia coli* K12 strains grown in a chemostat. *Molecular and General Genetics* **184**, 121–124.

NORDSTROM, K., MOLIN, S. & AAGAARD-HANSEN, H. 1980. Partitioning of plasmid RP1 in *Escherichia coli*: kinetics of loss of plasmid derivatives deleted from the *par* region. *Plasmid* **4**, 215–227.

NUGENT, M.E., PRIMROSE, S.B. & TACON, W.C.A. 1983. The stability of recombinant DNA. *Developments in Industrial Microbiology* **24**, 271–285.

O'CALLAGHAN, C.H., KIRBY, S.M. & MORRIS, A. 1972. Correlation between hydrolysis of the β-lactam bond of the cephalosporin nucleus and expulsion of the 3-substituents. *Journal of Bacteriology* **110**, 988–991.

PRIMROSE, S.B., DERBYSHIRE, P., JONES, I.M., NUGENT, M.E. & TACON, W.C.A. 1983. Hereditory instability of recombinant DNA molecules. In *Bioactive Microbial Products 2: Development and Production*. eds Nisbet, L.J. & Winstanley, D.J. pp. 63–77. London: Academic Press.

PROJAN, S.J., CARLETON, S. & NOVICK, R.P. 1983. Determination of plasmid copy number by fluorescence densitometry. *Plasmid* **9**, 182–190.

REEVE, J. 1979. Use of minicells for bacteriophage-directed polypeptide synthesis. *Methods in Enzymology* **68**, 493–503.

RUSH, M., NOVICK, R. & DELAP, P. 1975. Detection and quantitation of *Staphylococcus aureus* penicillinase plasmid deoxyribonucleic acid by reassociation kinetics. *Journal of Bacteriology* **124**, 1417–1423.

SANCAR, A., WHARTON, R.P., SELTZER, S., KACINSKI, B.M., CLARKE, N.D. & RUPP, W.D. 1981. Identification of the *uvr*R gene product. *Journal of Molecular Biology* **148**, 45–62.

SAYADI, S., BERRY, F., NASRI, M., BARBOTIN, J.N. & THOMAS, D. 1988. Increased stability of pBR322-related plasmids in *Escherichia coli* W3101 grown in carrageenan gel beads. *FEMS Microbiology Letters* **56**, 307–312.

SHEPARD, H.M., GELFAND, D.H. & POLISKY, B. 1979. Analysis of a recessive copy number mutant. *Cell* **18**, 267–275.

STÜBER, D. & BUJARD, H. 1982. Transcription from efficient promotors can interfere with plasmid replication and diminish expression of plasmid specified genes. *EMBO Journal* **1**, 1399–1402.

SUMMERS, D.K. & SHERRATT, D.J. 1984. Multimerization of high copy number plasmids causes instability: ColE1 encodes a determinant essential for plasmid monomerization and stability. *Cell* **36**, 1097–1103.

SUTCLIFFE, J.G. 1979. Complete nucleotide sequence of the *Escherichia coli* plasmid pBR322. *Cold Spring Harbour Symposium on Quantitative Biology* **43**, 77–90.

THOMAS, C.M. 1987. Plasmid replication. In *Plasmids: A Practical Approach*. ed. Hardy, K.G. pp. 7–36. Oxford: IRL Press.

TIMMIS, K.N., CABELLO, F. & COHEN, S.N. 1978. Cloning and characterization of *Eco*RI and *Hin*dIII restriction endonuclease-generated fragments of antibiotic resistance plasmids R6-5 and R6. *General Genetics* **162**, 121–137.

TIMMIS, K.N., DANBARA, H., BRADY, G. & LURZ, R. 1981. Inheritance functions of group IncFII transmissible antibiotic resistance plasmids. *Plasmid* **5**, 53–75.

TWIGG, A.J. & SHERRATT, D. 1980. Transcomplementable copy number-mutants of plasmid ColE1. *Nature* **238**, 216–218.

UHLIN, B.E. & NORDSTROM, K. 1977. R plasmid gene dosage effects in *Escherichia coli* K12. *Plasmid* **1**, 1–7.

WAHLE, E. & KORNBERG, A. 1988. The partitioning locus of plasmid pSC101 is a specific binding site for DNA gyrase. *EMBO Journal* **7**, 1889–1895.

WARNES, A. & STEPHENSON, J.R. 1986. The insertion of large pieces of foreign genetic material reduces the stability of bacterial plasmids. *Plasmid* **16**, 116–123.

WILLETS, N. & CROWTHERS, C. 1981. Mobilization of the non-conjugative IncQ plasmid RSF1010. *Genetics Research, Cambridge* **37**, 311–316.

WILLIAMS, N. & SCURRAY, R. 1980. The conjugation system of F-like plasmids. *Annual Review of Genetics* **14**, 41–76.

WOUTERS, J.T.M., DRIEHUIS, F.L., POLACZEK, P.J., VAN OPPENRAAG, M.L.H.A. & VAN ANDEL, J.G. 1980. Persistence of the pBR322 plasmid in *Escherichia coli* K12 grown in chemostat cultures. *Antonie van Leeuwenhoek* **37**, 311–316.

YOUNG, I.G. & POULIS, M.I. 1978. Conjugal transfer of cloning vectors derived from ColE1. *Gene* **4**, 175–179.

Large-scale Production of Proteins from Recombinant DNA in *Escherichia coli*

R.F. Sherwood, R. Plank, J. Baker and T. Atkinson
Division of Biotechnology, Public Health Laboratory Service, Centre for Applied Microbiology and Research, Porton Down, Salisbury, Wiltshire SP4 0JG, UK

A key to successful large-scale manufacture of protein products from recombinant DNA technology is to identify the fermentation and downstream recovery strategies from the outset to take full advantage of genetic design in the protein expression system. With the current range of vectors, hosts, gene promoters and other control elements, manipulation at the DNA level can determine overall product expression level, as well as location, timing of expression and the ease with which the product can be isolated and purified.

Escherichia coli is at the centre of gene manipulation, as a host for primary cloning prior to transfer of the recombinant DNA to other micro-organisms, or as the production host for protein expression and recovery. Many of the proteins derived from recombinant DNA which are today in the clinical market or undergoing trial are produced in *E. coli*: insulin, human growth hormone, interleukins, granulocyte macrophage colony stimulating factor and HIV antigens. The limitation of *E. coli* as a production host is its inability to perform the complex post-translational processing required for full biological activity of some proteins: glycosylation or specific amino acid modifications.

Current examples of protein production from cloned bacterial and mammalian genes enable the authors to demonstrate how genetic engineering complements the more traditional process of engineering aligned to computer control and data logging.

Genetic Design

Host strain

Before considering the detail of vector and gene construction it is worth noting that protein product expression level and recovery can be significantly

Genetic Manipulation
0–632–02926–9

affected by the choice of the *E. coli* host strain. There are a number of factors involved, including efficiency of transcription/translation, susceptibility of product to degradation by cellular proteases and, as will be described later, co-purification of protein contaminants.

Differences in product expression level between *E. coli* hosts can be demonstrated by small-scale (8 litre fermentation volume) studies (Lancaster *et al.* 1989) with carboxypeptidase G_2 (CPG_2), an enzyme with applications in cancer therapy (Sherwood *et al.* 1985). The enzyme specifically removes alpha-amino linked glutamate from a variety of compounds, notably folates, and the gene was cloned from *Pseudomonas* sp. strain RS-16 into *E. coli* by insertion into pBR322 within the region coding for tetracycline resistance (*tet*^{r}) yielding pNM21 (Minton *et al.* 1983). This gave a relatively low expression system based on the *tet*^{r} promoter, and pNM21 was tested in four different *E. coli* hosts for efficiency of production, measured as CPG_2 level as the percentage of total soluble protein (Table 1).

TABLE 1. *Effect of* E. coli *host strain on expression of cloned carboxypeptidase G_2 measured as percentage of total soluble protein*

Host strain	CPG2 as % soluble protein
JM83	0.2
W5445	0.4
MC1061	1.2
RV308	1.8

This effect is not predictable and a different pattern emerges with other cloned gene products, such as recombinant Protein A, an IgG binding protein cloned from *Staphylococcus aureus* and cloned human interferons.

Vector construction

Plasmid copy number

The majority of plasmid vectors used in *E. coli* are derived from pBR322 and genes cloned directly into pBR322, e.g. CPG_2 to form pNM21, give low to intermediate yield based on an average 50 plasmid copies per cell. A non-mobilizable derivative of pBR322 termed pAT153 yields about 150 plasmid copies per cell through deletion of the *rop* function which codes for a 63 amino acid polypeptide that accelerates and stabilizes the formation of RNAI and RNAII transcript hybrids, which in turn influences the rate of DNA synthesis from the plasmid origin of replication. A further series of

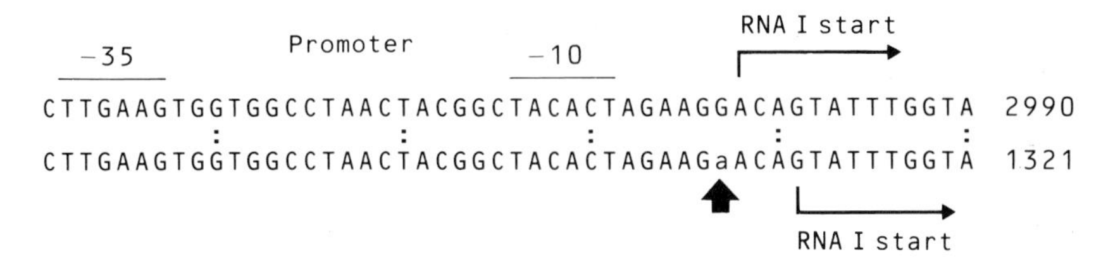

FIG. 1. Base change G to A adjacent to the RNAI transcription start site. Upper sequence = pBR322. Lower sequence = pUC19.

plasmids, designated pUC derived from pBR322 also lack *rop* and contain a single base change G to A adjacent to the RNAI transcription start site (Chambers *et al.* 1988). This affects the length of the RNAI transcript (Fig. 1), which in turn reduces the inhibitory effect of RNAI on the action of RNAII in promoting plasmid replication. The practical result of using pUC derived plasmids as vectors can be demonstrated again by comparing expression levels of CPG_2 in the *E. coli* host strain RV308 (Table 2).

TABLE 2. *Effect of plasmid copy number on expression of cloned carboxypeptidase G_2. The gene coding for CPG_2 was cloned into the ampicillin resistance gene of three related plasmids with varying copy number per cell (Chambers* et al. *1988)*

Plasmid designation	Origin	Average copy number per cell	CPG2 as % soluble protein
pCPM 1	pBR322	52 ± 6	2.3
pCPM 2	pAT153	152 ± 12	7.1
pCPM 3	pUC8	630 ± 10	26.9

Plasmid stability

Despite pBR322 being a standard plasmid vector for insertion of recombinant DNA, it exhibits instability when cultured in the absence of antibiotic selection pressure (tetracycline or ampicillin). Stability can be conferred by introducing the *par* locus, for example from pSC101 (Meacock & Cohen 1980). This function is concerned with partitioning of plasmids at cell division and its effect can be demonstrated by using pNM21 carried by *E. coli* strain W5445 in continuous fermentation (Fig. 2). The introduction of *par* does not influence growth rate of the *E. coli*, assessed by 'wash-out' in continuous culture, but the normally high plasmid copy number associated with pUC based plasmids is reduced 100-fold (Chambers 1989).

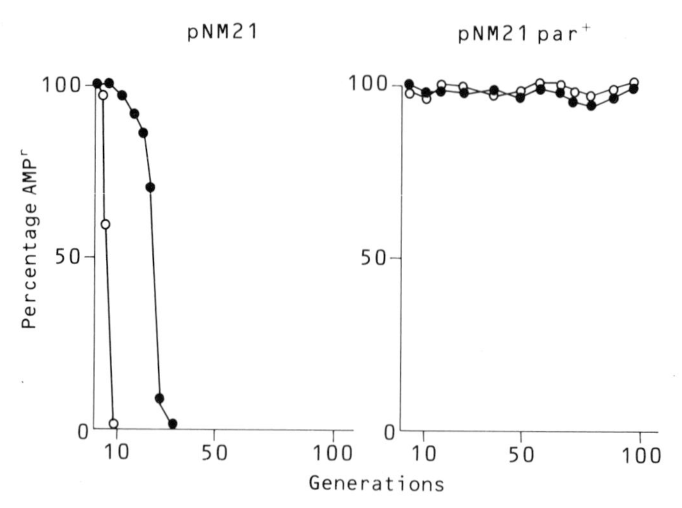

FIG. 2. Effect of introducing the *par* locus into pNM21 on stability, measured as percentage cell population retaining ampicillin resistance, during continuous culture of *E. coli* strain W5445. Chemostat cultures were maintained under glucose (●) or phosphate (○) limitation and grown at a dilution rate $D = 0.1$ at pH 7 and 37°C (Chambers 1989).

Gene promoters and expression

There are now a large number of high efficiency promoters in *E. coli* and choice depends largely on the degree of control required for gene expression. Switch on is normally 'controlled' by the level of particular medium ingredients, e.g. promoters derived from genes involved with *trp* (tryptophan), *lac* (sugar) or *pho* (phosphate) metabolism, which can be enhanced by using gratuitous inducers (IPTG for *lac* and IAA for *trp*), but these can be expensive commodities in large-scale fermentation. Promoter hybrids such as *tac* (*trp/lac*) can also provide consensus promoter sequences for *E. coli* transcription (Table 3).

TABLE 3. *Nucleotide sequences at the DNA dependent RNA-polymerase binding site (−35) and Sigma initiation site (−10) of promoters commonly used in* E. coli

Promoter	−35	−10
Consensus sequence	TTGACA	TATAAT
lac UV5	TTTACA	TATAAT
trp	TTGACA	TTAACT
tac	TTGACA	TATAAT
tc	TTGACA	TTTAAT
P_L	TTGACA	GATACT

TABLE 4. *Comparative efficiency of various promoters for the expression of chloramphenicol acetyl transferase (CAT) in* E. coli *strain TG-1*

Promoter	CAT as % soluble protein
lac	12
tac	20
tlc	11
trp	25
P_L	45

Temperature-sensitive promoters such as lambda P_L can be activated by a temperature increase to destroy repressor proteins, although it is normally necessary to clone additionally the repressor protein gene to ensure that sufficient levels are present to repress the promoter until required.

Variation in gene expression levels at comparable copy number per cell obtained with different promoters can be demonstrated with cloned chloramphenicol acetyl transferase (CAT) in *E. coli* strain TG-1 (Table 4).

Efficient translation of genes in *E. coli* depends on ribosomal recognition of the initiation codon and commences after the ribosome binds to a complementary site on the mRNA upstream of this codon. The mRNA usually contains part of the sequence 5′−AGGAGGTG−3′ which is complementary to the 3′ terminus of 16S ribosomal RNA (−CACCICCIAOH). The spacing between this sequence and the initiation codon of the cloned gene is critical to achieve high expression in *E. coli* where the limits are normally 6−11 base pairs (bp). Changes to this spacing are not always predictable as a reduction to 7 bp for cloned human growth hormone nearly halves expression level (Goeddel *et al.* 1979), whilst giving a 100-fold increase for β-interferon (Shepherd *et al.* 1982). It is worth noting that, in addition to providing the necessary sequences to promote gene expression, it is important to have strong termination sequences to prevent continued readthrough beyond the cloned gene.

Product localization

Secretion and fusion proteins

One of the more recent and interesting developments in genetic design has been the introduction of nucleotide sequences which can dictate product localization within the cell and subsequently aid recovery and purification. *E. coli* is rarely known to secrete proteins into the culture medium, but proteins can be transported across the cytoplasmic membrane into the cell's periplasmic space. Transport is directed by gene leader sequences coding for signal peptides which are cleaved from the mature protein within the membrane. The leader sequence may occur naturally on the cloned gene; e.g. CPG_2 from *Pseudomonas*, Protein A from *Staphylococcus* and mammalian insulin, or can be introduced upstream of the gene start codon by cloning the sequence from a naturally occurring periplasmic protein such as *E. coli* alkaline phosphatase. The features of a typical signal peptide, 20–30 amino acids with hydrophilic and hydrophobic regions is demonstrated by CPG_2, isolated from *Pseudomonas*, but working efficiently in *E. coli* (Fig. 3). Periplasmic location in *E. coli* can provide particular advantages for ease of product release from the cell by means of hot/cold or osmotic shock and selectively reduce the background of contaminating proteins requiring removal during purification.

Selective release of periplasmic proteins has been taken one stage further by Nilsson *et al.* (1985) who produced a protein fusion comprising Protein A and alkaline phosphatase in *E. coli* which could be purified in a one-step

Hydrophilic region | Hydrophobic

Met Arg Pro Ser Ile His Arg Thr Ala Ile Ala Ala Val Leu Ala

Region | Mature protein

Thr Ala Phe Val Ala Gly Thr | Ala Leu Ala Gln Lys Arg Asp Asn

Percentage enzyme activity

Fraction	CPG2	AP	GAPDH	NADHOX
Periplasmic	97.0	97.1	6.8	0.3
Cytoplasmic	2.6	2.3	93.0	8.4
Membrane-bound	0.4	0.6	0.2	89.1

AP = alkaline phosphatase;
GAPDH = glyceraldehyde-3-phosphate dehydrogenase;
NADHOX = NADH.O_2 reductase

FIG. 3. Signal peptide sequence of carboxypeptidase G_2 and effectiveness in promoting transport into the periplasmic space in *E. coli*. *E. coli* cells were suspended in 0.9 ml of 0.58 M sucrose–0.2 mM dithiothreitol–30 mM Tris–HCl, pH 8.0. Conversion to spheroplasts was by addition of 20 µl of lysozyme (2 mg/ml) and 40 µl of 0.1 M EDTA at 23°C for 10 min. The spheroplasts were held on ice and 0.1 ml of 30 % w/v BSA added, followed by 5 ml of sucrose–Tris buffer and then sedimented by centrifugation at 5 000 *g* for 10 min. The supernatant was retained as the periplasmic fraction. The pellet was suspended in 5 ml of 10 mM Tris–HCl–0.2 mM dithiothreitol, pH 7.0 and sonicated at 20 Kc/s, 2 A for 15 s. Centrifugation at 100 000 *g* for 1 h at 4°C separated soluble (cytoplasmic) from particulate (membrane-bound) proteins. Known marker enzymes for cytoplasmic (GAPDH), periplasmic (AP) and membrane (NADHOX) location were used as controls.

procedure based on the binding of Protein A to the Fc region of immobilized IgG (dissociation constant 2×10^{-8}). This technique can be applied to the production of polypeptide hormones such as insulin-like growth factor I and the natural product specifically cleaved from the fusion, on the basis of amino acid sequence, by chemical (e.g. hydroxylamine ASP:GLY) or enzymic (e.g. enterokinase ASP–ASP–ASP–LYS:) methods.

Manipulations of DNA in which a small number of additional codons are placed downstream of the C-terminal codon of the cloned gene have been used to confer specific properties on the protein formed to aid purification. The classic example was the use of a polyarginine tail on recombinant urogastrone to alter its binding characteristics on a cation ion-exchange resin (Sassenfeld & Brewer 1984). The tail was subsequently removed by carboxypeptidase B cleavage. In a similar manner, other poly(amino acid) tails have been used to purify galactokinase (cysteines) specifically on thiopropyl–Sepharose; galactosidase (phenylalanines) by increased retention on phenyl-Sepharose (Persson *et al.* 1988) and dihydrofolate reductase (histidines) by metal chelate chromatography (Hochuli *et al.* 1988). Analogous to this approach, Hopp *et al.* (1988) have fused a short, antigenic peptide at the N-terminus of a number of lymphokines which can be purified on an immobilized monoclonal antibody, specific for the peptide, from *E. coli* extracts or the culture medium following fermentation of *Saccharomyces cerevisiae.*

Inclusion bodies

Another type of localization is the formation of product into inclusion bodies, microcrystalline or amorphous bodies containing product in biologically inactive form. These are usually protein-dependent (e.g. hydrophobicity of the protein), although fermentation conditions, particularly temperature (low temperature reduces formation) and pH (low pH increases formation) influence their production. Examples of proteins which can occur in this form in *E. coli* are bovine growth hormone, prochymosin, insulin and interleukin-2. Such proteins have tertiary structures dependent on the formation of disulphide bridges and it appears that the redox system of *E. coli* is incapable of full oxidation, since many (70%) of the sulphydryl groups in these proteins are found in the reduced form. Inclusion bodies can protect the product from proteolytic attack within the cell and provide an efficient means of isolating product from the cell by differential centrifugation, but conditions for solubilization and renaturation to biologically active structure are stringent and not applicable to many proteins.

A good example of inclusion body formation and recovery in *E. coli* is provided by cloned calf prochymosin (Marston *et al.* 1984; Kawaguchi *et al.* 1984). Inclusion bodies are isolated from lysed cells by centrifugation and

TABLE 5. *Steps for isolation, solubilization and renaturation of cloned prochymosin inclusion bodies from* E. coli

Step	Procedure
Isolation from the cell	Mechanical disruption (sonication at 3 × 20 s) or enzymic lysis (lysozyme at 20 μg/ml at 37°C for 10 min in 50 mM Tris–HCl, pH 8.0 plus 50 mM NaCl, 1 mM EDTA and 0.1 mM PMSF)
Separation from cell debris	Sedimentation at 12 000 ***g*** for 5 min at 4°C
Washing to remove contaminants	Wash with a mixture of 0.5% v/v Triton X-100 and 10 mM EDTA in the absence of Mg^{2+}
Solubilization under alkaline conditions	Suspend in isolation buffer containing 8 M urea for 1 h at RT and then add 9 volumes of 50 mM KH_2PO_4, pH 10.7 with 1 mM EDTA and 50 mM NaCl for 30 min
Neutralization and diafiltration	Adjust pH to 8.0 with HCl and diafilter in an Amicon hollow fibre concentrator (10 000 MW cut off) to reduce conductivity
Purification by ion-exchange chromatography	Batch binding on DE-52 equilibrated in 20 mM Tris–HCl pH 8.0 with 1 mM EDTA and 50 mM NaCl and elution with buffer supplemented with 350 mM NaCl, followed by chromatography on DE-52 using a gradient elution of 100–500 mM NaCl
Activation to chymosin	Autocatalytic by incubation at acid pH. Prochymosin in 50 mM Tris–HCl pH 7.5 adjusted to 0.2 M glycine–HCl pH 2.3 and held for 15 min at RT before neutralization with 0.1 M sodium phosphate, pH 6.3

then solubilized and renatured to give active product which can be purified by ion-exchange chromatography (Table 5).

Fermentation

Culture conditions

Fermentation of organisms carrying recombinant DNA does not significantly differ from traditional processes in which a wide range of micro-organisms have been used as sources of specialized enzymes or other proteins. In such practice, development of growth conditions is often simplified as the host, *E. coli*, is common to several products and the expression control system is defined from the outset. This removes the need to tailor growth medium; for example, to a selected and high producing 'natural' isolate, which may have stringent requirements within a complex induction/repression control system.

Even after extensive strain selection and fermentation development, such isolates rarely yield more than 5% of their soluble protein as the product of interest. In contrast, recombinant systems in *E. coli* can yield between 10 and 50% soluble protein as a single product and this reduces the fermentation volumes required to fulfill production requirements. A typical growth medium comprises:

Casamino acids	20 g/litre
Yeast extract	3
Phosphate salts	5
Potassium salts	2
Ammonium salts	1
Glycerol	5% v/v

The fermentations are rapid (6–12 h with a generation time of about 60 min) and initially proceed by utilization of the amino acids as primary carbon and nitrogen sources with a switch to glycerol (or glucose) later in the culture and with subsequent batch feeding, cell yields of 50–100 g wet weight cells per litre are attainable. Coupled with an expression level of 20% soluble protein, product yields can be in excess of 1 g per litre.

Fermentations of *E. coli* are normally at pH 7.0, but growth temperature ranges from 28 to 37°C dependent on a number of factors, including promoter used and tendency of product to form inclusion bodies. Aeration is at 1 vvm although it is often necessary to increase aeration rate during periods of product induction. A major factor in determining the precise fermentation detail is the promoter used. Promoters based on *lac* can be induced by IPTG (isopropylthiogalactopyranoside) during the fermentation, typically added to a concentration of 2 mM at a culture optical density of 12 and allowing growth to continue for a further 4 h. *Trp* promoters can be induced with IAA (indole acrylic acid), which can be added (20 μg/ml) early in the culture (2 h), allowing growth to continue to stationary phase. In the case of temperature-sensitive promoters, notably the lambda P_L promoter, growth temperature is controlled in the range 28–32°C until late logarithmic phase and then raised rapidly to above 40°C and held for about 2 h. Some practical examples of product expression in *E. coli* are shown in Fig. 4 and are representative of results obtained from 10–500 litre scale fermentation. Controlled timing of product expression during fermentation is particularly important where the protein is susceptible to proteolysis in the cell or where high levels of the protein interfere with cell metabolism through specific toxicity or more general effects such as membrane 'jamming' leading to cell fragility.

Computer control system

It is important to supplement the design of the biological system used for production of recombinant DNA derived proteins with a data logging and

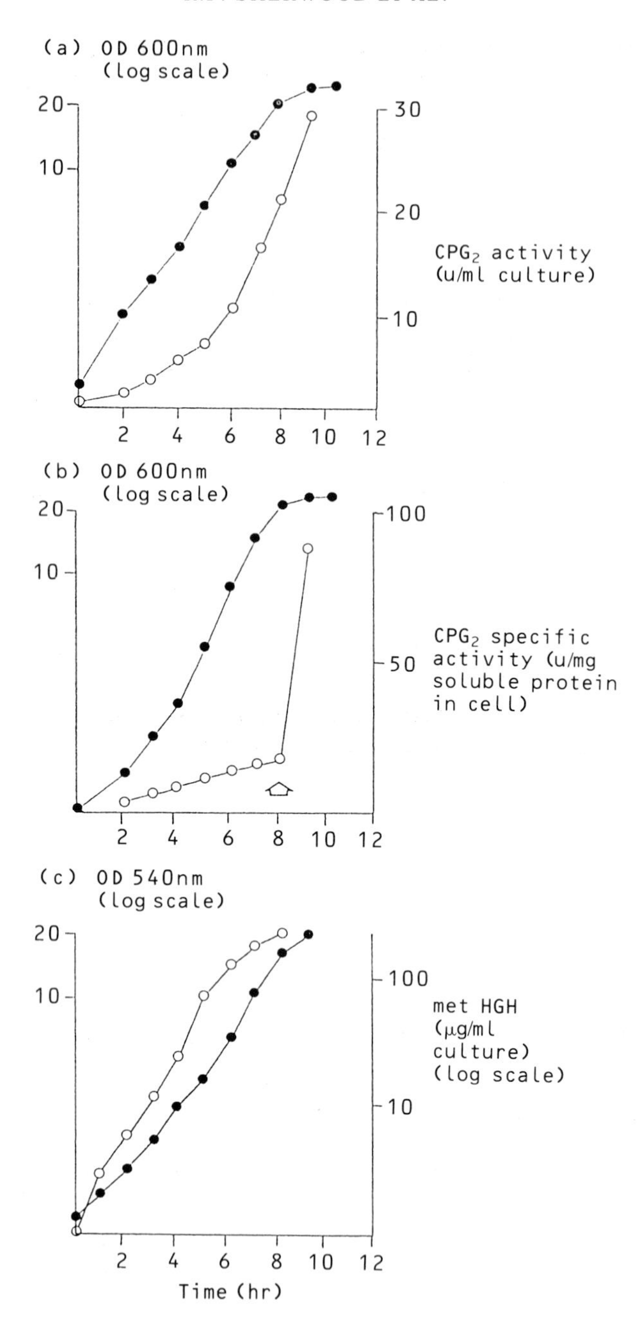
(a) OD 600nm
(log scale)
20
10
30
20
10
CPG2 activity
(u/ml culture)
2 4 6 8 10 12
(b) OD 600nm
(log scale)
20
10
100
50
CPG2 specific
activity (u/mg
soluble protein
in cell)
2 4 6 8 10 12
(c) OD 540nm
(log scale)
20
10
100
10
met HGH
(μg/ml
culture)
(log scale)
2 4 6 8 10 12
Time (hr)

feedback control system during fermentation. It is standard practice to control cell culture conditions precisely and expert systems are being increasingly used for discrete product purification stages. The Biotechnology Division, CAMR, uses a centralized control system for fermentation which has the following general features:

Hardware

The main computer is a Hewlett Packard 1000 Series A900 on which the control software resides. It has two HP2250 monitoring and control processors connected to it via an HPIB cable. Terminal access is via normal terminals or Fluke 1780A touchscreens in the process areas where a keyboard is not desired.

The sensors, valves, motors, and other controllers and transducers are connected to HP2250s via twisted pair cables. The control output signals are mainly 4–20 mA loops to reduce the effects of noise in the process environment, but there are a few 0.5 V connections. The input signals from the sensors/transducers are generally 4–20 mA loops, with valve feedback signals giving digital logic levels of 0.5 V.

Also connected to the A900 is a VG8-80 gas mass spectrometer which is used to analyse the off-gas from the fermentations. These measurements are used with an analysis of the feed air to calculate respiratory quotients, oxygen uptake rates, and carbon dioxide production rates.

Software

The operating system of the HP1000 Series A900 computer is real time executive version A (RTEA) which underpins all other software.

FIG. 4. Expression of recombinant DNA derived proteins during fermentation of *E. coli*. Medium in general contains (g/litre): casamino acids, 20; $(NH_4)_2SO_4$, 5.4; $K_2HPO_4.2\ H_2O$, 2.0; yeast extract, 1.5; $NaH_2PO_4.2H_2O$, 1.0; $MgSO_4.7\ H_2O$, 0.3; $FeSO_4.7\ H_2O$, 0.1; $CaCl_2.6\ H_2O$, 0.03; $MnSO_4.4\ H_2O$, 0.02; $ZnSO_4.7\ H_2O$, 0.02; $CuSO_4.5\ H_2O$, 0.001; $CoCl_2.6\ H_2O$, 0.001; glucose, 25. Antibiotics used were either ampicillin at 50 μg/ml or tetracycline at 5 μg/ml. Temperature was controlled at 37°C ± 0.2°C (except for cultures exploiting P_L promoter grown at 28°C and raised to 42°C) and pH at 7.0 ± 0.02 using automatic addition of 10 M NaOH or 3.2 M H_3PO_4 solution. Dissolved oxygen was maintained above 30% saturation by automatic adjustment of stirrer speed with air supply kept constant at 1 vvm. Foaming was controlled by addition of PPG2000 antifoam.
(a) Carboxypeptidase G_2 expressed from pNM21 (*lac* promoter) in *E. coli* RV308; (b) carboxypeptidase G_2 expressed from pNM501CP (P_L promoter in *E. coli* MC1061. Effect of temperature induction (arrow) at 42°C shown by increase in specific activity of the enzyme in the cell; (c) methionyl human growth hormone expressed from a pHGH107 derivative (*lac/trp* tandem promoter) in *E. coli* RV308. In each case, cell density = (•) and product = (○).

The software that controls and monitors the fermentations is a PMC1000 package which accesses the HP2250 control processors to monitor and control the vessels. PMC1000 provides 3-term PID control if required to any control loop, and being entirely software driven and configured, is extremely flexible and easy to use.

The process operators access PMC1000s facilities through applications software written by inhouse programmers to the operator's specifications predominantly in Pascal 1000.

Data is logged by the applications software into databases created and managed by an HP Image II database. Data held consist of measured values of monitored parameters logged at intervals determined by the operators, offline data entered following laboratory analysis, and controlled parameter setpoints. Monitored parameters are logged in 'real time' on the running fermentations and can be used as required for control loop feedback.

Operation

Process operators can schedule fermentations via terminals and select the control parameters and logging intervals required. The logging interval can be dynamic, if required, to provide faster data capture in periods when the measured value is changing rapidly. It is possible to interrogate the databases of logged data for both past and current fermentations to compare trends and output can be presented in tabular or graphical form.

Using the touchscreens, the operators are presented with menus of possible actions. These include the ability to start a process presterilization cycle, inoculate and terminate the culture, as well as change parameter setpoints and monitor vessel performance during the fermentations. The vessel control menus permit the operators to access any vessel from any one touchscreen.

System utilities permit normal archive and restoration of databases and packing of databases to conserve disk space.

Downstream Processing

Product recovery

How genetic design can aid product recovery by dictating the location and form of the protein has already been described. Intracellular proteins are still most often released by mechanical shear equipment. Separation of soluble protein product from cell debris is critical as insufficient removal of fine particulate matter can dramatically effect subsequent purification steps, acting as centres for protein aggregation/precipitation and causing filter blocking and deterioration in column chromatography performance. Chemical lysis by acid/alkali is attractive on a large-scale, but care has to be taken to avoid deamidation of protein (e.g. asparagine residues at high pH) and oxidation. Enzymic treatment of *E. coli*, with lysozyme presents problems on a large-scale and regulatory requirements for clinical products would entail positive proof that any enzyme added to the process was removed at a subsequent purification stage. For proteins located in the periplasmic space, temperature and osmotic shock can be used to remove selectively the highly expressed product which is often the predominant protein in the fraction. This also reduces viscosity of solutions and membrane association and keeps purification protocols to a minimum.

A general scheme for isolation and purification of proteins from *E. coli* is shown in Fig. 5.

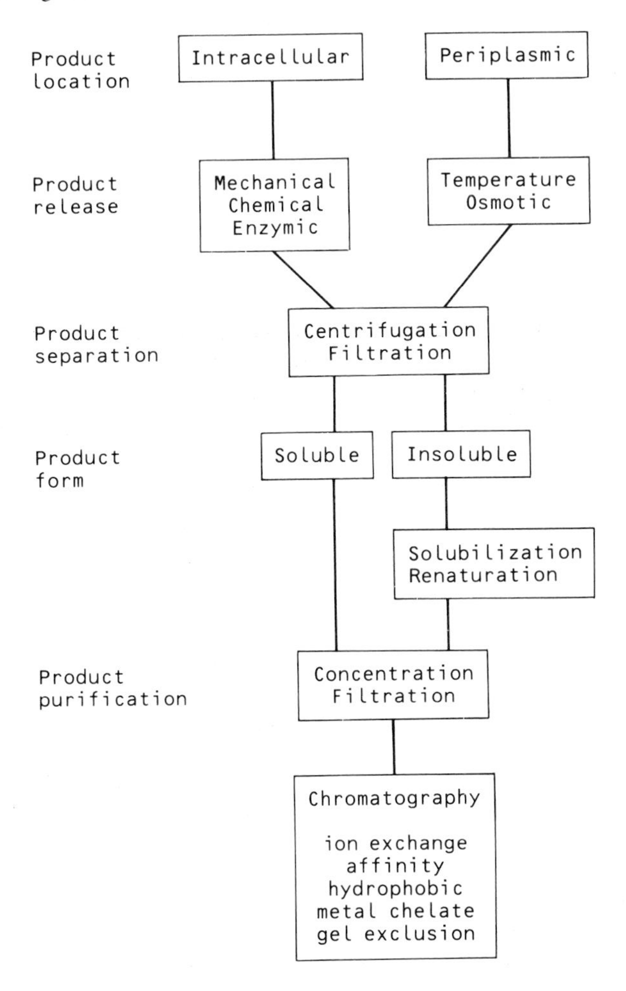

FIG. 5. General scheme for isolation and purification of proteins from *E. coli*.

Protein purification

Given the high expression levels of most recombinant DNA-derived proteins, the purification protocols are usually confined to only a few stages of which the final one or two are 'polishing' stages with little increase in protein-specific activity. The first step would normally involve ion-exchange column chromatography, although a precipitation by means of heat or ammonium sulphate can be employed either side. There may be a second ion-exchange

TABLE 6. *Large-scale isolation and purification of carboxypeptidase G_2 from* E. coli

Step	Procedure	Enzyme-specific activity (U/mg)	Yield
Cell harvest	Centrifugation in Westfalia KA25 six chamber disc bowl centrifuge at 6000 *g* and 200 litre/h flow rate. Capacity 25 kg cells		
Cell breakage	Mechanical shear disruption in a Manton-Gaulin 15 M8BA homogenizer. Pressure set at 33 MPa and flow rate at 25 litre/h. Two passes of a suspension containing 10 kg cells in 30 litre 10 mM Na acetate pH 5.5 with 0.5 mM EDTA, held at 4°C		
Cell debris removal (clarification)	Centrifugation in a Sharples AS-26 tubular bowl centrifuge at 12000 kg and 35 litre/h flow rate. Capacity 8 g cell debris	18	100
Ion-exchange chromatography I	Binding to SP-Sephadex C-50 equilibrated in 10 mM Na-acetate pH 5.5 (10 litres settled volume) at 4 litres/h flow rate. Two buffer washes followed by elution with 25 mM Tris–acetate pH 9.0. Active fractions diluted to 2 mS conductivity	90	80
Ion-exchange chromatography II	Binding to DEAE-Cellulose DE52 equilibrated in 20 mM Tris–acetate pH 9.0 (3 litres settled volume) at 0.4 litres/h flow rate. Two buffer washes followed by elution with an 8 litre gradient of 20 to 500 mM Tris–acetate pH 9.0. Active fractions pooled and adjusted to pH 7.3 with 0.1 M acetic acid	400	30
Gel exclusion chromatography	Concentration of active fractions to 50 ml and passage through Ultrogel ACA44 column (2 litre volume) equilibrated in 50 mM Tris–HCl pH 7.3 with 0.2 mM $ZnCl_2$	500	25

chromatography followed by a final step involving gel exclusion, hydrophobic or affinity chromatography. Ion-exchange chromatography serves not only product purification, but also provides a key step in removal of pyrogens originating from the *E. coli* host, particularly relevant to proteins destined for therapeutic use. The development of fast-flow chromatographic matrices has provided the opportunity to design single bed large-scale columns which can be operated automatically by a variety of expert control systems.

The detail of typical methods employed to isolate and purify proteins from *E. coli* can best be described by reference to two cloned products, carboxypeptidase G_2 (Table 6) and methionyl human growth hormone (Table 7).

There were some interesting host strain-related effects on binding of carboxypeptidase G_2 to the SP-Sephadex ion-exchanger described in Table 6. These took the form of variability in percentage enzyme that could be bound

TABLE 7. *Isolation and purification of methionyl human growth hormone from* E. coli

Step	Procedure	Hormone-specific activity (mg/mg)	Yield
Release from cells	Adjust cell culture to pH 12.0 by addition of 10 M NaOH at 20 ml per min		
Cell debris removal (clarification)	Centrifuge extract in Westfalia KA25 and Sharples AS-26 as described in Table 6	0.05	100
Ammonium sulphate precipitation	Precipitate clarified lysate in a 20–50% $(NH_4)_2$ SO_4 cut, leaving for 1 h at 4°C before centrifugation. Resuspend pellet in 50 mM Tris–HCl pH 8.5 and dialyse overnight against the same buffer	0.65	92
Ion exchange chromatography	Bind dialysate on DEAE-Sepharose CL-6B equilibrated in 50 mM Tris–HCl pH 8.5. Wash with 2 volumes of same buffer containing 50 mM NaCl and elute with a gradient of 50–150 mM NaCl at 2 column volumes per h	0.65	62
Gel exclusion chromatography	Chromatography on Sephadex G-100 equilibrated in 50 mM Tris–HCl pH 8.0 plus 100 mM NaCl	0.86	40
Hydrophobic chromatography	Adjust active fractions to 1 M with NaCl and bind to phenyl-Sepharose CL-4B equilibrated in 50 mM Tris–HCl pH 8.0 containing 1 M NaCl. Elute with deionized water at 1.5 column volumes per h	1.0	37

TABLE 8. *Comparison of the binding and elution of carboxypeptidase G_2 extracted from 4* E. coli *host strains on SP-Sephadex*

	Conductivity (mS)	Percentage enzyme from strains			
		RV308	W5445	MC1061	JM83
Enzyme bound	4	5	18	23	78
	2	80	85	71	98
	1	62	86	94	98
Enzyme eluted	4	ND*	ND	ND	80
	2	82	82	100	100
	1	68	100	92	35

* Not determined.

and eluted at different conductivity values (Table 8). This was thought to reflect differences in the profile of contaminating proteins from the four *E. coli* strains competing for ion-exchange sites. Some of these differences could be identified by analysis on a gel exclusion column, notably with strain RV308 where a second major, low molecular weight, contaminating protein peak was eluted prior to carboxypeptidase G_2 (Fig. 6).

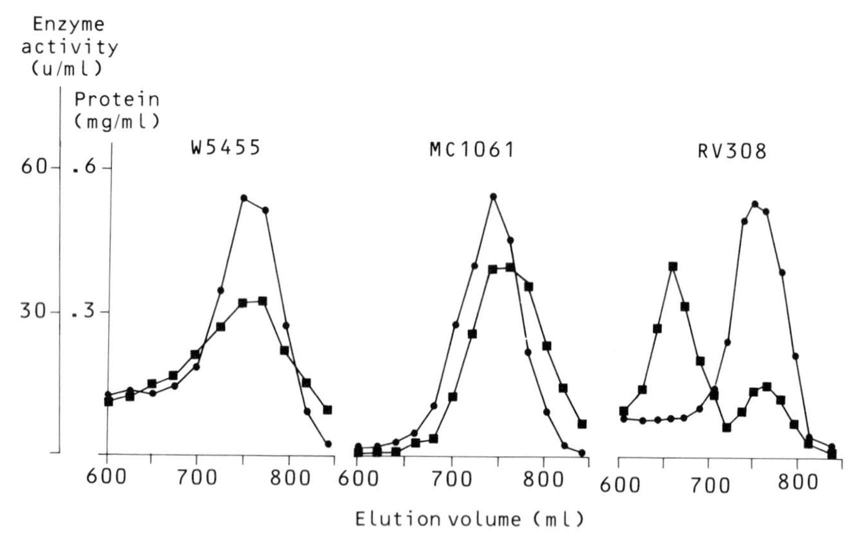

FIG. 6. Gel-exclusion analysis of carboxypeptidase G_2 eluting from SP-Sephadex. Proteins were chromatographed on an Ultrogel ACA44 column (4.4 × 90 cm) equilibrated in 50 mM K-phosphate pH 7.5 and run at 40 ml per h. Fractions (10 ml) were collected and assayed for enzyme (●) activity (Sherwood *et al.* 1985) and protein (■) by the Coomassie blue G250 binding assay (Bradford 1976).

The effectiveness of producing proteins in *E. coli* carrying recombinant DNA can be demonstrated by the example of the cloned pituitary hormone, methionyl human growth hormone (*met* HGH). *E. coli* fermentations yield in excess of 300 mg per litre of culture compared with an average content of 4 mg per pituitary gland. The high purity of recombinant *met* HGH that can be attained is demonstrated when comparing product from *E. coli* (Fig. 7a) with a standard HGH preparation from pituitaries (Fig. 7b). The latest forms

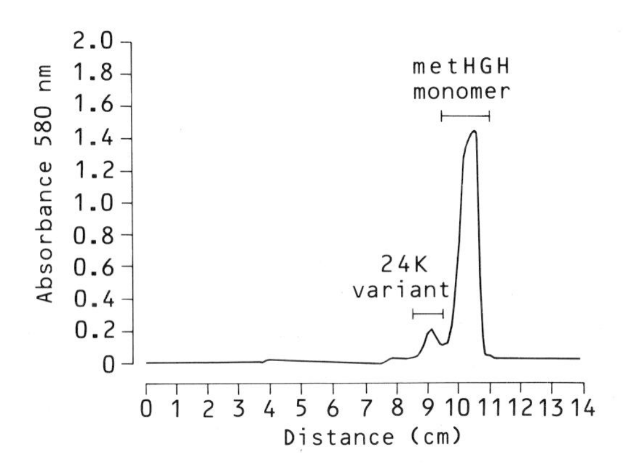

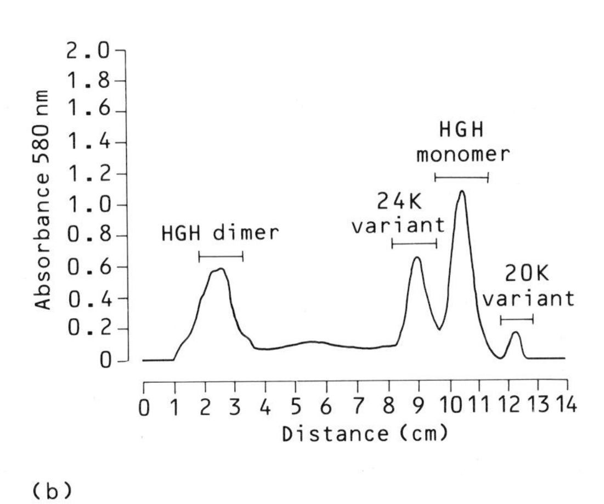

FIG. 7. Gel electrophoresis comparing human growth hormone produced from *E. coli* and pituitary glands. Methionyl HGH from *E. coli* (a) and HGH from pituitaries (b) were electrophoresed on 12 % polyacrylamide disc gels (14 cm) under non-reducing conditions at a constant current of 2.5 mA per gel. Gels were stained with Coomassie blue and scanned at 580 nm using a Gelman scanner.

of recombinant HGH now lack the N-terminal methionine through insertion of a gene leader sequence, coding for a signal peptide, which is cleaved from the mature protein in the *E. coli* cell membrane to leave the natural N-terminal amino acid, phenylalanine (Becker & Hsiung 1986; Chang *et al.* 1989).

Quality Assurance

Quality assurance applies to all stages of the production process, starting with a definition of the recombinant DNA and the host into which it is inserted and ending with a fully tested specification for the final product in its formulated state.

With reference to the major divisions used in this chapter, key tests can be summarized as follows:

Genetic design

Defined genotype of the host cell tested over multiple subculturing to ensure conformity.
Plasmid vector defined by restriction mapping and sequencing across key control elements on the plasmid (many plasmid vectors are now fully sequenced).
Gene sequence defined.

Fermentation

Raw materials analysed to defined specification.
Inoculum conforming to phenotype of host cell.
End of fermentation sample conforming to phenotype.
Plasmid vector isolated from end of fermentation sample and tested to correct restriction fragment pattern (plasmid stability).
Partial sequencing of plasmid across inserted gene sequence to ensure no sequence deletion/rearrangement.
Culture parameters within limits during fermentation

Downstream processing

Raw materials analysed to defined specification.
Product levels within limits at each stage of purification.
Final product methods of analysis may include:
Biological activity;
Electrophoretic purity;

HPLC/FPLC analysis of protein and peptide fragments;
Partial amino acid sequencing;
Cross-reactivity with antibody;
Pyrogens below set limits.

The source of final product contamination can vary and it is important to identify and remove contaminants as early as possible in the process.

Product-related contaminants

Fragments can arise by proteolytic 'nicking' of protein in the cell or during purification and can be further complicated for secreted proteins when the pre-protein is incorrectly processed to the mature form as it is transported across the cell membrane. Protein aggregates (particularly dimers of small polypeptides) can also be encountered. Modified forms of the protein can arise through proteolytic clips; for example, between amino acids 142 and 143 in the primary sequence of *met* HGH to give a molecule which is still disulphide bridged, but with a different conformation. Deamidation of asparagines and glutamines can occur at alkaline pH; the combination ASP–GLY or ASP–SER provide 'hot-spots' for deamidation and can result in the formation of an isopeptide via a cyclic imide intermediate. This alters the isoelectric point of the protein and can affect folding. Oxidation of methionines (e.g. met^{125} in *met* HGH) can form a sulphoxide, which requires analysis of tryptic fragments by mass spectrometry for its detection. Cyanates formed in urea solutions used for inclusion body solubilization can lead to carbamoylation of primary amines.

Given a combination of the above events, it is easy to see why product-related contaminants are the most difficult to deal with as many of the modifications give forms which exhibit very similar biological activity and may co-purify with the desired 'native' protein.

Host-related contaminants

Soluble cell proteins are the most common contaminants, but with the high levels of product expression achieved, these can be removed with minimum purification steps. Plasmid encoded proteins, other than the product, can appear at high level during product purification. These include β-lactamase, expressed when ampicillin is used as the selective antibiotic for amp^r plasmid vectors.

Nucleic acids are generally removed at early purification stages, but there are stringent limits set for any carryover into proteins to be used in therapy (less than 10 pg per therapeutic dose).

Endotoxins, which in *E. coli* are predominantly cell wall lipopolysaccharides, are strongly negatively charged and are efficiently removed by anionic ion exchangers, but it is necessary to introduce specific removal steps, such as filtration through positively charged filters or binding to Alhydrogel, at the end of the purification process prior to formulation and freeze-drying of clinical products. Such products have to pass rigorous pyrogenicity tests in rabbits, which in general equates to less than 1 endotoxin unit per mg protein as defined by the limulus amaebocyte lysate assay.

Process-related contaminants

The general rule for process-related contaminants is that 'anything deliberately added to the process must be shown to be absent in the final protein product'. Chemical protease inhibitors can protect the product at early process stages, but choice depends on perceived toxicity and ability to remove later. Enzymes used for cell lysis or specific cleavage of fusion proteins must also be shown to have been removed. Affinity ligands, whether chemical or a monoclonal antibody, which come into contact with the protein product during purification, must be characterized and the degree of 'leaching' into product streams defined.

It is perhaps quality assurance of proteins derived from recombinant DNA that has most preoccupied biotechnology and pharmaceutical companies in recent years as products have come through clinical trials into the market place.

Conclusion

In conclusion, it is perhaps sufficient to re-emphasize that each protein or polypeptide expressed in *E. coli* presents its own particular problems and that genetic design linked to controlled fermentation and efficient protein isolation needs to accommodate them. This should not be surprising as the products have evolved and are of value on the basis of their often unique specificity.

References

Becker, G.W. & Hsiung, H.M. 1986. Expression, secretion and folding of human growth hormone in *Escherichia coli*. *FEBS Letters* **204**, 145–150.

Bradford, M.M. 1976. A rapid and sensitive method for the quantitation of microgram quantities of protein utilising the principle of protein-dye binding. *Analytical Biochemistry* **72**, 248–254.

Chambers, S.P. 1989. Construction and characterisation of a novel *Escherichia coli* expression vector. PhD. Thesis, University of Warwick.

Chambers, S.P., Prior, S.E., Evans, R.A., Sherwood, R.F. & Minton, N.P. 1988. Plasmid pMTL153: a high copy number version of pAT153 and its use to obtain high expression of the *Pseudomonas* carboxypeptidase G_2 gene. *Applied Microbiology and Biotechnology* **29**, 572–578.

CHANG, J.Y-H., PAI, R-C., BENNETT, W.F. & BOCHNER, B.R. 1989. Periplasmic secretion of human growth by *Escherichia coli*. *Biochemical Society Transactions* **17**, 335–337.

GOEDDEL, D.V., HEYNEKER, H.L., HOZUMI, T., ARENTZEN, R., ITAKURA, K., YANSURA, D.G., ROSS, M.J., MIOZARRI, G., CREA, R. & SEEBURG, P.H. 1979. Direct expression in *E. coli* of a DNA sequence coding for human growth hormone. *Nature* (*London*) **281**, 544–548.

HOCHULI, E., BANNWARTH, W., DOBELI, R.G. & STUBER, D. 1988. Genetic approach to facilitate purification of recombinant proteins with a novel metal chelate adsorbent. *Bio/Technology* **6**, 1321–1325.

HOPP, T.P., PRICKETT, K.S., PRICE, V.L., LIBBY, R.T, MARCH, C.J., CERRETTI, D.P., URDAL, D.L. and CONLON, P.J. 1988. A short polypeptide marker sequence useful for recombinant protein identification and purification. *Bio/Technology* **6**, 1204–1210.

KAWAGUCHI, Y., SHIMIZU, N., NISHIMORI, K., UOZOMI, T. & BEPPU, T. 1984. Renaturation and activation of calf prochymosin produced in an insoluble form in *Escherichia coli*. *Journal of Biotechnology* **1**, 307–315.

LANCASTER, M.J., SHARP, R.J., COURT, J.R., MCENTEE, I., MELTON, R.G. & SHERWOOD, R.F. 1989. Production of cloned carboxypeptidase G_2 by *Escherichia coli*: Genetic and environmental considerations. *Biotechnology Letters* **11**, 699–704.

MARSTON, F.A.O., LOWE, P.A., DOEL, M.T., SCHOEMAKER, J.M., WHITE, S. & ANGAL, S. 1984. Purification of calf prochymosin (prorennin) synthesised in *Escherichia coli*. *Bio/Technology* **2**, 800–804.

MEACOCK, P.A. & COHEN, S.N. 1980. Partitioning of bacterial plasmids during cell division: a *cis*-acting locus that accomplishes stable plasmid inheritance. *Cell* **20**, 529–543.

MINTON, N.P., ATKINSON, T. & SHERWOOD, R.F. 1983. Molecular cloning of the *Pseudomonas* carboxypeptidase G_2 gene and its expression in *Escherichia coli* and *Pseudomonas putida*. *Journal of Bacteriology* **156**, 1222–1227.

NILSSON, B., ABRAHMSEN, L. & UHLEN, M. 1985. Immobilisation and purification of enzymes with staphylococcal protein A gene fusion vectors. *EMBO Journal* **4**, 1075–1080.

PERSSON, M., BERGSTRAND, M.G., BULOW, L. & MOSBACH, K. 1988. Enzyme purification by genetically attached polycysteine and polyphenylalanine affinity tails. *Analytical Biochemistry* **172**, 330–337.

SASSENFELD, H.M. & BREWER, S.J. 1984. A polypeptide fusion designed for the purification of recombinant proteins. *Bio/Technology* **2**, 76–81.

SHEPHERD, H.M., YELVERTON, E. & GOEDDEL, D. 1982. Expression of beta-interferon in *E. coli*: The effect of altering the distance between the ribosome binding site and the initiation codon. *DNA* **1**, 125–131.

SHERWOOD, R.F., MELTON, R.F., ALWAN, S.M. & HUGHES, P. 1985. Purification and properties of carboxypeptidase G_2 from *Pseudomonas sp.* RS-16: Use of a novel triazine dye affinity method. *European Journal of Biochemistry* **148**, 447–453.

Stability and Copy Number of Yeast Cloning Vectors

E.B. Gingold[1], M.J. Kleinman[2] and Virginia Bugeja[3]
[1]*Department of Biotechnology, South Bank Polytechnic, London SE1 0AA, UK;*
[2]*Department of Biotechnology, Ngee Ann Polytechnic, Singapore, 2159, Singapore;*
[3]*Department of Biological Sciences, Hatfield Polytechnic, Hatfield, Hertfordshire AL10 9AB, UK*

Gene Cloning in Yeasts

The yeast *Saccharomyces cerevisiae* has become an important host for work in gene cloning. This well-studied single-celled microbe can be manipulated by most of the standard techniques that have been developed for bacteria but nonetheless has a full eukaryotic biology. Hence *S. cerevisiae* has often served as a model system for the study of eukaryotic process such as the control of the cell cycle, the sexual cycle and the organization and properties of chromosomes. The introduction of cloning technologies has proven particularly advantageous in these studies.

The aspect of yeast cloning that has been of greatest commercial interest, however, has been the use of yeast as a host for the production of protein products from the higher eukaryotes. The eukaryotic nature of this species would make it appear a more attractive host than *Escherichia coli* for products which require eukaryotic post-translational modification, or which are unstable in bacterial backgrounds. Indeed, after unsuccessful attempts to express the hepatitis B virus surface antigen in *E. coli*, the immunologically active product was finally produced in yeast (Valenzuela *et al.* 1982), a result that did much to encourage interest in its potential as a production host.

There have been many reviews on the theory and practice of yeast cloning including Beggs (1981), Rothstein (1985), Gingold (1987), and Kingsman *et al.* (1985, 1987).

The range of yeast cloning vectors

One of the most outstanding features of cloning in yeast has been the wide variety of vectors that have been developed. What distinguishes these classes

0–632–02926–9

of vectors from each other are the differing mechanisms by which they replicate inside yeast cells and by which they are transmitted to daughter cells at cell division. As a consequence of this variation, there are large differences in the stability levels of these vectors and in their copy number within cells.

The simplest vectors are the so-called integrating plasmids which in reality are merely bacterial plasmids that can replicate inside a yeast cell only by integration into a yeast chromosome. Despite their low transformation efficiency and their copy number of, in general, one per cell, they have proven of immense usefulness in the development of unique techniques for the study of yeast genes. For a discussion of the use of these techniques see the reviews cited above.

Far more complex are the vectors built from elements obtained from yeast chromosomes. Initially, these simply consisted of a chromosomal DNA replication origin. Such vectors were highly unstable, but this was found to be much improved by the addition of a centromeric sequence. The further addition of telomeric sequences to these plasmids has enabled the construction of artificial linear yeast chromosomes and this process has given much insight into our understanding of chromosome organization (Kingsman & Kingsman 1988). It should be noted, however, that such artificial chromosomes are generally only present at a single copy per cell.

For practical purposes such as expression of foreign genes, by far the most commonly used vectors are those based on the native yeast 2μ plasmid. The reasons for this choice are twofold: these vectors give high copy numbers, and can be of at least moderately high stability. High copy number is often desirable as this can be a major factor in determining the levels of product produced (although one cannot assume a linear relationship). The importance of plasmid stability is, of course, obvious. It must be realized, however, that a laboratory worker's idea of stability is often very different from that of an industrial biologist. Thus many plasmids described as stable on the basis of 90% retention after an overnight growth would be totally lost in large-scale culture. Furthermore, the selective media used in laboratories to overcome problems of instability would generally be economically unsuited to industrial processes. Very few yeast plasmids are stable by the more demanding criterion of commercial systems and hence the problem of plasmid stability is a serious one.

The authors have been conducting a study into the factors influencing plasmid stability and copy number for a range of 2μ based plasmids. The methods that are described in this chapter were developed for this study, but should prove of general use in the evaluation of the properties of yeast vector systems under a range of conditions.

Plasmid Stability

Plasmid stability is a measure of the maintainence by cells in a culture of a plasmid, in its original form, during growth of that culture. There are two distinct ways in which instability can arise. Firstly, the plasmid could be lost completely from a proportion of the cells in the culture. This failure of plasmid to be be transmitted to daughter cells is referred to as *segregational instability*. Secondly, the plasmid may become rearranged, or lose sequences during growth of the culture, that is, undergo *genetic instability*. In each case plasmid-borne characteristics of importance to the investigator may be progressively lost from the culture.

Factors affecting the degree of instability have been found to include the nature of the plasmid, its copy number, the genetic characteristics of the host cell, selective effects of the plasmid on host cells including the consequences of plasmid gene expression, and the environmental conditions in which the culture is grown (Walmsley *et al.* 1983; Futcher & Cox 1983, 1984; Cashmore *et al.* 1986, 1988; Kleinman *et al.* 1986; Mead *et al.* 1986; Murray *et al.* 1987; Caunt *et al.* 1988; Bugeja *et al.* 1989). What has become clear from these studies is that each coding region of the native yeast 2μ plasmid plays a role in maintaining the plasmid and hence vectors which do not include the complete native sequence are likely to be unstable. The absence of some regions can, however, be at least partially complemented by the presence of native 2μ plasmid in the cell along with the chimeric plasmid. On the other hand, there is as yet no clear picture of the complete requirements of a stable plasmid and hence each new construction must be tested empirically.

Retention and loss of a phenotypic characteristic

The simplest approach to the determination of plasmid stability is to look for the loss of a plasmid-borne characteristic. In the authors' work, all plasmids carried markers that enabled simple selection for their loss. Hence the plasmid pJDB248 (all plasmids are shown in Fig. 1) carried the yeast *LEU*2 gene and was used in host strains which were *leu*2, that is, auxotrophic for leucine. Other yeast plasmids carry genes such as *TRP*1 and *HIS*3 that can be used in a similar fashion. The percentage of cells in any culture that have lost the plasmid-borne marker can be determined using the replica plating technique of Lederberg & Lederberg (1952).

Replica plating procedure

Spread 0.1 ml of a diluted sample of the yeast culture onto five complete medium YEPD (1% yeast extract, 2% bactopeptone, 2% glucose, 2%

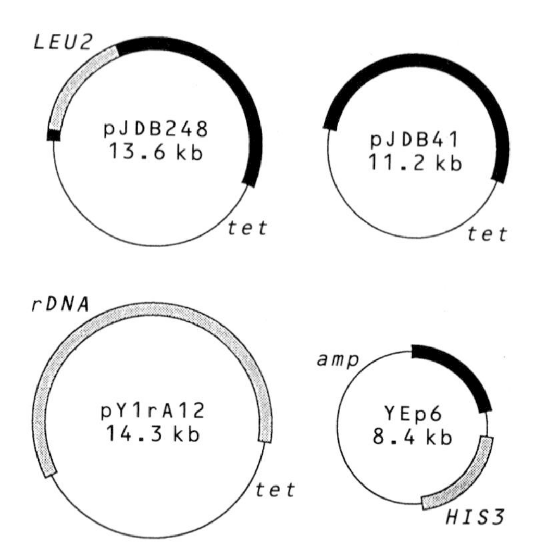

FIG. 1. The plasmids used in this chapter. Segments derived from bacterial DNA are shown as ———, yeast chromosomal DNA as ▭, and yeast 2μ DNA as ▬. The bacterial ampicillin resistance and tetracycline resistance genes are shown as *amp* and *tet*, yeast histidine and leucine biosynthetic genes as *His*3 and *Leu*2, while rDNA represents the yeast ribosomal RNA genes. Plasmids pJDB248 and pJDB41 are described in Beggs (1978), pY1rA12 in Petes *et al.* (1978), and YEp6 in Struhl *et al.* (1979).

agar) plates to give 100–200 colonies per plate. Incubate the plates at 30°C for 2–3 days. Replica plate onto defined medium (Wickerham 1946) with the appropriate amino acid selection marker for the plasmid (*non-selective medium*) and defined medium without this supplement (*selective medium*). Incubate the plates at 30°C for a 2–3 days. All cells are able to grow on YEPD and non-selective medium, whereas only those containing the plasmid are able to grow on *selective medium*. The proportion of plasmid containing cells can then be calculated as:

$$\% \text{ plasmid containing cells} = \frac{\text{number of colonies on selective medium}}{\text{number of colonies on non-selective medium}} \times 100$$

Results and comments

The results of an experiment to measure the stability of the plasmid pJDB248 under different conditions of continuous growth in a chemostat are shown in Fig. 2. The experimental approach is described more fully in Kleinman *et al.* (1986). As can be seen from Fig. 2, we were able to demonstrate that the *Leu*2

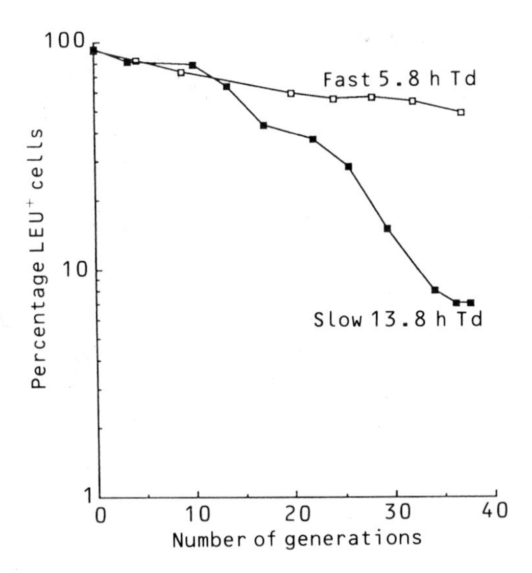

FIG. 2. Kinetics of loss of leucine prototrophy from *S. cerevisiae* S150–2B (*cir*^{0}) (pJDB248) in glucose-limited continuous culture maintained at dilution rates of 0.12/h (□) (5.8 h generation time, 'fast') and 0.05/h (■) (13.8 h generation time, 'slow').

marker was lost more rapidly when a glucose-limited culture was grown at a slow rate than when it was grown at a higher rate.

It should be noted that this method worked well in our experiments as the *LEU2* allele carried by pJDB248 produces sufficient product to overcome the auxotrophy even if present only as a single copy per cell. Results obtained with plasmids that have a defective form of the *LEU2* gene, e.g. pJDB219 (Beggs 1978), may be less clearcut as these require high copy number to produce sufficient leucine for growth and hence may lead to false negatives from colonies in which copy number is low, but not zero.

One must be aware that this approach will only determine loss of the plasmid-borne phenotype and not necessarily of the plasmid itself. Clearly, genetic instability leading to deletion or inactivation of the test marker would have the same result and hence further work is needed to distinguish between plasmid loss and modification.

Detection of plasmid by colony hybridization

The authors have tested directly for the presence of the plasmid in individual colonies by a variation of the standard bacterial colony hybridization technique of Grunstein & Hogness (1975) as described in Sherman *et al.* (1982). The choice of a probe is important. One cannot generally use yeast vectors as they contain regions homologous to the yeast chromosome and hence would give a signal against all yeast colonies whether or not they carry the plasmid. The authors have used the plasmid pJDB41 (Fig. 1) as the probe in all their work as this includes both the complete 2μ sequence and the bacterial plasmid regions carried by yeast vectors, but no yeast chromosomal regions. Any colony containing the vector even in a deleted or rearranged form should give a positive hybridization signal when tested against this probe.

Yeast colony hybridization procedure

Replicate plate colonies grown on YEPD plates onto nitrocellulose filters (Millipore HATF 0.45 μm). Place the nitrocellulose filters onto fresh YEPD plates and allow them to grow for 4–12 h. Remove the filters, air-dry briefly, then place onto several layers of 3MM papers with the following solutions, using fresh 3MM filter paper in petri dishes. Treat the filters by saturating the 3MM papers for each solution change:

1 1 M sorbitol, 20 mM EDTA, 50 mM dithiothreitol (10 min).

2 1 M sorbitol, 20 mM EDTA, 1 mg/ml zymolase (2–3 h, 37°C).

(a) Lysis should be checked by picking cells onto a microscope slide with 5% SDS;

(b) Any enzyme or combination of enzymes which produce yeast protoplasts can be used, e.g. 1% glucuronidase, 1% SP234 (Novozyme).

3 0.5 M NaOH (7 min, room temperature).

4 0.5 M Tris–HCl pH 7.5, 10 × SSC (standard or 1 × SSC is 0.15 M NaCl, 0.015 M sodium citrate) (4 min, room temperature).

5 0.5 M Tris–HCl pH 7.5, 10 × SSC (4 min, room temperature).

6 2 × SSC (2 min, room temperature).

Then place the filters on dry 3MM filter paper and allow to air dry (0.5 h) before baking in a vacuum oven at 80°C for 2 h.

Label the probe with ^{32}P or ^{35}S using a standard nick translation kit (Amersham International). Hybridization to filters is carried out as follows:

1 The filters are placed in a polythene bag and pre-wet with 6 × SSC.

2 Hybridization is carried out in 10 ml of fluid (6 × SSC, 0.1 mg/ml denatured calf thymus DNA, 0.5% SDS, 0.01 M EDTA, 5 × Denhardt's solution (0.02% Ficoll, 0.02% polyvinylpyrrolidone, 0.02% bovine serum albumin)) with 0.01–0.1 μg of labelled denatured DNA as probe. Seal the bag, place in a second sealed bag and incubate for 18–24 h at 68°C. (Note that a prehybridization incubation for several hours in the hybridization fluid without the probe can be included to reduce non-specific binding).

3 Open the bag and dispose of liquid radioactive waste into a designated waste receptacle.

4 Wash filters in a large beaker as follows:
2 × SSC, 0.5% SDS (5 min, room temperature);
2 × SSC, 0.1% SDS (15 min, room temperature);
0.1 × SSC, 0.5% SDS (2–3 h, 68°C).
During this final wash carry out several changes of buffer.
5 Dry the filters at room temperature.
6 Wrap the filters in Saranwrap (Dow Ltd)
7 The filters are then exposed to X-ray film for autoradiography by standard techniques.

Results

This method was applied to a sample from the previously described experiment, consisting of about 600 colonies of which over 500 had lost the *LEU2* marker. Each of these leucine-requiring colonies gave a negative response in this test. The colonies that retained this marker gave, as expected, a positive hybridization signal. The authors were thus able to conclude that the loss of the *LEU2* marker was, in this case, a direct measure of plasmid loss and that it was segregational rather than genetic instability. It must be borne in mind, however, that this result is not general and good evidence exists for genetic rearrangements in other plasmid-host systems.

It should be noted that this method gives good results when testing for the stability of the native 2μ plasmid. Here, of course, phenotypic scoring cannot be used due to the lack of a detectable phenotype. If the aim of this testing is to obtain plasmid-free cells, however, it would be wise to confirm that the negative colonies do not retain a low level of plasmid by testing individual colonies with the more sensitive DNA isolation and Southern blotting test described below.

Copy Number Determination

The determination of copy number is of importance in understanding the reasons for instability of plasmids, and also in its own right as an indication of the gene dosage per cell. A number of different methods exist, but all are based on the isolation of total DNA from the yeast culture followed by comparison of the levels of plasmid DNA with that of a recognizable chromosomal species.

Isolation of total yeast DNA

This method involves physical cell disruption by glass beads. It is based on the method of Piper *et al.* (1986) and is suitable for extracting total DNA from

20–100 ml of an exponentially growing culture (5×10^6–2×10^7 cells per ml) or from yeast pellets frozen at −20°C.

Method

1 Add 0.5 ml cold buffer (10 mM EDTA, 1% SDS, 20 mM Tris–HCl pH 8.5), 0.5 ml cold phenol/chloroform (50/50, pre-equilibrated with 0.1 M Tris) and glass beads (0.45–0.5 mm diameter, same volume as pellet) to the pelleted cells in an Eppendorf tube (1.5 ml capacity).
2 Vortex for about 5 min.
3 Spin in a microcentrifuge at 12 000 rpm for 10 min.
4 Transfer the aqueous (upper) layer to a fresh Eppendorf tube, note volume and add 1/5 volume cold 6 M ammonium acetate (pH 6) and twice the total volume of cold ethanol. Leave on ice for 30 min to precipitate the DNA.
5 Spin in a microfuge for 15 min.
6 Resuspend the pellet in 0.4 ml TE buffer (1 mM EDTA, 10 mM Tris–HCl pH 8) and add 1 μl pancreatic RNase (20 mg/ml). Leave at 37°C for 1 h.
7 Add 0.3 ml phenol/chloroform. Vortex and spin in microfuge for 10 min.
8 Remove aqueous layer (top) to a fresh Eppendorf tube and add 0.4 ml diethyl ether. Vortex and spin in microfuge for 10 min.
9 Discard ether layer (top). Add 100 μl cold 6 M ammonium acetate and 1 ml cold ethanol. Leave at −20°C for 15 min to precipitate DNA.
10 Spin in microfuge for 15 min.
11 Discard supernatant. Add 0.5 ml cold 70% ethanol to pellet. Spin in microfuge for 10 min.
12 Discard ethanol, dry pellet under vacuum and resuspend in 20 μl TE buffer.

Analysis of plasmid levels in the isolated DNA

Perhaps the simplest method of analysis was that described by Futcher & Cox (1984) who cut the total cellular DNA with a suitable restriction enzyme (*Eco*RI), ran the products on a gel which was then stained and scanned with a densitometer. It was possible to distinguish bands above the general smear which came from the multicopy plasmid and ribosomal RNA genes. As an estimate of the number of copies of the ribosomal genes was available, the relative levels of these two species was used to determine copy number of the plasmid.

The authors preferred to base their analysis on hybridization of a Southern blot of the gel with a suitable probe as this eliminated the background of non-specific chromosomal fragments. The choice of probe that is to be used is particularly important. Clearly, the probe must have a region unique to the plasmid that is being investigated, and a second region found only in the chromosomal DNA. But as the analysis is based on a comparison of the relative intensity of hybridization of the probe to these two regions, it is clearly advantageous if the plasmid bands and the chromosomal bands give signals of the same order of magnitude. This will not be the case if a single copy chromosomal gene is being compared to a multicopy plasmid and hence, if possible, the chromosomal region should be one that is multicopy. The ribosomal RNA coding sequences (rDNA) are the most obvious candidates here. The authors have mostly used the plasmid pYlrA12, which carries the yeast ribosomal genes on the bacterial vector pMB9. The pMB9 region was homologous to the bacterial sequences on the yeast plasmid under investigation, whilst the ribosomal genes provided homology with a multicopy region of the chromosome.

Southern blotting procedure

Five microlitres of the DNA preparation is digested with a suitable restriction endonuclease (*Eco*RI was generally used in our work), electrophoresed on an agarose gel, and transferred to a nylon membrane (Genescreen Plus, Dupont) according to the manufacturer's instructions (Capillary Blot Procedure, Genescreen Plus Catalog No. NEF-976 (1985) Dupont) and based on the method of Southern (1975) as follows:

1 Incubate the agarose gel in 0.4 M NaOH, 0.6 M NaCl for 30 min followed by 1.5 M NaCl, 0.5 M Tris–HCl (pH 7.5) for a further 30 min.

2 Transfer DNA to the nylon membrane (pre-wet in 10 × SSC) using 10 × SSC in a standard blotting apparatus. Allow transfer to continue for 16–24 h.

3 Immerse the membrane in 0.4 M NaOH for 30–60 s, followed by 0.2 M Tris–HCl (pH 7.5), 2 × SSC for a further 30–60 s.

4 Allow the membrane to dry at room temperature, then place it in a polythene bag and pre-wet with 0.1 × SSC.

5 Add 10 ml hybridization fluid (1 M NaCl, 10% (w/v) dextran sulphate, 1% SDS, 100 μg/ml denatured salmon sperm DNA) to bag and seal. Incubate at 65°C for at least 15 min.

6 Add radioactive probe (labelled by means of a standard nick translation kit) to bag. Final concentration of the probe in the bag should be $\leq$ 10 ng/ml. Reseal bag, place in second bag and incubate for 18–24 h at 65°C.

7 Open bag and dispose of radioactive waste in designated waste receptacle.

8 Wash membrane as follows:

2 × 100 ml of 2 × SSC (5 min, room temperature);

2 × 200 ml of 2 × SSC, 1% SDS (30 min, 65°C);
2 × 100 ml of 0.1 × SSC (30 min, room temperature).

9 Wrap membrane in Saranwrap (Dow Ltd).

10 The membranes are then exposed to X-ray film for autoradiography using standard techniques.

Analysis of copy number data

A typical autoradiograph obtained in our investigation is shown in Fig. 3. The bands corresponding to the chromosome ribosomal RNA genes and the plasmid sequences can be clearly seen. It will be noted that as the probe used was homologous to the multi-copy yeast chromosomal rDNA, similar levels of intensity are found for the plasmid and chromosomal bands.

A densitometer scan of this autoradiograph is shown in Fig. 4. A quantitative estimation of the hybridization to each band can be arrived at by measuring

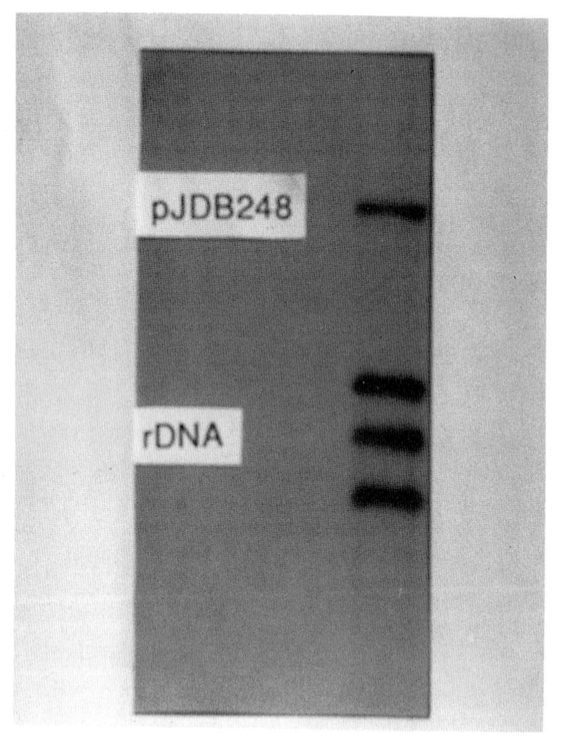

FIG. 3. Autoradiograph of total yeast DNA from *S. cerevisiae* S150–2B (*cir*^{0}) (pJDB248) digested with *Eco*RI and hybridized to the rDNA probe pY1rA12.

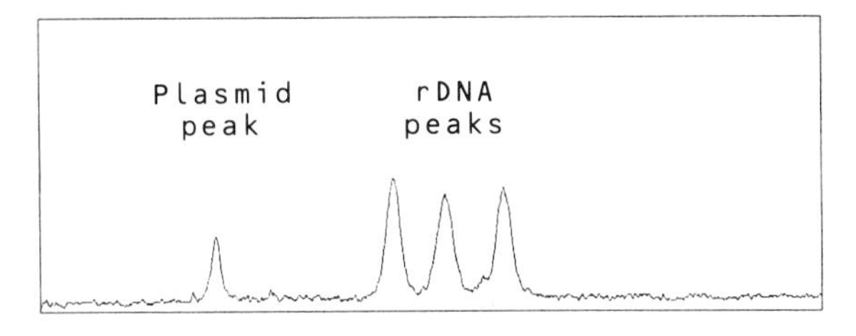

FIG. 4. Densitometer scan (Bio-Rad vd620 densitometer) of the autoradiograph in Fig. 3. The areas under the peaks represent the amount of hybridization of the probe to the plasmid and rDNA bands.

the area under the peaks. The plasmid copy number can thus be determined relative to the internal control of the rDNA repeat unit, estimated at 100 copies per haploid genome (Petes *et al.* 1978). Taking into account the degree of homology of the probe to the plasmid and ribosomal DNA (that is, the sizes of the sequences carried on the probe homologous to the individual bands), the average copy number in the total cell population can be calculated as:

$$\begin{array}{l}\text{Average plasmid}\\ \text{copy number}\end{array} = \begin{array}{l}\text{Copy number of}\\ \text{chromosomal gene}\end{array} \times \frac{\text{Hybridization to plasmid}}{\text{Hybridization to chromosome}} \times \frac{\text{Homology to chromosome}}{\text{Homology to plasmid}}.$$

To obtain the copy number in cells actually containing plasmid, a correction must be made for the proportions of plasmid containing cells as follows:

$$\begin{array}{l}\text{Average plasmid copy number}\\ \text{in plasmid containing cells}\end{array} = \frac{\text{Average plasmid copy number}}{\text{\% plasmid containing cells}} \times 100.$$

The results from one of the authors' stability experiments is shown in Fig. 5. This shows that at slow growth rates, as the number of plasmid carrying cells decreased, the plasmid copy number in the remaining cells increased dramatically. This provided an indication that it was a partitioning breakdown. Further details of this work are reported in Bugeja *et al.* (1989).

In other work, the authors have used a probe, YEp6, containing a single copy chromosomal gene, *HIS*3, and a section of the 2μ plasmid plus a pBR322 sequence. When analysis of a densitometer tracing of the Southern blot autoradiograph was attempted, the differences in intensity between the single copy *HIS*3 band and the multicopy plasmid bands made accurate analysis by densitometry difficult. To avoid this problem it was found necessary to cut the bands from the membrane filters after they were located by autoradiography and determine activity by liquid scintillation counting.

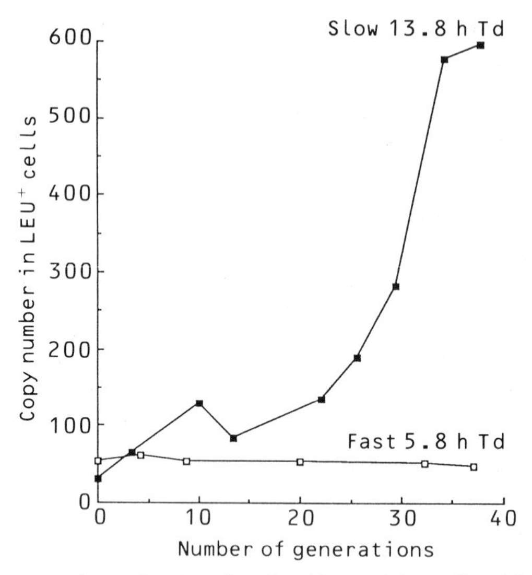

FIG. 5. Plasmid copy number estimations for plasmid containing cells of *S. cerevisiae* S150–2B (*cir*^{0}) (pJDB248) growing in the 'fast' (□) and 'slow' (■) continuous cultures described in Fig. 2. The rDNA probe pY1rA12 was used for the copy number estimations.

References

BEGGS, J.D. 1978. Transformation of yeast by a replicating hybrid plasmid. *Nature* **275**, 104–109.

BEGGS, J.D. 1981. Gene cloning in yeast. In *Genetic Engineering*, Vol. 2. ed. Williamson, R. pp. 175–203. London: Academic Press.

BUGEJA, V.C., KLEINMAN, M.J., STANBURY, P.F. & GINGOLD, E.B. 1989. The segregation of the 2μ-based yeast plasmid pJDB248 breaks down under conditions of slow, glucose limited growth. *Journal of General Microbiology* **135**, 2891–2897.

CASHMORE, A.M., ALBURY, M.S., HADFIELD, C. & MEACOCK, P.A. 1986. Genetic analysis of partitioning functions encoded by the 2μ circle of *Saccharomyces cerevisiae*. *Molecular and General Genetics* **203**, 154–162.

CASHMORE, A.M., ALBURY, M.S., HADFIELD, C. & MEACOCK, P.A. 1988. The 2μ D region plays a role in yeast plasmid maintenance. *Molecular and General Genetics* **212**, 426–431.

CAUNT, P., IMPOOLSUP, A. & GREENFIELD, P.F. 1988. Stability of recombinant plasmids in yeast. *Journal of Biotechnology* **8**, 173–192.

FUTCHER, A.B. & COX, B.S. 1983. Maintenance of the 2μ circle plasmid in populations of *Saccharomyces cerevisiae*. *Journal of Bacteriology* **154**, 612–622.

FUTCHER, A.B. & COX, B.S. 1984. Copy number and the stability of 2μ circle-based artificial plasmids of *Saccharomyces cerevisiae*. *Journal of Bacteriology* **157**, 283–290.

GINGOLD, E.B. 1987. Cloning in yeast. In *Techniques in Molecular Biology*, Vol. 2. eds Walker, J.M. & Gaastra, W. pp. 140–158. London: Croom Helm.

GRUNSTEIN, M. & HOGNESS, S.D. 1975. Colony hybridisation: A method for the isolation of cloned DNAs that contain a specific gene. *Proceedings of the National Academy of Sciences of the United States of America* **72**, 3961–3965.

KINGSMAN, S.M., KINGSMAN, A.J., DOBSON, M.J., MELLOR, J. & ROBERTS, N.A. 1985. Heterologous gene expression in *Saccharomyces cerevisiae.* In *Biotechnology and Genetic Engineering Reviews.* ed. Russell, G.E. pp. 377–416. Newcastle upon Tyne: Intercept.

KINGSMAN, S.M., KINGSMAN, A.J. & MELLOR, J. 1987. The production of mammalian proteins in *Saccharomyces cerevisiae. Trends in Biotechnology* **5**, 53–57.

KINGSMAN, S.M. & KINGSMAN, A.J. 1988. *Genetic Engineering.* pp. 320–348. Oxford: Blackwell Scientific Publications.

KLEINMAN, M.J., GINGOLD, E.B. & STANBURY, P.F. 1986. The stability of the yeast plasmid pJDB248 depends on growth rate of the culture. *Biotechnology Letters* **8**, 225–230.

LEDERBERG, J. & LEDERBERG, E. 1952. Replica plating and indirect selection of bacterial mutants. *Journal of Bacteriology* **63**, 399–406.

MEAD, D.J., GARDNER, D.C.J. & OLIVER, S.G. 1986. The yeast 2μ plasmid: strategies for the survival of a selfish DNA. *Molecular and General Genetics* **205**, 417–421.

MURRAY, J.A.H., SCARPA, M., ROSSI, N. & CESARENI, G. 1987. Antagonistic controls regulate copy number of the yeast 2μ plasmid. *The European Molecular Biology Organisation Journal* **6**, 4205–4212.

PETES, T.D., HEREFORD, L.M. & SKRYABIN, K.G. 1978. Characterization of two types of yeast ribosomal DNA genes. *Journal of Bacteriology* **134**, 295–305.

PIPER, P.W., CURRAN, B., DAVIES, M.W., LOCKHEART, A. & REID, G. 1986. Transcription of the phosphoglycerate kinase gene of *Saccharomyces cerevisiae* increases when fermentative cultures are stressed by heat shock. *European Journal of Biochemistry* **161**, 525–531.

ROTHSTEIN, R. 1985. Cloning in yeast. In *DNA Cloning,* Vol. 2. ed. Glover, D.M. pp. 45–66. Oxford: IRL Press.

SHERMAN, F., FINK, G.R. & HICKS, J.B. 1982. *Methods in Yeast Genetics Laboratory Manual.* p. 118. Cold Spring Harbor, New York: Cold Spring Harbor Laboratory.

SOUTHERN, E.M. 1975. Detection of specific sequences among DNA fragments separated by gel electrophoresis. *Journal of Molecular Biology* **98**, 503–517.

STRUHL, K., STINCHCOMBE, D.T., SCHERER, S. & DAVIS, R.W. 1979. High-frequency transformation of yeast: Autonomous replication of hybrid DNA molecules. *Proceedings of the National Academy of Sciences of the United States of America* **76**, 1035–1039.

VALENZUELA, P., MEDINA, A., BUTLER, W.J., AMMERER, G. & HALL, B.D. 1982. Synthesis and assembly of Hepatitis B virus surface antigen particles in yeast. *Nature* **298**, 347–350.

WALMSLEY, R.M., GARDNER, D.C.J. & OLIVER, S.G. 1983. Stability of a cloned gene in yeast grown in chemostat culture. *Molecular and General Genetics* **192**, 361–365.

WICKERHAM, L.J. 1946. A critical evaluation of the nitrogen assimilation tests commonly used in the classification of yeasts. *Journal of Bacteriology* **52**, 293–301.

Designer Genes — Design and Synthesis Illustrated for the Gene Coding for Human Macrophage Colony Stimulating Factor

LINDA DAVIES, I.D. JOHNSON AND J.A. DAVIES
British Biotechnology Ltd, Watlington Road, Cowley, Oxford OX4 5LY, UK

Synthetic genes have an advantage over natural cDNA in that the sequence can be manipulated to afford greater control over restriction sites and codon usage within the DNA sequence. The greater flexibility and modularity of the gene facilitates its use for structure–function studies and protein engineering research.

The usefulness of this approach has only been realized as confidence has grown in the ability to design and synthesize fully functional genes. During the last three years, the authors have been able to rationalize their design and synthesis strategy to an extent where the synthesis of genes from 500–2 000 base pairs (bp) is rapid and routine.

The following presentation details the current strategy for the synthesis of 'designer genes', with the gene coding for human macrophage colony stimulating factor (MCSF) as an example.

Gene Design

Figure 1a shows the steps involved for the design of a typical synthetic gene. Two approaches are possible: one being to take the amino acid sequence of the protein and to choose a DNA sequence to code for this, based on the favoured codon usage for expression in a particular host. Restriction enzyme sites can then be introduced at intervals throughout the gene by making use of the degeneracy of the genetic code. Another approach is to base the sequence on that of the natural cDNA and to introduce a minimum number of changes to the DNA sequence for addition or deletion of restriction enzyme sites.

The latter approach was adopted for the design of the MCSF gene. The natural sequence was obtained (Wong *et al.* 1987) and analysed for actual and

0–632–02926–9

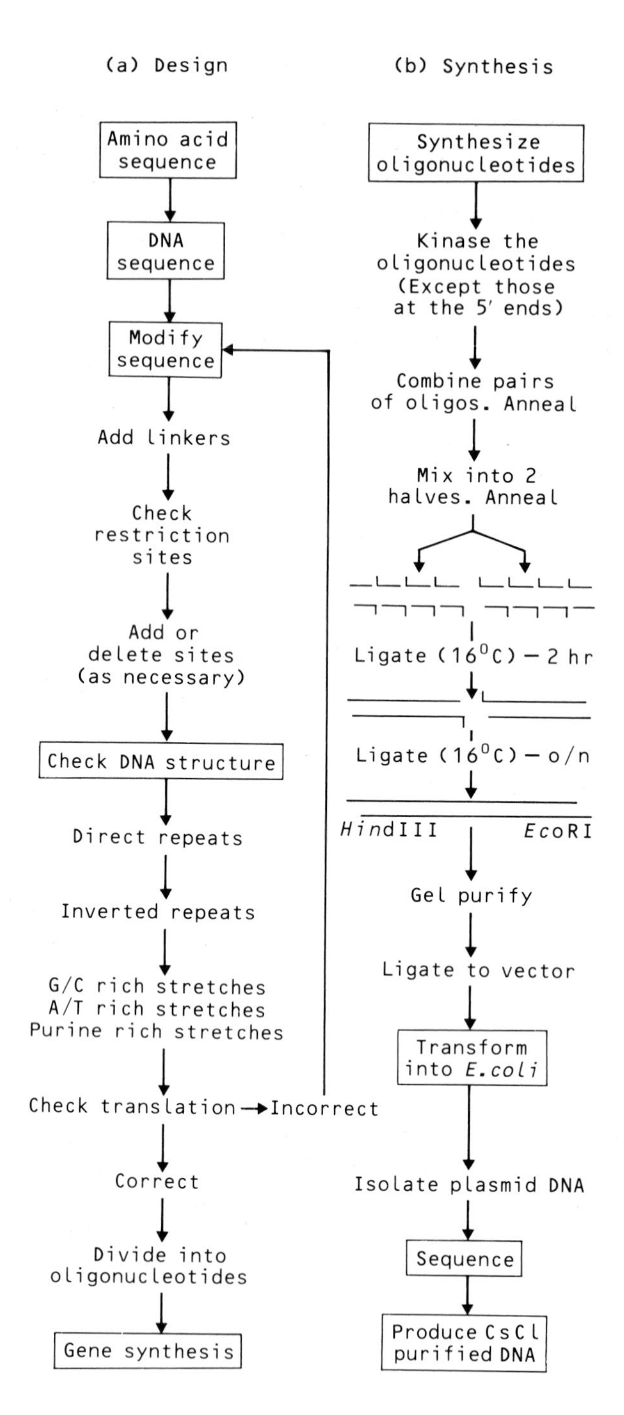

FIG. 1. Flow diagrams showing the steps involved in the design and synthesis of a synthetic gene.

potential restriction enzyme sites. Five base changes were made to the DNA sequence whilst still maintaining the correct amino acid translation. Two *Pst*I sites, a *Bsp*MI and a *Stu*I site were removed, whilst a *Sal*I site was introduced. Restriction sites were added at the 5′, 3′ ends of the gene to facilitate manipulation between different vectors. The sequence of the synthetic gene is shown in Fig. 2.

Finally, the sequence was divided into oligonucleotides such that unwanted homology between sticky ends of non-adjacent oligonucleotides was minimized. It was decided to build the gene in two parts and assemble the entire gene, once correct clones had been identified for these two fragments. This strategy is adopted for any gene greater than approximately 500 bp, as it facilitates the identification of mutation-free clones.

Materials and Methods

Bacterial strains and plasmids

Escherichia coli strain HW87 (*ara* D139 (*ara–leu*) del 7697 (*lac* IPOZY) del 174 *gal*U *gal*K *hsd*R *rps*L *srt rec*A56) was used for plasmid cloning and preparation. Recombinant *E. coli* cultures were grown in L-broth containing carbenicillin at 100 μg/ml. Plasmid pUC18 (Pharmacia) was used for cloning the synthetic gene fragments.

Chemicals

Restriction enzymes, T4 DNA ligase and *Hae*III-digested φX174 DNA were purchased from New England Biolabs. T4 polynucleotide kinase was from BRL, Klenow polymerase from Boehringer Mannheim and [α-^{35}S]dATP from Amersham Corp.

Oligonucleotide synthesis

The oligonucleotides were synthesized by means of cyanoethyl phosphoramidite chemistry on an Applied Biosystems 380B DNA synthesizer. All syntheses were performed at the 0.2 μmol scale and, after removal from the solid support, purified on denaturing polyacrylamide gels. Each oligonucleotide was dissolved in 1 ml sterile distilled water and its concentration was determined by measuring the absorbance at 260 nm.

Agarose gel electrophoresis

DNA fragments were analysed by electrophoresis in 1–2% agarose gels

MCSF (natural) catalogue gene (front half)

```
                                  A1
AAGCTTACCT  GCCATGGAGG  AGGTGTCGGA  GTACTGTAGC  CACATGATTG  GGAGTGGACA
TTCGAATGGA  CGGTACCTCC  TCCACAGCCT  CATGACATCG  GTGTACTAAC  CCTCACCTGT
        10          20  A2      30          40          50          60
    A3                                          A5
CCTCCAGTCT  CTGCAACGGC  TGATTGACAG  TCAGATGGAG  ACCTCGTGCC  AAATTACATT
GGAGGTCAGA  GACGTTGCCG  ACTAACTGTC  AGTCTACCTC  TGGAGCACGG  TTTAATGTAA
    A4  70          80          90         100  A6     110         120
                        A7                                  A9
TGAGTTTGTC  GACCAGGAAC  AGTTGAAAGA  TCCAGTGTGC  TACCTTAAGA  AGGCATTTCT
ACTCAAACAG  CTGGTCCTTG  TCAACTTTCT  AGGTCACACG  ATGGAATTCT  TCCGTAAAGA
       130         140  A8     150         160         170  A10    180
                                    A11
CCTGGTACAA  GACATAATGG  AGGACACCAT  GCGCTTCAGA  GATAACACCC  CCAATGCCAT
GGACCATGTT  CTGTATTACC  TCCTGTGGTA  CGCGAAGTCT  CTATTGTGGG  GGTTACGGTA
       190         200         210  A12    220         230         240
            A13
CGCCATTGTG  CAGCTGCAGT  TGAATTC
GCGGTAACAC  GTCGACGTCA  ACTTAAG
       250  A14    260
```

MCSF (natural) catalogue gene (back half)

```
                        B1
AAGCTTTTCT  GCAGGAACTC  TCTTTGAGGC  TGAAGAGCTG  CTTCACCAAG  GATTATGAAG
TTCGAAAAGA  CGTCCTTGAG  AGAAACTCCG  ACTTCTCGAC  GAAGTGGTTC  CTAATACTTC
        10          20  B2      30          40          50          60
B3                                              B5
AGCATGACAA  GGCCTGCGTC  CGAACTTTCT  ATGAGACACC  TCTCCAGTTG  CTGGAGAAGG
TCGTACTGTT  CCGGACGCAG  GCTTGAAAGA  TACTCTGTGG  AGAGGTCAAC  GACCTCTTCC
B4      70          80          90         100  B6     110         120
                        B7                                  B9
TCAAGAATGT  CTTTAATGAA  ACAAAGAATC  TCCTTGACAA  GGACTGGAAT  ATTTTCAGCA
AGTTCTTACA  GAAATTACTT  TGTTTCTTAG  AGGAACTGTT  CCTGACCTTA  TAAAAGTCGT
       130         140  B8     150         160         170  B10    180
                                    B11
AGAACTGCAA  CAACAGCTTT  GCTGAATGCT  CCAGCCAAGG  CCATGAGAGG  CAGTCCGAGG
TCTTGACGTT  GTTGTCGAAA  CGACTTACGA  GGTCGGTTCC  GGTACTCTCC  GTCAGGCTCC
       190         200         210  B12    220         230         240
            B13                                             B15
GATCCTCCAG  CCCGCAGCTC  CAGGAGTCTG  TCTTCCATCT  GCTGGTGCCC  AGTGTCATCC
CTAGGAGGTC  GGGCGTCGAG  GTCCTCAGAC  AGAAGGTAGA  CGACCACGGG  TCACAGTAGG
       250  B14    260         270         280         290  B16    300
                                    B17
TGGTCTTGCT  GGCTGTCGGT  GGCCTCTTGT  TCTACAGGTG  GAGGCGGCGG  AGCCATCAAG
ACCAGAACGA  CCGACAGCCA  CCGGAGAACA  AGATGTCCAC  CTCCGCCGCC  TCGGTAGTTC
       310         320         330  B18    340         350         360
B19                                             B21
AGCCTCAGAG  AGCGGATTCT  CCCTTGGAGC  AACCAGAGGG  CAGCCCCCTG  ACTCAGGATG
TCGGAGTCTC  TCGCCTAAGA  GGGAACCTCG  TTGGTCTCCC  GTCGGGGGAC  TGAGTCCTAC
B20    370         380         390         400  B22    410         420
                        B23
ACAGACAGGT  GGAACTGCCA  GTGTAGTAAA  GATCTGAATTC
TGTCTGTCCA  CCTTGACGGT  CACATCATTT  CTAGACTTAAG
       430         440  B24    450         460
```

FIG. 2. Sequence of the synthetic gene fragments for MCSF.

(Sigma Type I–A) and isolated from 1–2% low melting point (LMP) agarose gels (BRL Ultrapure grade) by standard procedures (Sambrook *et al.* 1989).

Sequence analysis

DNA sequencing of the DNA was by the Sanger dideoxy chain termination method modified for alkali denatured plasmid DNA (Chen & Seeburg 1985). 'Universal' and 'reverse' M13/pUC primers were used, together with other synthetic primers.

Gene Synthesis

Assembly of oligonucleotides into fragments A and B

Figure 1b shows the steps involved in the synthesis of a gene sequence. Each of the two fragments for the MCSF gene were produced in this way before joining them together to give the full length sequence. Fragment A was assembled from 14 oligonucleotides whilst fragment B was assembled from 24 (see Fig. 2). The 5′,3′ ends of each fragment were compatible with *Hin*dIII and *Eco*RI cohesive ends respectively.

The next step was to dry 250 pm of each oligonucleotide, except those forming the 5′-overhangs (A1, A14, B1 and B24). They were then resuspended in 20 μl kinase buffer (50 mM Tris pH 7.6, 10 mM $MgCl_2$, 2 mM ATP, 0.1 mM spermidine, 5 mM dithiothreitol, 10 U T4 polynucleotide kinase). These were incubated at 37°C for 30 min followed by heat inactivation of the kinase at 85°C for 15 min. The two 5′-overhang oligonucleotides for each fragment were not phosphorylated, so as to prevent concatamer formation. Instead, 250 pmole were dried and resuspended in 20 μl sterile distilled water.

Then 6 μl of each oligonucleotide was taken and mixed with its complementary pair. These were then annealed by heating at 90°C for 10 min followed by slow cooling to room temperature. Ten microlitres of each pair was then mixed for each half of the fragment—the volume then being made up to 100 μl with sterile distilled water, 10 μl 10 mM ATP and 10 μl 10 × ligase buffer (250 mM Tris pH 7.8, 200 mM DTT, 50 mM $MgCl_2$). Then 100 units of T4 ligase were added and left at 16°C for a minimum of 2 h. Then 5 μl of each half fragment were analysed on a 2% agarose gel to check the ligation had gone to completion.

Finally 50 μl of each half fragment were mixed together, a further 100 U of T4 ligase added, and the ligation left to proceed at 16°C overnight. Again, a 2% agarose gel was used to check the extent of ligation. Figure 3 shows the check gel for the two half length and the full length products of the assembly

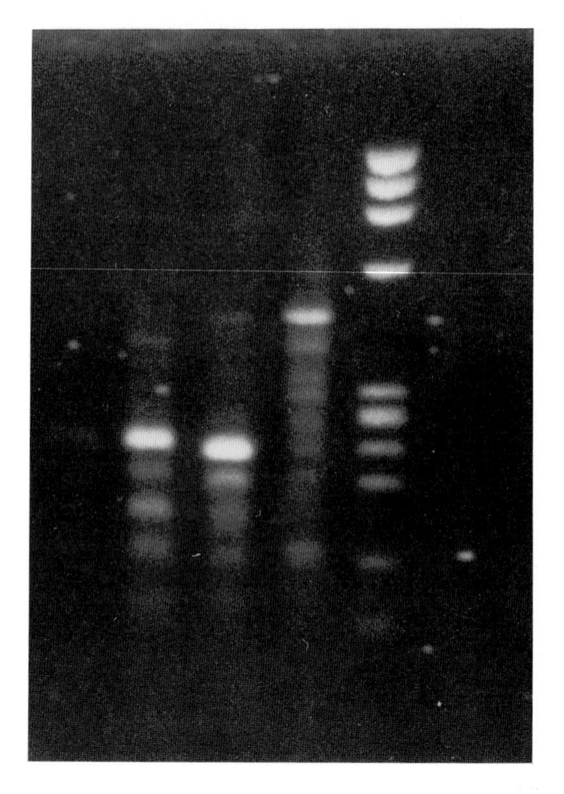

FIG. 3. Agarose gel showing the assembly products for fragment B. From the left to right the lanes are: 1st and 2nd half length products (240 bp and 220 bp respectively), full length product (460 bp) and φX174/*Hae*III maker DNA (giving bands of 72, 118, 194, 234, 271, 310, 603, 872, 1078 and 1353 bp).

of fragment B. The fully assembled fragments were then purified by running the ligation mixtures on a 2% LMP agarose gel—the band corresponding to the full length product (as compared to φX174/*Hae*III marker DNA run simultaneously) was cut out and extracted by standard procedures (Sambrook *et al.* 1989).

Cloning of synthetic fragments A and B

Each purified fragment was ligated to pUC18 linearized with *Hin*dIII and *Eco*RI, the ligation mixtures were then used to transform $CaCl_2$-treated *E. coli* strain HW87. Both vector and fragment concentrations were 20 ng/μl—the ligation mixtures being set up as stated in Table 1, together with the resulting numbers of colonies obtained.

TABLE 1. *Ligation mixtures (in μl) for the transformation of fragments A and B*

	Vector	10 × ligase buffer	10 mM ATP	T4 ligase	H_2O	No. of colonies
Fragment A						
—	1	2	2	—	15	5
—	1	2	2	1	14	47
0.5	1	2	2	1	13.5	180
1	1	2	2	1	13	152
2	1	2	2	1	12	456
4	1	2	2	1	10	320
Fragment B						
—	1	2	2	—	15	1
—	1	2	2	1	14	39
0.5	1	2	2	1	13.5	140
1	1	2	2	1	13	456
2	1	2	2	1	12	560
4	1	2	2	1	10	304

Plasmid DNA was prepared from 10 ml overnight culture using the 'boiling lysis' approach (Holmes & Quigley 1981). These were then used to check for the presence of insert DNA (*Hin*dIII/*Eco*RI digests will release the appropriate size band; 261 bp and 455 bp for fragments A and B respectively).

Sequencing of the synthetic fragments

The plasmids which contained inserts were then sequenced with 'Universal' and 'Reverse' M13/pUC primers by the Sanger dideoxy chain termination method. The double-stranded DNA was treated with sodium hydroxide/EDTA to denature the strands prior to annealing of the sequencing primer (Chen & Seeburg 1985). After checking the autoradiographs, the areas which could not be read were sequenced. The appropriate oligonucleotides which had been used to construct the gene were used as sequencing primers. Four clones were fully sequenced for fragment A. Of these, one was found to be correct. The other three each contained single base mutations (T → G at position 15, G → C at position 67 and A → G at position 111 respectively).

For fragment B, three clones were fully sequenced. All were correct except for a C → G mutation at position 454 which occurred in all three clones. As this mutation occurred 3′ to the gene coding sequence, it was decided to continue with one of these clones instead of sequencing further clones to obtain one without the mutation.

For each of the chosen clones, large-scale preparations of plasmid DNA were carried out by alkaline lysis of 500 ml cultures (containing 100 μg/ml carbenicillin). These were then purified by equilibration centrifugation in CsCl–ethidium bromide gradients (Sambrook *et al.* 1989)—the DNA being finally dissolved in 500 μl 1 × TE, and its concentration determined by measuring the absorbance at 260 nm.

Joining of fragments A and B to form the complete MCSF gene

One microgram of each of the two fragments was digested with the restriction enzymes *Pst*I and *Eco*RI and then run on a 1% low melting point agarose gel, together with a track containing φX174/*Hae*III marker DNA. The 442 bp band for fragment B was cut out and extracted, whereas for fragment A it was the 2893 bp vector band. Extractions were carried out by standard procedures (Sambrook *et al.* 1989) and the DNA was dissolved in 50 μl 1 × TE.

The 442 bp insert (containing fragment B) was ligated to the vector (containing fragment A) at 16°C overnight. This ligation mixture was then used to transform $CaCl_2$-treated *E. coli* strain HW87. From the resulting colonies, four were chosen to prepare plasmid DNA. A restriction digest with *Hin*dIII and *Eco*RI showed all four to contain the full length MCSF gene. One of these was then taken and plasmid DNA prepared by alkaline lysis of a 500 ml overnight culture (containing 100 μg/ml carbenicillin) followed by purification by equilibration centrifugation in a CsCl–ethidium bromide gradient.

Final confirmatory sequencing of the caesium-purified DNA was carried out by using primers for both strands (the oligonucleotides made for assembly of the gene were used as sequencing primers, together with the 'Universal' and 'Reverse' M13/pUC primers).

Discussion

The gene synthesis protocol described in this chapter has been successfully applied to the production of over 70 synthetic genes in the authors' laboratory. These have ranged from 150 bp up to some reaching 2 000 bp. Different assembly strategies, though, have been used by other groups—some of them designed to minimize the amount of oligonucleotide synthesis required. One such approach involves the use of DNA polymerase to fill in single-stranded regions created by annealing long oligonucleotides with complementary 3′ ends (Rossi *et al.* 1982). The synthesis of a gene for Eglin C has been successfully accomplished by this approach (Rink *et al.* 1984) though the method is limited because it is particularly prone to deletions and rearrangements caused by the polymerase step.

Another approach adapted from this was in the synthesis of the HIV transactivator protein TAT (Adams *et al.* 1988). Again long oligonucleotides were used, this time covering just one strand of the desired gene with small complementary adapter oligonucleotides to direct the assembly and cloning. After ligation to the cloning vector the partially single-stranded intermediate is transformed directly into the recipient host where the plasmid is repaired *in vivo*. This approach avoids the problems associated with the polymerase step because repair occurs after ligation to the vector. The use of longer oligonucleotides, however, increases the tendency for one base pair deletions to occur, as it becomes increasingly difficult to separate the full length oligonucleotides from those one base shorter during the purification step.

References

ADAMS, S.E., JOHNSON, I.D., BRADDOCK, M., KINGSMAN, A.J., KINGSMAN, S.M. & EDWARDS, R.M. 1988. Synthesis of a gene for the HIV transactivator protein TAT by a novel single stranded approach involving in vivo gap repair. *Nucleic Acids Research* **16**, 4287–4298.

CHEN, E.Y. & SEEBURG, P.H. 1985. Supercoil sequencing: A fast and simple method for sequencing plasmid DNA. *DNA* **4**, 165–170.

HOLMES, D.S. & QUIGLEY, M. 1981. A rapid boiling method for the preparation of bacterial plasmids. *Analytical Biochemistry* **114**, 193.

RINK, H., LIERSCH, M., SIEBER, P. & MEYER, F. 1984. A large fragment approach to DNA synthesis: Total synthesis of a gene for the protease inhibitor eglin C from the leech *Hirudo medicinalis* and its expression in *E. coli*. *Nucleic Acids Research* **12**, 6369–6387.

ROSSI, J.J., KIERZEK, R., HUANG, T., WALKER, P.A. & ITAKURA, K. 1982. An alternative method for synthesis of double-stranded DNA segments. *Journal of Biological Chemistry* **257**, 9226–9229.

SAMBROOK, J., FRITSCH, E.F. & MANIATIS, T. 1989. *Molecular Cloning: A Laboratory Manual.* Cold Spring Harbor, New York: Cold Spring Harbor Laboratory.

WONG, G.G., TEMPLE, P.A., LEARY, A.C., WITEK-GIANNOTTI, J.S., YANG, Y., CIARLETTA, A.B., CHUNG, M., MURTHA, P., KRIZ, R., KAUFMAN, R.J., FERENZ, C.R., SIBLEY, B.S., TURNER, K.J., HEWICK, R.M., CLARK, S.C., YANAI, N., YOKOTA, H., YAMADA, M., SAITO, M., MOTOYOSHI, K. & TAKAKU, F. 1987. Human CSF-1: Molecular cloning and expression of 4-kb cDNA encoding the human urinary protein. *Science* **235**, 1504–1508.

Monitoring Safety in Process Biotechnology

A.M. BENNETT, S.E. HILL, J.E. BENBOUGH AND P. HAMBLETON
Division of Biologics, Public Health Laboratory Service, Centre for Applied Microbiology and Research, Porton Down, Salisbury, Wiltshire SP4 0JG, UK

The recently implemented Control of Substances Hazardous to Health (COSHH) legislation stresses the need to carry out assessments on risks associated with hazardous substances, to control employees' exposure to such substances and to monitor the workplace if a problem exists (Health and Safety Executive 1988). The objective of this chapter is to provide a preliminary survey of the potential hazards associated with processes involved in biotechnology and to review methods of monitoring these hazards.

Generally, the biotechnology industry has an excellent health and safety record but incidents such as those detailed in Table 1 warn against complacency. All these incidents resulted from exposure to high airborne concentrations of 'non-pathogenic' micro-organisms or their products. It is important to realize that any micro-organism or any biological product is capable of producing allergic symptoms, including asthma and dermatitis, in workers exposed to them (Maroni *et al.* 1987). Lacey & Crook (1988) reported that in extrinsic allergic alveolitis, a chronic lung disease, acute symptoms occurred after exposure to concentrations of $10^8/m^3$ fungal or actinomycete spores but then could recur at exposure to lower concentrations of $10^6/m^3$. Gram-negative bacterial lipopolysaccharide seems capable of causing non-allergic respiratory health effects as well as kidney and stomach pains in exposed workers (Donham *et al.* 1989; Dunnil 1982). Other health problems found in the biochemical industry include shifts in the natural microbial flora of workers caused by antibiotic exposure resulting in increased incidence of candidiasis amongst workers in the antibiotic industry (Maroni *et al.* 1987), use of improperly identified pathogenic strains causing disease in screening of strains for production of enzymes (Schlech *et al.* 1981) and toxic effects caused by a specific pharmaceutical product (Poller *et al.* 1979).

The health risks associated with microbiological aerosols have been shown by Druett *et al.* (1953) to depend not only on the actual concentration of

0-632-02926-9

TABLE 1. *Outbreaks of serious illness in pilot to large scale biotechnology*

Incident	Organism	Process	Symptoms	Reference
UCL Biochemical Engineering Dept 1960s	*Pseudomonas aeruginosa*	Centrifugation of cell debris	Stomach and kidney pains. Influenza-like symptoms	Dunnil (1982)
Citric acid plant, Czechoslovakia	*Aspergillus Penicillium*	Open pan fermentation	40% incidence of bronchitis 23 retired on health grounds	Horejsi *et al.* (1960)
Citric acid plant, Britain	*Aspergillus niger*	Fermentation and aerosolization of contaminated washing water	4.9% incidence of occupational asthma	Topping *et al.* (1985)
Single-cell protein production, ICI	*Methylophilus methylotrophus*	Fermentation and finishing	Conjunctivitis and fever	Mayes (1982)
Single-cell protein production, Sweden	*Methylomonas methanolica*	Spray drier producing respirable dust	Conjunctivitis and fever	Ekenvall *et al.* (1983)
Single-cell protein production, USSR 1980s	Various yeasts	Fermentation and finishing	Bronchitis Dermatitis Asthma	Rimmington (1989)

micro-organisms but also on the proportion of the aerosol in smaller sized airborne particles. They showed that smaller sized airborne particles containing pathogenic micro-organisms pose a significantly greater threat because they can penetrate deeper into the human respiratory tract whereas larger particles are trapped in the upper respiratory system where they are normally less infectious. Therefore, to assess the hazard, airborne microbial samplers designed to determine the airborne concentration of micro-organisms are used in conjunction with samplers which can fractionate the aerosol into separate sizes.

Methods of Microbial Air Monitoring

The information obtained by using the samplers described here could give an idea of the total operator exposure to the organism and indicate where in the respiratory tract the organisms would be deposited. The samplers used should ideally cause minimal damage to the collected organisms and enable assay of the micro-organisms collected to be done with the minimum of sample manipulation. Unfortunately no air sampling device or technique fulfills all these functions, as is shown in Table 2. The three most commonly used types of air sampler are discussed below.

TABLE 2. *Advantages and disadvantages of commonly used air sampling devices*

Sampler	Flow rate (litre/min)	Advantages	Disadvantages
Impactors		*No sample manipulation*	*Overcrowded plates*
Casella slit	30–700	Efficient Easy to use	Static Bulky
Andersen	27	Particle size information	Difficult to assemble quickly
SAS Cherwell	180	Portable Easy to use Pointable	Inefficient at low particle size ranges
Impingers		*Ability to measure high concentrations*	*Dilution needed. Ineffective at low concentrations*
Porton	11	Easy to use	Evaporation
Three stage	55	Particle size information	Manufacturing difficulties
Other sampling methods			
Personal filter samplers	1–4	Measures high concentrations	Interference from filter material
Cyclone	750	Measures very wide range of concentration	
Settle plates		Very easy to use	Inefficient Qualitative

Impactors

Impaction samplers are devices which accelerate air through some form of flow constriction onto solid surfaces, such as agar plates, on which the particles containing the micro-organisms are deposited. These samplers measure colony forming units (cfu) derived from particles containing viable micro-organisms and not necessarily the true airborne concentration of microbes, i.e. if a particle containing 10 viable organisms impinges on the agar plate one colony and not 10 is formed. Two commonly used impactors are the Casella (Casella London Ltd) slit sampler which accelerates known volumes of air (30–700 litres/min) through a slit onto a rotating nutrient agar plate and the Andersen (Analysis Automation Ltd, Eynsham) sampler which can separate the particles of different size ranges on six separate agar plates. Other impactors are the Cherwell (Cherwell Laboratories Ltd, Bicester) Surface to Air Sampler

(SAS), Casella cascade impactor and the Biotest RCS (Biotest-Folex Ltd, Birmingham) sampler. In the cascade impactor, organisms are collected on glass slides covered in gelatin. The micro-organisms are subsequently washed off the slide by a suitable buffer and the microbial count is determined. In this case the true airborne concentration of micro-organisms is obtained.

Impingers

Impingers collect airborne particles by accelerating air into a fluid, normally phosphate buffer containing additives such as antifoam and anti-evaporation agents. The fluid is then diluted and plated out on agar to determine the airborne concentration of micro-organisms. The main types of impinger that are often used are the Porton raised impinger (May & Harper 1957), which operates at 11 litres/min, and the May three-stage impinger (May 1966) which separates the aerosol into particles of more than 6 μm, those between 3 and 6 μm and less than 3 μm. This impinger normally operates at a flow rate of 55 litres/min. Another type of impinger is the cyclone sampler which is often used to concentrate micro-organisms from large volumes of air. The cyclone sampler is an inexpensive sampler which can sample air at up to 750 litres/min into a small quantity of swirling buffer which can be readily removed from the cyclone and assayed without interrupting the sampling process.

Other sampling methods

Sampling devices with filters are often used to capture particles containing non-viable biological matter likely to be retained in the human respiratory tract and can be easily attached to workers' clothing. Settle plates are often used for monitoring air in pharmaceutical manufacturing units because they tend to be non-intrusive. Settle plates, however, are not generally useful for monitoring hazards because the method tends to select for very large aerosol particles which have been shown to be less hazardous (Muir 1973) and the technique is qualitative not quantitative.

Sampling strategy

Before any microbial aerosol monitoring is carried out it should be decided what sort of information is needed. It is important to consider the activities carried out by the facilities to be assessed before deciding on the monitoring procedures to be used. In facilities having large numbers of personnel involved in processing there will be fairly high levels (up to 300 cfu/m^3) of background organisms in the atmosphere. These organisms will mainly be Gram-positive organisms shed by the workforce. In addition there will be large and highly variable populations of other microbes (fungi or yeast) derived from vegetation and drawn in from the outside air. Hence it is necessary to discriminate between the production strain used in the monitored process and these large and variable numbers of background organisms. The most convenient way of doing this is to use a general nutrient agar plate for sampling, to limit the amount of stress caused by impingement or impaction, and then either replica plate or pick colonies onto a selective plate for positive identification. The background count can be minimized by filtration of the inlet air and careful design of the ventilation and by ensuring that staff wear specialized clothing. It is then likely that most of the airborne organisms would be the process organism.

In deciding where to position samplers, the need to determine concentrations of airborne microbes close to the breathing zone of any operators present should be considered. Also, for accurate estimation of the extent of the hazard, sampling methods which yield particle size information are needed. For the sensitive assay of biological products released into the air, preliminary concentration of the material in a large volume cyclone sampler or prolonged filtration is necessary. The sensitivity of such methods can be determined by incorporating both negative and positive controls.

Microbial Monitoring of a Bioprocessing Unit

A systematic programme of air monitoring of a bioprocessing plant was carried out with the principles described above. Since the monitored plant was a commercial enterprise, the microbial aerosol monitoring programme was constrained by the operations being carried out. Therefore, the data obtained might not be as detailed and controlled as would have been the case in a laboratory situation. For example, for many of the processes assessed only one sampler could be used due mainly to constraints of space. Therefore decisions had to be made on whether the priority was to record the time of emission from a process, to determine particles size distributions of the emissions, concentrations of product emitted or the amount of material to which the operator is exposed. The samplers used in these studies were the Casella slit sampler, Andersen multistage, Porton all glass and the Cherwell SAS samplers. The relative merits of these samplers are discussed below.

Casella slit sampler

The Casella slit sampler was operated at 30 litres/min. It was normally placed at operator head height approximately 1 m from the process being undertaken and was operated for periods of 0.5, 2 or 5 min. The collecting medium used was Tryptone Soya Broth Agar (TSBA) with identification of colonies carried out by replica plating onto selective media after 24 h incubation. This sampler has been shown to be 98% efficient for particles retained by the human respiratory system (Lach 1985).

Andersen sampler

The Andersen sampler is a six stage impactor that allows the particle size distribution of an aerosol within six ranges from less than 0.5 to more than 7 μm to be determined. This sampler was operated at 27 litres/min with a vacuum pump and a critical orifice. The incubated plates of TSBA were counted and, by using the statistical tables provided, the airborne microbial concentration was presented as total cfu and as a cumulative percentage particle size distribution.

Cherwell surface to air sampler (SAS)

The Cherwell SAS is basically an air pump which pulls air through a perforated plate and impacts micro-organisms on a 55 mm diameter RODAC plate. The sampler is relatively inefficient for collecting particles of less than 2 μm diameter (Lach 1985). However, it has the advantage of being portable, having a rechargeable powerpack, and can be easily carried and easily directed to the desired location.

Porton all glass impinger (AGI)

The AGI is a glass device in which particles in air are accelerated through a sonic nozzle and impinge in 10 ml of a collection fluid, normally phosphate buffer containing antifoam and anti-evaporants. This type of sampler has the disadvantage of subjecting the micro-organism to high levels of stress so it should only be used for hardy organisms or stable products. The advantage of this sampler is that it allows the dilution of samples and so is effective at high aerosol concentrations. In this study the sampler was used to measure product concentrations only.

Results and Discussion

The bioprocessing plant environment was monitored to assess the background microbial levels found in the production facility and then individual steps in the processes were monitored for microbial aerosol generation.

Background aerosol in the fermenter facility

Figure 1 shows the background level of micro-organisms found close to a fermenter situated in the main entrance to the production facility as determined

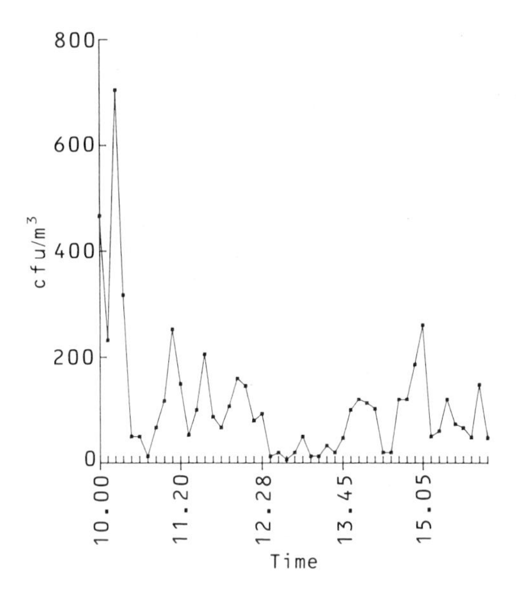

FIG. 1. Background microbial aerosol count, measured by the Casella slit sampler, at the fermenter.

with a Casella slit sampler. No production strain organisms were positively detected around the fermenter during operation. The fluctuations in airborne microbial concentration were thought to be due either to microbial shedding from operators or to deposited microbe-containing particles being re-aerosolized as a result of human activity. As can be seen from Fig. 1, the lowest levels of airborne micro-organisms were found at times of least human activity, i.e. during tea (1 030–1 100 h) and lunch breaks (1 230–1 330 h). The highest levels found were achieved when three workers were present to repair a leak in a stream line close to the fermenter at approximately 1 000 h. Curtailment of non-essential worker access to the main production plant reduced the background aerosol level before individual stages of the process were monitored.

Operation of fermenter sampling valves

Fermenters are basically pressure vessels and are subject to integrity tests. Therefore it is unlikely that aerosols are generated from these vessels during normal operation. The main dangers of release of aerosols occur when the physical integrity of the pressure vessel is, of necessity, breached at points to allow intrusions such as the stirrer shaft, outlet air discharge, probes and the sampling valve etc. To prevent emissions occurring, both the stirrer shaft and probes tend to have double seals to prevent them leaking and the exhausted air is usually filtered, at least in fermenters of under 1 000 litres capacity. The main potential for the generation of aerosolized production strain arises from the operation of the fermenter sampling valve. Cameron *et al.* (1987) have shown that some sampling valves do not produce aerosols if they are carefully designed and are maintained and used correctly but can do in the event of component failure or incorrect operation. This was assessed by carrying out microbial monitoring of the routine sampling procedures in a small bio-processing facility.

Two different sampling valves were tested for aerosol production. Valve 1 was a simple conventional sampling valve which was steamed prior to use. The sample was then taken and any excess of culture fluid or condensate was poured into a tub of disinfectant and the pipework was sterilized by steaming to atmosphere. During this operation the total airborne microbial level increased from 100 cfu/m^3 to between 330 and 500 before falling rapidly to the initial level. Selective media were not used in this case and therefore it was difficult to identify whether this increase represented release of production strain or was due to the operator shedding micro-organisms.

Valve 2 was operated in a similar way to valve 1 except that it was internally sterilized by steam so there was no steam leak to atmosphere. In this case there was no increase in background micro-organisms or process strain levels during operation. This suggests that the increase noted with valve 1 was due to a release from the sampling valve. However the increase in concentration was minimal and the aerosol dispersed so quickly that the operator was probably not exposed to significant levels of airborne micro-organisms.

Disc bowl centrifugation of fermentation broth

The disc bowl centrifuge was situated in a ventilated room (22 air changes each hour) operating under negative pressure with all operators wearing laboratory coats, gowns and rubber boots. The monitored procedure consisted of the continuous centrifugation of fermentation broth, followed by dismantling of the parts, and then by the removal, weighing and bagging of cell paste. This procedure was monitored with both the Casella slit sampler and the SAS sampler during two different process cycles. The Casella slit sampler was placed at head height approximately one metre away from the centrifuge while the SAS sampler was placed as close to the position of the operator as possible.

The results of monitoring, during operation, of the disc bowl centrifugation of a fermentation broth containing Product A are shown in Figs 2 and 3. Both samplers seem to give the same levels of airborne microbes even though the SAS (Fig. 2) was always closer to the aerosolization process than the Casella (Fig. 3). This may reflect difference in the efficiency of the

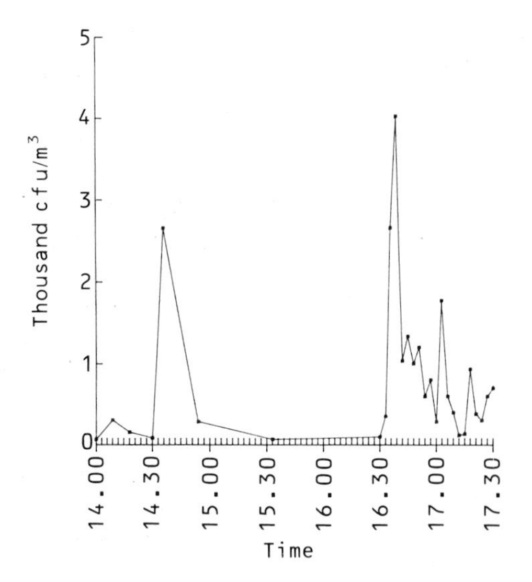

FIG. 2. Microbial aerosol produced by disc bowl centrifugation and subsequent removal of cell paste measured by the Casella slit sampler (Product A).

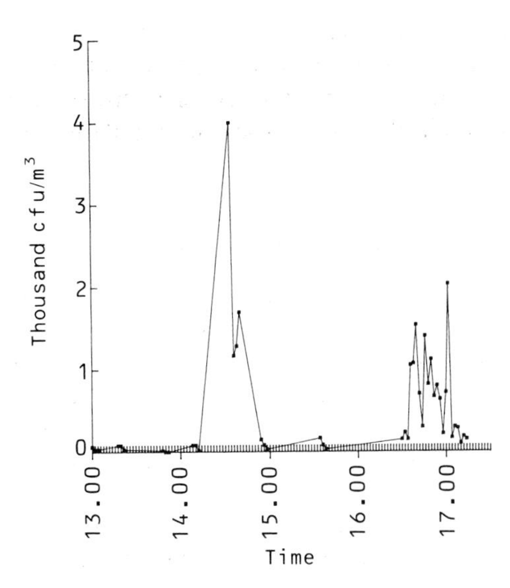

FIG. 3. Microbial aerosol generation by disc bowl centrifugation and subsequent removal of cell paste measured by the Cherwell SAS sampler (Product A).

samplers or may indicate that the air in the room was well and rapidly mixed. Significant amounts of microbial aerosol were produced during centrifugation (1400–1630 h). This was identified as being due to to a leak in the inlet tube to the disc bowl centrifuge. The airborne microbial concentration produced was shown to be in excess of 4 000 cfu/m^3 by the SAS pointed at the leak and 2 583 cfu by the Casella. During the cleaning of the centrifuge parts, bagging and weighing of the cell paste the airborne process microbial levels were found to be 2 034 cfu/m^3 by the SAS and 4 012 cfu/m^3 measured by the Casella. At the end of this process the inlet tubing which generated the aerosol was replaced.

Monitoring was carried out in the same way as described above during a second process involving two consecutive runs of the disc bowl centrifuge with product B fermentation broth. The results of these processes are shown in Figs 4 and 5. The results show firstly that no detectable leak occurred during centrifugation (1100–1245 h) confirming that the repairs carried out after the last centrifugation were effective. Secondly, the microbial aerosol levels detected during removal of cell paste were much higher than for the disc bowl centrifugation of product A (more than 22 000 cfu/m^3). The high levels were due to the cell paste being less viscous and of a higher loading than with product A. The lower viscosity means that less energy was needed for aerosolization of the microbes. The results show that the amount of aerosol produced is not only dependent on the machinery used but on the nature of the material processed. The data again suggest that the SAS sampler is less effective at sampling small particles.

From these data an assessment of the dose which operators receive can be made. If the average microbial aerosol concentration during the first cell paste removal between 1300 and 1430 h is taken as 6 830 cfu/m^3 over 1.5 h and the operators' breathing rate is estimated to be 1 m^3 per h and the micro-organism retention factor is taken as 0.5 (Muir 1973) then it can be estimated that more than 5 000 cfu of the production strain may have been retained in their lungs.

Homogenization and subsequent centrifugation

A Casella slit sampler was used to monitor these processes (Fig. 6). Low levels of airborne micro-organisms were produced during homogenization and centrifugation (between 0000 and 0300 h). When cell debris was removed the levels of airborne micro-organisms increased to a maximum of 3 000 cfu/m^3 suggesting that a significant fraction of the microbiological population was not inactivated by homogenization. The processes steps were also monitored for aerosols of product using the AGI sampler with distilled water as the collecting fluid. Collecting fluid was assayed by a product-specific

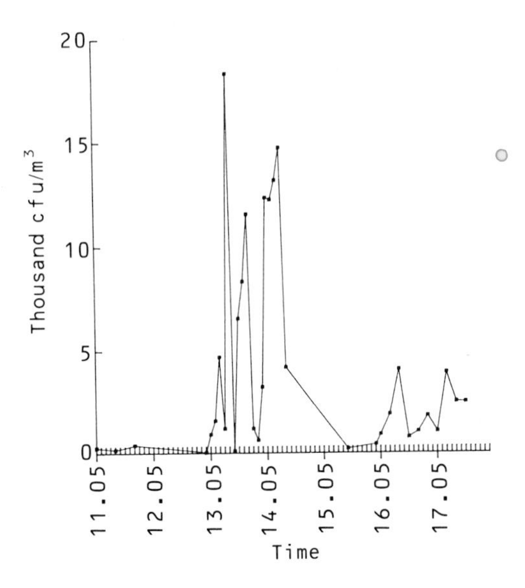

FIG. 4. Microbial aerosol generation by disc bowl centrifugation and subsequent removal of cell paste measured by the Casella slit sampler (Product B).

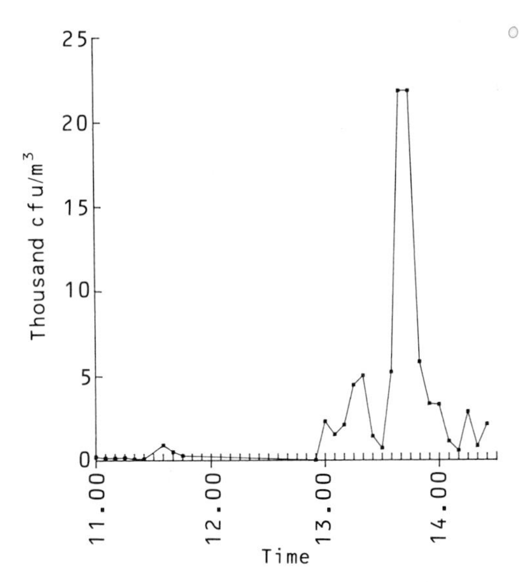

FIG. 5. Microbial aerosol generation by disc bowl centrifugation and subsequent removal of cell paste measured by the Cherwell SAS sampler (Product B).

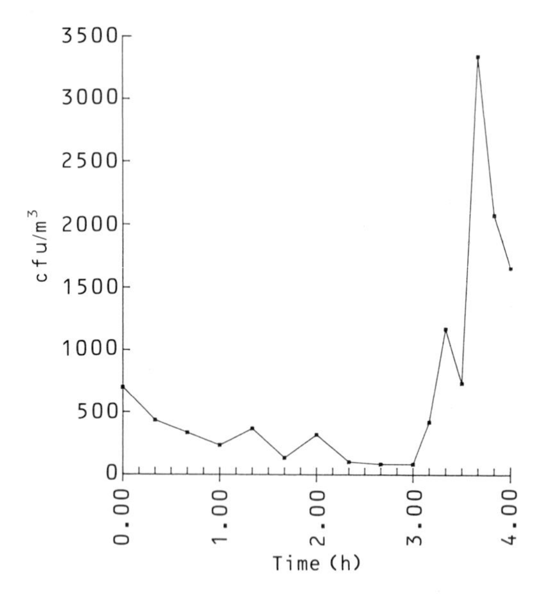

FIG. 6. Microbial aerosol generation during homogenization and the subsequent centrifugation of the homogenate measured by the Casella slit sampler.

radio-immunoassay. This showed that the airborne concentration of product was in the range of 30 to 70 ng/m^3 during the process.

Tubular bowl centrifugation of fermentation broth

The centrifugation of the fermentation broth containing a product C took place in a room where the exhaust air was filtered and was operating at negative pressure. This room was unoccupied during the centrifugation (except for the sampler operator) but an operator was present during the cell paste removal. During the centrifugation step no smell of the micro-organism or the broth was noted but during cell paste removal a pungent smell was present. The concentration of airborne microbes released by this process was monitored by the particle sizing Andersen sampler.

The results of this monitoring are shown in Fig. 7 and the cumulative percentage particle size distribution is shown in Fig. 8. During centrifugation (between 0000 and 0320 h), micro-organisms were aerosolized in concentrations of 50 000–90 000 cfu/m^3. Figure 8 shows that 90% of these particles were of particle diameter of below 3 μm, because they were collected in the first three stages of the Andersen sampler. The microbial levels detected during cell paste removal was much lower, at less than 1 000/m^3, and showed

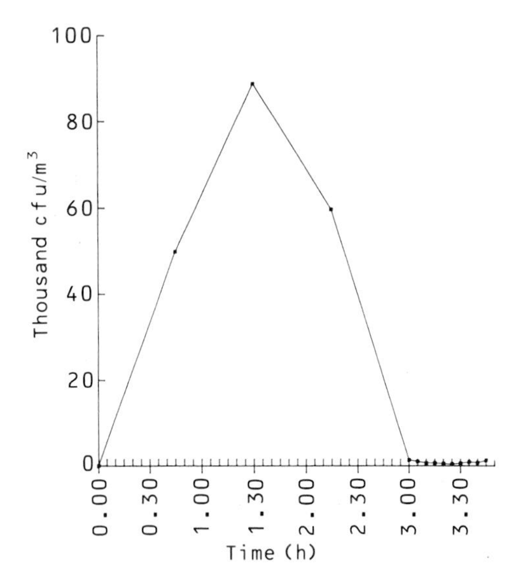

FIG. 7. Microbial aerosol generation during tubular bowl centrifugation and the subsequent removal of the cell paste measured by the Andersen sampler.

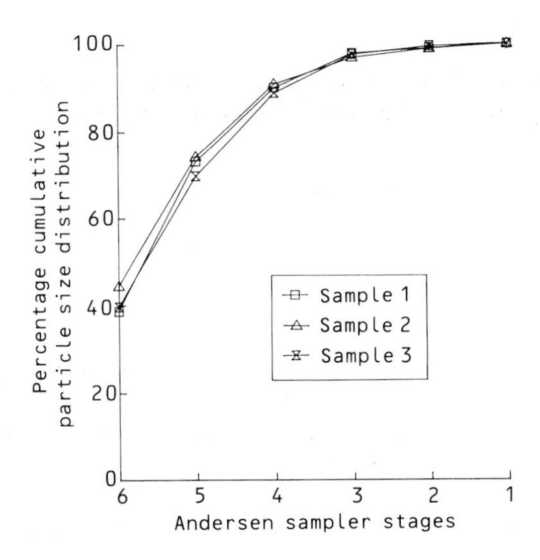

FIG. 8. Percentage cumulative particle size distribution of the microbial aerosol generated by the tubular bowl centrifuge.

no consistent particle size distribution. The lower aerosol production during removal of cell paste from the tubular bowl compared to that after disc bowl centrifugation reflected the high viscosity of the cell paste and the fact that weighing and bagging of the product was not involved in the former case. The smell detected during cell paste removal and not during centrifugation is probably related to the fact that only larger size particles (5–20 μm) are deposited in the nose (Muir 1973). Particles of this size were probably more prevalent during cell paste removal than during centrifugation.

Conclusions

This work was carried out to assess microbial aerosol levels produced in a normal pilot plant scale bioprocessing facility and also to determine the efficacy of a variety of sampling devices for process monitoring in biotechnology. The study was limited to microbial monitoring although future work will investigate the levels of product and endotoxin in the atmosphere of biotechnology plants during operations. It has to be stressed that any work carried out to monitor an industrial process generally has to take second place to the process itself, so that, except in extreme circumstances, completely controlled experimental work is impossible. The sampler(s) need to be positioned in such a way as allow repeated removal or replacement of samples without interfering with the process or with the operators. It is also important to ensure that monitored processes are being carried out in the normal way, i.e. the operators are not creating or preventing aerosol production for the monitor's benefit.

The results presented here show that the levels of aerosol produced by exhaust filtered fermenters with well designed sampling valves are negligible under normal circumstances. In the event of accident, however, it has been shown that sizeable aerosols can be formed (Ashcroft & Pomeroy 1983) and so it is important to have emergency plans in case of this circumstance. The normal operation of the disc bowl centrifuge also produced minimal levels of microbial aerosol except when there was an integrity failure. Thus, if such process equipment is carefully maintained and operated correctly, it is unlikely that significant microbial aerosol will be generated.

This study shows clearly, however, that care should be exercised during the handling of biomass paste recovered from the disc bowl centrifuge due to the poor dewatering associated with this device (Schmidt-Kastner & Golker 1987). If the cell paste produced had a lower water content, then aerosol generation would be minimized. Tubular bowl centrifuges should be contained if possible since they have been known to cause aerosolization of micro-organisms resulting in health problems in biotechnology workers (Gibson 1972; Dunnil 1982).

All of the samplers used in this study have advantages and disadvantages as monitors of biotechnology processes. The type of sampling device most likely to find favour for use in an industrial situation will be that which is easiest to use, involving the minimum of sample manipulation and does not take up too much space. In this context the SAS sampler is probably the simplest and most manoeuverable of the samplers used as it is compact, mobile and easily positioned. It is, however, inefficient in monitoring particles of the sizes most likely to be retained in the lung. The Casella slit sampler is less portable but is a more efficient device for collecting these particles and is the preferred sampler for this purpose.

Acknowledgements

This work was undertaken for the Department of Trade and Industry, Industrial Biosafety Project, Warren Spring Laboratory, Stevenage. We would also like to thank the management and production staff at the monitored facility for their help and patience.

References

ASHCROFT, J. & POMEROY, N.P. 1983. The generation of aerosols which may occur during plant scale production of microorganisms. *Journal of Hygiene* **91**, 81–91.

CAMERON, R., HAMBLETON, P. & MELLING, J. 1987. Assessing microbiological safety of bioprocessing equipment. In *Proceedings of the Fourth European Congress on Biotechnology* **1**, 139–142.

DONHAM, K., HAGLIND, P., PETERSEN, Y., RYLANDER, R. & BELIN, L. 1989. Environmental and health studies of farm workers in Swedish swine confinement buildings. *British Journal of Industrial Medicine* **46**, 31–37.

DRUETT, H.A., HENDERSON, D.W., PACKMAN, L. & PEACOCK, S. 1953. Studies on respiratory infection. I. The influence of particle size on respiratory infection with anthrax spores. *Journal of Hygiene* **51**, 359–371.

DUNNIL, P. 1982. Biosafety in the large scale isolation of intracellular microbial enzymes. *Chemistry and Industry* **22**, 877–879.

EKENVALL, L., DOLLING, B., GOTHE, C-J., EBBINGHAUS, L., VON STEDINGK, L-V. & WASSERMAN, J.M. 1983. Single cell protein as an occupational hazard. *British Journal of Industrial Medicine* **40**, 212–215.

GIBSON, D.E. 1972. *Hygiene and Safety in Biotechnology: A Case History At University College.* M.Sc. Thesis, London School of Hygiene and Tropical Medicine.

HEALTH AND SAFETY EXECUTIVE 1988. *Control of Substances Hazardous to Health.* London: Health and Safety Executive.

HOREJSI, M., SACH, J., TOMASIKOVA, A., MECL, A., BLAHNIKOVA, D., TUMOVA, M. & VALISOVA, A. 1960. A syndrome resembling farmers lung in workers inhaling spores of *Aspergillus* and *Penicillia* moulds. *Thorax* **15**, 212–217.

LACH, V.H. 1985. Performance of the surface air system sampler. *Journal of Hospital Infection* **6**, 102–107.

LACEY, J. & CROOK, B. 1988. Fungal and actinomycete spores as pollutants of the workplace and occupational allergens. *Annals of Occupational Hygiene* **32**, 515–533.

MARONI, M., COLOMBI, A., ALCINI, D. & FOA, V. 1987. Health risks in the biotechnology industry. *Medicina del Lavorna* **78**, 272–282.

MAY, K.R. 1966. Multistage liquid impinger. *Bacteriological Reviews* **14**, 37–43.

MAY, K.R. & HARPER, G.J. 1957. The efficiency of various liquid impinger samplers for sampling biological aerosols. *British Journal of Industrial Medicine* **10**, 142–151.

MAYES, R.W. 1982. Lack of allergic reaction in workers exposed to Pruteen (bacterial single-cell protein). *British Journal of Industrial Medicine* **39**, 183–186.

MUIR, D.C.F. 1973. *Airborne Allergens in Clinical Aspects of Inhaled Particles*. ed. Muir D.C.F. London: William Heinemann Medical Books Ltd.

POLLER, L., THOMSON, J.M., OTTERIDGE, B.W., YEE, K.F. & LOGAN, S.H.M. 1979. Effects of manufacturing oral contraceptives on blood clotting. *British Medical Journal* **1**, 1761–1762.

RIMMINGTON, A. 1989. *The Release of Microorganisms and Other Pollutants From Soviet Microbiological Facilities. The Political and Environment Fallout*. University of Birmingham Report.

SCHLECH, W.F., TURCHIK, J.B., WESTLAKE, R.E., KLEIN, G.C., BAND, J.D. & WEAVER, R.E. 1981. Laboratory-acquired infection with *Pseudomonas pseudomallei* (melidiosis). *New England Journal of Medicine* **305**, 1133–1135.

SCHMIDT-KASTNER, G. & GOLKER, C.F. 1987. Downstream processing in biotechnology. In *Basic Biotechnology* ed. Bu'Lock, J. & Kristiansen, B. pp. 173–196. London: Academic Press.

TOPPING, M.D., SCARISBRICK, D.A., LUCZYNSKA, C.M., CLARKE, E.C. & SEATON, A. 1985. Clinical and immunological reaction to *Aspergillus niger* among workers at a biotechnological plant. *British Journal of Industrial Medicine* **42**, 312–318.

Patenting in Biotechnology

D.L. Wood
The Patent Office, State House, 66–71 High Holborn, London WCIR 4TP, UK

It is probably true to say that the average research worker has a very limited knowledge of the whole subject of patents. This is partly due, particularly in the field of biotechnology, to the need to keep up to date with a rapid expansion of knowledge in their own subject matter area but also because the whole subject is perceived as being so complex as to be best left to those lawyers whose job it is to solve all the problems associated with the relevant legislation.

In the UK that legislation is primarily the Patents Act 1977 or, alternatively, the European Patent Convention which came into force in 1978 and since then has been of increasing importance. What follows is an attempt to give an overview of the essential features of the patent system as it applies to the UK. Since it used to be thought that 'life' could not be patented, a particular emphasis will be placed on those features as they apply to inventions in the field of biotechnology. In a world where research departments must increasingly look to gaining some economic return from their work it is hoped that some of the mystique will be removed from patents and an understanding of their importance in the process of bringing a product to the market place will be achieved.

Fundamentals of the Patent System

A patent is a form of intellectual property granted by the state and serves to prevent others from using or benefiting from the patented invention without permission from the patentee. Thus it confers a monopoly which, in the UK and most European countries, is limited to a period of 20 years from the date the application is made provided that annual renewal fees are paid.

Briefly, to obtain a patent, an application is filed with the Patent Office in each country where protection is required and after search and examination by the Patent Office Examiner the application may be refused or accepted. The search serves to provide information about all prior published inventions

Genetic Manipulation
0-632-02926-9

in the same area as that of the subject of the application and may indicate to the applicant that it is not worth pursuing the application or it may become a focal point of negotiations between the applicant (or agent) in the subsequent examination stage.

The significant advantage of the European Patent system is that a single application filed with the European Patent Office will serve as the appropriate application for all the European countries the applicant chooses to designate. Search and examination of this single application will result, if all the requirements are met, in a bundle of patents being granted, the number in the bundle corresponding to the number of countries designated.

Before a patent can be granted, the Patent Office Examiner must pay attention, in the main, to four basic requirements. Three of these concern the invention to which the application relates as, to be patentable, the invention must be (a) novel, (b) not obvious to a person skilled in the particular art and (c) applicable to industry. The fourth requirement concerns the patent specification filed as the major part of the application. In short the specification must present the invention with sufficient clarity and detail that those with ordinary experience in the relevant field following all the instructions will be enabled to achieve the promised results.

Once a patent has been granted it is still open to an interested third party to commence opposition proceedings if they feel that the grant should not have taken place because of, e.g. the existence of prior art undiscovered by the Patent Office Examiner during the search.

The four basic requirements referred to above will now be looked at in greater detail. Where appropriate, examples from the field of biotechnology will be used to illustrate a particular point, but it will still be necessary at a later point to look at some of the problems thrown up by the fact that in biotechnology we are dealing with living materials rather than inanimate matter.

Novelty

Under this requirement the invention must not have already been available to others by any form of public disclosure prior to the filing date of the application. The disclosure will most commonly be in the form of a publication in another patent or a journal reference but also extends to public use of the invention before the filing date. Prior experimental use within the confines of a research laboratory is not regarded as a disclosure so long as the details have been of restricted availability. At present, within the UK and the European system, prior disclosure of an invention by the inventor him or herself in, say, a learned journal before filing a patent application will mean that it will be impossible to secure the grant of a patent on that invention. Thus it is a cardinal rule to tell nobody about an invention until a patent application has been filed. Thereafter details of the invention can be made available at a scientific conference, in a learned journal or to those who might provide financial support for exploitation of the invention in the knowledge that none of this disclosure will ultimately affect the validity of the granted patent.

Whether something is regarded as novel or new will depend on how that thing, be it a product, process or apparatus, is defined in the patent claims. It is the function of claims to define the scope of the legal protection sought and as such they must set out the essential features of that product, process or apparatus in technical terms without reference to the advantages of the invention or the manner of carrying it out as these will occur in the descriptive part of the specification. Therefore to destroy the novelty of a claim it will be necessary to show that something disclosed in the prior art has all the same features in the same combination as those defined in the claim. Any difference, however small, will not destroy the novelty of an invention, but a small difference might well have relevance when the matter of inventive step is considered.

It follows from what has been said above that a prior disclosure of a product is sufficient to destroy the novelty of the same product even if it is determined to put that product to a different use than that originally disclosed. Where a product has biological utility, however, there is an exception to this general rule. Two situations generally occur: (a) a compound may be known in the chemical literature but for the first time a biological utility is discovered for that compound; or (b) a compound already known to have a biological utility is found to have another different biological utility. In both these situations 'purpose limited' claims of specific wording are allowed so as to provide a monopoly over what is already known about the compound.

Obviousness (inventive step)

The determination of novelty involves a precise comparison between what the inventor claims to have invented and all inventions in the same field in the prior art. No such precision exists when the matter of obviousness or lack of inventive step is to be determined. Under UK law an invention involves an inventive step if it is not obvious to a person skilled in the art in the light of the knowledge which forms the state of the art. Because of this lack of precision most Patent Offices have provided guidelines for the examiner to take account of in their determination of whether an invention is obvious or not. The European guidelines define 'obvious' in the following way:

> 'The term 'obvious' means that which does not go beyond the normal progress of technology but merely follows plainly or logically

> from the prior art, i.e. something which does not involve the exercise of any skill or ability beyond that to be expected of the person skilled in the art.'

Clearly then it is essential that a 'real-life' assessment is made in this area because once a new idea has been formulated it is all too easy to show theoretically how it might have been arrived at, starting from something known, by a series of apparently easy steps.

At the time of writing it has to be said that there are many unsolved problems relating to the issue of obviousness in respect of patents in the field of biotechnology. Many of these problems are associated with the perceived complexity of the technology, and the ability of Patent Offices and the Courts to cope in this area is being severely stretched. Compared to any other subject matter area it is probably more difficult to determine what 'follows plainly or logically from the prior art'. Thus, the danger exists, on the one extreme, of seeing very little inventive merit in any biotechnological invention because the prior art, though complex, describes well understood techniques and, on the other extreme, of seeing inventive merit in almost any invention since the art is so complex, and thus to grant patents which are for no more than routine developments.

The difficulties associated with assessing obviousness are perhaps best illustrated at present by the recent judgment delivered by the Court of Appeal in the case of Genentech versus Wellcome. The patentees, Genentech, had been granted a patent (UK 2119804) relating to human tissue plasminogen activator (TPA), a natural protein occurring in the human body, but in their case produced by recombinant DNA technology. Without going into detail, what Genentech were essentially claiming was human TPA prepared by any recombinant DNA method, a claim which they felt to be justified because they were the first to prepare such a product. In the Patents Court, Genentech's claims were rejected chiefly along the lines that TPA was a known protein and Genentech had only given limited guidance as to how to prepare it by recombinant means. Thus it was the breadth of Genentech's claims that were their downfall. In the Court of Appeal the legal argument against Genentech's claims took a different turn, two of the judges deciding that what Genentech had done, whilst being an excellent piece of work involving a combination of tenacity, skill and managerial efficiency, did not involve an inventive step. The third judge thought otherwise but nevertheless revoked the patent for different reasons. What this decision has done is to show that work of the type undertaken by Genentech should not be rewarded by a patent and obviously it will have a serious impact upon the biotechnology industry if followed elsewhere.

Before leaving the area of obviousness it is perhaps instructive to look also at the field of monoclonal antibodies. The basic Milstein/Köhler technique is so well known that it is exceedingly difficult to obtain patent protection for antibodies against well-established antigens. In most cases, having decided on the antigen, the production of the corresponding monoclonal antibody follows by routine application of the basic technique. Only if the basic technique had to be specifically adapted or if, say, some of the produced hybridomas were shown to have unexpected advantages might there be the possibility of patent protection in this area. Of course, if the antigen against which the antibody was raised had not previously been recognized, patent protection would again be possible.

Industrial applicability

In general, this criteria for patentability throws up fewer problems than the areas of novelty and inventive step previously discussed. The basic test is whether the invention is capable of being made or used in any kind of industry: 'industry', being interpreted in a broad sense as including any physical activity of a technical character rather than an activity which belongs to the aesthetic arts.

It is sufficient to point out in an article such as this that, both under UK and European law, methods for treatment of the human or animal body by surgery, therapy or diagnosis are not regarded as inventions which are capable of industrial application. Products such as substances, compositions or apparatus for use in such methods may, of course, be patented as long as the other patentability criteria are met.

Clarity and sufficiency of the description

Whilst to relate to a patentable invention, that invention must show novelty, inventive step and industrial applicability a patent will still not be granted if the specification describing it does not do so in terms that are clear, so as to be understood by a man skilled in the art, and sufficient so that he has all the details necessary to reproduce the work carried out by the inventor. A specification that contains insufficient detail will be fatally flawed for it is not possible to add new matter to an application once it has been filed.

There are obvious problems here for patents in the area of biotechnology for very often an invention will require the use of, for example, a microorganism, a cell line or plasmid which is incapable of being described with adequate precision. Thus the man skilled in the art faced only with the shorthand name of the organism, is unable to reproduce what the inventor has shown he has achieved.

The problem has been solved by depositing new strains in recognized culture collections from which they can be made available to those wishing to carry out a process involving their use. Thus, under UK law, when the invention involves the use of a new strain it is necessary for the patent specification to disclose the date and place of deposit together with the accession number and as much relevant information as is available to the applicant on the characteristics of the strain. This requirement obviously only applies to organisms not already available to the public and to those which cannot be described with precision by means of a verbal description. In cases of doubt it is well worth making the required deposit rather than to run the risk of providing an insufficient description. Whether a description is considered sufficient or not also depends on the breadth of the monopoly that is sought by the inventor. The wider the monopoly the more incumbent upon the inventor it is to provide a sufficient description to show that the invention will work across the range contemplated by the patent claims.

Taking a simple example, it may be that an inventor has carried out a recombinant DNA process in which a culture of *E. coli* has been transformed but which, in their opinion, could equally well be performed with other host cell cultures. In these circumstances the resulting patent claim may be drafted so as to relate to the use of any host cell culture and may well be a valid claim if there is no good reason to suppose that the inventor was wrong in his thinking. However, if there were clear problems in moving to the use of other cultures and, for example, it could be established that the process would not work with a mammalian cell line, then clearly the inventor has been too greedy and we are dealing with an invalid claim. In an ideal world the inventor would provide a number of examples to support the breadth of a wide claim but often, because of time and financial constraints, a compromise has to be struck and careful thought has to be given to the exemplification necessarily commensurate with the monopoly sought.

So much for the basic legal requirements necessary before any patent can be granted. What then is patentable in the field of biotechnology?

Biotechnological Inventions

As has been hinted at already, the problem for the patent system when the new art of biotechnology burst onto the scene was that we were dealing with 'living' material and previously it had been thought that such material could not be patented. The 1977 UK Patents Act came into force at just about the time inventions in biotechnology began to gather in momentum so even this Act was not able to take into account the problems that this art was to throw up. In many ways the problems surrounding the art have been solved by reference to what was already considered patentable in biotechnology prior to 1977, i.e. in those areas involving fermentation techniques, and to the whole area of chemical inventions where case law is well developed.

As well as making this kind of extrapolation, regard also had to be had to one particular subsection of the 1977 Act which is of prime importance in deciding what is or is not patentable. Subsection 1(3)(b) is worded as follows:

> 'A patent shall not be granted—for any variety of animal or plant or any essentially biological process for the production of animals or plants, not being a microbiological process or the product of such a process.'

Clearly then the first part of this subsection does not allow the patenting of plant and animal varieties, or the biological processes for preparing them, and we shall look at this area later, but it appears from the second part that what are termed 'microbiological processes' and their products may well be valid subjects for patent protection. In general this has been taken to be the case, with 'microbiological process' being interpreted as being any process involving the use of production of a micro-organism and 'micro-organism' being interpreted as broadly as possible so as to include not only bacteria and fungi but also other replicable material such as viruses, plasmids and cell lines. It will be my purpose in what follows to set out what may be patented in certain well-defined areas without necessarily going into detail as to how such products or processes should be characterized for the purpose of a patent claim. The latter often involves complex arguments which are best left to patent practitioners to resolve.

Patents involving micro-organisms

To all intents and purposes it is possible to distinguish six patentable areas under this section as follows:

1 A process of producing a new micro-organism.

2 The new micro-organism produced by the process of 1.

3 The new micro-organism *per se* (i.e. without reference to the means of its production).

4 A process of preparing an end product by cultivating or otherwise using a new or known micro-organism.

5 The end products prepared by the process of 4. If the end product is new then a *per se* claim is permissible otherwise a product-by-process claim would be appropriate.

6 A formulation relating to a particular strain or culture thereof.

A particular problem associated with the above has centred on whether a claim to a new micro-organism *per se* is in any way allowable. To the extent that technical intervention is involved in the production of the micro-organism it is generally accepted that such a claim is permissible, but the problem is really acute in the instance that the micro-organism, or for that matter a plasmid or gene, is a product that exists already in nature. This is because most patent laws exclude from patentability anything which consists of a discovery as opposed to being an invention. The argument therefore goes that the isolation of a product existing in nature is mere discovery and patent protection ought not to extend to that product. In another sense it is also argued that because of its pre-existence the product cannot be regarded as being new. If, however, the product had no previously recognized existence, and can be adequately identified without reference to the process by which it is obtained, then it may be patentable *per se*.

Patents involving proteins

In many ways it is possible to argue that a protein is no different to any other polymer which may be claimed by reference to, say, its average molecular weight and repeating structural unit. This is, however, rarely an adequate or even a viable definition, particularly when the protein has biological activity and is in the form of an enzyme, antigen or antibody. A convenient way of defining a biologically-active protein is therefore in terms of its biological properties, most particularly by reference to its activity, specificity on a substrate and optimum and stable pH range.

Patents involving genes and gene products

What may be patented in this area is still the subject of considerable debate particularly as it can be argued that a gene may be regarded as a product of nature with a certain definable biological property. For this reason a broad claim of the type 'A nucleotide sequence coding for polypeptide *x*' would not be allowed where *x* is known, but would probably be allowed where *x* is new. Specifying the base sequence of the nucleotide or the amino acid sequence of the resulting polypeptide would make the claim more acceptable.

Patents involving vectors

Plasmids are the most common form of vector used in biotechnology and may of course be naturally occurring, in the sense that they are contained in a micro-organism, or prepared by recombinant means. In patent claims they may be defined:

1 By reference to their code name together with a restriction map.
2 In terms of the DNA segments making up the whole.
3 In terms of the added DNA, i.e. the gene insert or its expression product.

Patents involving hybridomas

As has previously been pointed out, the knowledge of the basic Milstein/Köhler technique has made it possible to produce thousands of hybridomas which may be novel but can hardly be said to involve an inventive step. Any modification of the basic technique involving an unexpected advantage will, however, probably result in a patentable invention. Thus the use of a novel parent myeloma cell line may lead to patentability, as may the use of the resulting monoclonal antibodies in therapy or diagnosis.

Plant and animal varieties

It will be recalled that under Section 1(3)(b) of the Patents Act 1977 it is not possible to protect plant and animal varieties or, essentially, biological processes for producing them. In the case of plant varieties this exclusion from patentability is justified by the existence of plant variety rights legislation, the intention being that it should not be possible to gain protection via two different systems of enacted legislation. In the case of animal varieties the reason for the exclusion is not quite so clear since there is not an alternative system and the whole issue, in any case, is clouded by the uncertainty surrounding the definition of 'animal variety'.

Although the plant variety right will cover the new variety it does not extend to the process of preparing that variety and so where this process is of a technical rather than a biological nature it ought to be protectable under the patent system. This is particularly true in the case where a foreign gene is inserted into a plant cell so as to confer a desirable property on the resulting plant although there is considerable debate over whether claims to the plant *per se* should be covered by patents or plant variety rights. A commonly held position is that patents should be allowed for new plants produced by a process involving technical intervention provided the claims are not directed to specific varieties.

The situation concerning the protection of animals is interesting if only because the US Patent Office has recently granted patents for transgenic animals, the most notable being the transgenic mouse reported by most of the British press. It has been argued that Section 1(3)(b) of the 1977 Act excludes the protection of animals or, if it does not, that animal patents are contrary to morality, but it will probably be necessary to get a case through the Courts to determine the law on this matter.

Before closing this chapter it is worth spending some time describing the procedure related to the obtaining of a patent since this whole area is often a mystery to the uninitiated. What follows is a brief description of the practice before the UK Patent Office. Practice before the European Patent Office is similar but there are significant differences.

Obtaining a Patent

The whole process is commenced by filing at the Patent Office an application together with a fee which is currently £15 (1990 price). The application at this stage need be no more than a *full* description of the invention. Patent claims may be filed at the same time or within one year of the filing date of the basic application. Once the claims have been filed and a fee of £95 (1990 price) for preliminary examination paid, the application will be referred to a Patent Office Examiner for a search to be performed. This search will cover inventions in the same area as that of the invention in question and will be conducted through Patent Office files and extended through on-line databases to provide effectively a world-wide search covering patents and journal references.

On receipt of the search report the applicant then has the opportunity to withdraw the application if the results appear to be unfavourable or, within 6 months of publication of the application, to request that the application proceed to substantive examination. At 18 months after the application date, unless it has been previously withdrawn, the application will be published together with the results of the search. Soon after the fee for substantive examination (£110—1990 price) has been paid the application will again be referred to an examiner for full examination particularly in the four areas referred to above. In determining patentability, the results of the search report will be taken into account. Communication between the examiner and the applicant, or the patent agent, then takes place until, in the majority of cases, the application is regarded as being in order for grant. In cases where agreement cannot be reached the applicant may take advantage of a hearing before a senior Patent Office official who may uphold the examiner's decision or order that a patent should be granted. If the hearing goes against the applicant he may take the matter to the Patents Court and ultimately right up to the House of Lords. Once a patent has been granted, the payment of annual renewal fees will keep it in force up to a maximum of 20 years from the date of filing the application.

Conclusions

From what has been said it should be apparent that, with few exceptions, there is no bar to the patenting of the products of biotechnological processes, the processes themselves and apparatus used in those processes provided that the normal criteria for patentability, i.e. novelty, inventive step and industrial applicability, are met. If there is uncertainty this generally arises because there are as yet few precedent cases to establish what is the law in this complex area. It is to be hoped that as precedents emerge, inventors will be encouraged to protect their inventions by means of patents to their own benefit and that of the country as a whole.

Index

THE SOCIETY FOR APPLIED BACTERIOLOGY TECHNICAL SERIES

General Editor: F.A. Skinner

1 Identification Methods for Microbiologists Part A *1966*
Edited by B.M. Gibbs and F.A. Skinner
Out of print

2 Identification Methods for Microbiologists Part B *1968*
Edited by B.M. Gibbs and D.A. Shapton
Out of print

3 Isolation Methods for Microbiologists *1969*
Edited by D.A. Shapton and G.W. Gould

4 Automation, Mechanization and Data Handling in Microbiology *1970*
Edited by Ann Baillie and R.J. Gilbert

5 Isolation of Anaerobes *1971*
Edited by D.A. Shapton and R.G. Board

6 Safety in Microbiology *1972*
Edited by D.A. Shapton and R.G. Board

7 Sampling—Microbiological Monitoring of Environments *1973*
Edited by R.G. Board and D.W. Lovelock

8 Some Methods for Microbiological Assay *1975*
Edited by R.G. Board and D.W. Lovelock

9 Microbial Aspects of the Deterioration of Materials *1975*
Edited by R.J. Gilbert and D.W. Lovelock

10 Microbial Ultrastructure: The Use of the Electron Microscope *1976*
Edited by R. Fuller and D.W. Lovelock

11 Techniques for the Study of Mixed Populations *1978*
Edited by D.W. Lovelock and R. Davies

12 Plant Pathogens *1979*
Edited by D.W. Lovelock

13 Cold Tolerant Microbes in Spoilage and the Environment *1979*
Edited by A.D. Russell and R. Fuller

14 Identification Methods for Microbiologists (2nd Edn) *1979*
Edited by F.A. Skinner and D.W. Lovelock

15 Microbial Growth and Survival in Extremes of Environment *1980*
Edited by G.W. Gould and Janet E.L. Corry

16 Disinfectants: Their Use and Evaluation of Effectiveness *1981*
Edited by C.H. Collins, M.C. Allwood, Sally F. Bloomfield and A. Fox

17 Isolation and Identification Methods for Food Poisoning Organisms *1982*
Edited by Janet E.L. Corry, Diane Roberts and F.A. Skinner

18 Antibiotics: Assessment of Antimicrobial Activity and Resistance *1983*
Edited by A.D. Rusell and L.B. Quesnel

19 Microbiological Methods for Environmental Biotechnology *1984*
Edited by J.M. Grainger and J.M. Lynch

20 Chemical Methods in Bacterial Systematics *1985*
Edited by M. Goodfellow and D.E. Minnikin

21 Isolation and Identification of Micro-organisms of Medical and Veterinary Importance *1985*
Edited by C.H. Collins and J.M. Grange

22 Preservatives in the Food, Pharmaceutical and Environmental Industries *1987*
Edited by R.G. Board, M.C. Allwood and J.G. Banks

23 Industrial Microbiological Testing *1987*
Edited by J.W. Hopton and E.C. Hill

24 Immunological Techniques in Microbiology *1987*
Edited by J.M. Grange, A. Fox and N.L. Morgan

25 Rapid Microbiological Methods for Foods, Beverages and Pharmaceuticals *1989*
Edited by C.J. Stannard, S.B. Petitt and F.A. Skinner

26 ATP Luminescence: Rapid Methods in Microbiology *1989*
Edited by P.E. Stanley, B.J. McCarthy and R. Smither

27 Mechanisms of Action of Chemical Biocides *1990*
Edited by S.P. Denyer and W.B. Hugo